PAGAN

THE COLD WARRIOR'S TALE

ROBERT MANTEUFFEL

TACTICAL16 PUBLISHING

PAGAN

The Cold Warrior's Tale

First Edition

Published by Tactical 16 Publishing

Colorado Springs, Colorado

www.Tactical16.com

ISBN: 978-1-966413-04-2 (paperback)

ISBN: 978-1-966413-90-5 (hard back)

Contents

Pagan was their call sign –
Pagan was their creed –
Pagan was their religion

He fought like a pagan who defends his religion. - Stephen Crane

For the women and men who served the United States of America in uniform when no one else would - who fought and bled and died in the war that never happened.

Chapter 1
Welcome

The stewardess announced the descent into Rhein-Main Air Force Base, and he nudged the shoulder of his best friend, a short, blonde, all-state wrestler from western Pennsylvania who had served with him for the last year and half. A practical joker, with a wry sense of humor and a ready wit, came to quickly, shook his head, and prepared to land. They were both Lieutenants commissioned in the armor branch of the Army, who had spent a year at Fort Knox, Kentucky, as executive officers in basic training companies, turning raw recruits into light infantry soldiers and walking mile after mile on the "trail" as the drill instructors referred to the blacktop roads on hilly terrain that wound through the post.

More importantly, they were cavalrymen, "pony soldiers" who had earned their spurs in the Third Armored Cavalry Regiment under the watchful eyes and none-too gentle voice of Major Charles F. Riley, known as "Chuck the Fucker," "Uncle Chuck," and "Papa Gator." He had received a battlefield commission, promoting him from First Sergeant to Lieutenant in Vietnam when all of the officers in his troop of the Eleventh Armored Cavalry Regiment were killed. The Lieutenant's buddy, Scottie, who had a fierce temper, was known as "Baby Gator," but was also called, "Gator Bait," because of his propensity to get chewed on by Papa Gator. Ten months with "Uncle Chuck," in the deserts of Texas and New Mexico that surrounded Fort Bliss, had taught them to work around difficulties and persevere in the face of an impatient Congress, hellish weather, AWOL troopers, and broken equipment.

Over the preceding Christmas holiday, he struck up a relationship with an intrepid, petite red head who was formerly his former best friend's girlfriend and was now engaged to her. So, besides being involuntarily transferred to Europe and extended for a three-year tour, he would also have to arrange for leave to get married in the States and bring her to West Germany.

Descending from the plane, they shielded their eyes from the sun and walked over to an olive drab bus and stood outside, watching as the luggage from the plane, including their duffle bags and suitcases, was transferred. Then they climbed up and sat side-by-side once again, as the bus made its way out of the airport and on to the autobahn. Neither of them had been to Germany before so they did not know what to expect, still this wasn't it.

For one thing, the land was flat as Indiana and the soil appeared to be mostly sand, with broken wooded areas and some fields. Both had orders to tank battalions in the Eighth Infantry "Pathfinder" Division – the Crazy Eight or Shaft of Gold. The Lieutenant had been looking forward to making his application for the Army's law school program after two years and Scotty was planning on spending another year or two at Fort Knox, between a staff assignment in one of the battalions there followed by advanced course and assignment to a line tank battalion or cavalry squadron.

However, both had been diverted for immediate assignment to Europe because in conjunction with the Shah falling in Iran and the Russians massing on the border with Afghanistan, President Carter was informed of the bad news: the units defending West Germany were Category III – combat ineffective due to deficiencies in maintenance, training, and officer strength; the result of the destruction of the Army post Vietnam, recruiting difficulties posed by transition to the volunteer Army, and cost constraints imposed by the administration. Of the line divisions, the Eighth was stationed the farthest to the rear and was arguably in the worst shape since it took transfers of problem soldiers from the forward divisions and the two armored cavalry regiments that patrolled the "Iron Curtain" in the Fulda Gap along the border with East Germany and Czechoslovakia. So, the two lieutenants were shoved into the breach to shore up the defense of the west. The bus left the autobahn and rolled up to the transient billets where newly arrived and departing soldiers stayed while other arrangements were made for them. The Lieutenant shook hands with his buddy, walked to the door of the bus, and jumped down to the ground, carrying his American Tourister attaché case and a suitcase with his duffel bag over his shoulder. Then he went into the transient billets, checked in at the front desk, and dropped the duffle bag and suitcase inside his room. Then he took his attaché, he went out onto the street and walked to the entrance to Sullivan Barracks, where his battalion was stationed.

HELL IN A VERY SMALL PLACE

He walked past the guards at the gate and turned right into the very first building. The heavy oak door of the building that had once served as barracks for a German air defense unit during the war creaked slowly open for him. The Lieutenant peered into the dark entry way. The floors were black, and the walls were as well. At the far end, beside the light from a single dim bulb, sat a blonde gap-toothed Specialist 4 who was apparently the CQ, or charge of quarters. At his feet was an Army cot chained to the desk where he sat. Hand cuffed to the cot was a slender black man wearing a practically transparent black negligée. "Good morning, Specialist," said the Lieutenant, "I just arrived from CONUS and need to sign in. Is the First Sergeant around?"

The specialist looked up dully and then disappeared down a dark set of stairs to his left, remarking over his shoulder, "I'll check with Top and see what he wants you to do." The newly arrived officer peered down at the man in the cot and said, "Pardon me, soldier. I assume that you are a soldier in the United States Army?"

"Yes, sir," the soldier replied.

"Then if I might be so bold as to inquire – what in the name of heaven or hell is going on around here?" demanded the Lieutenant.

"What do you mean, sir"?

"For one, why are you dressed in a black negligée during duty hours?"

"Because I am gay, sir."

"All right – why are you handcuffed to that cot?"

"Well, sir, you see, they are chaptering me out. I'm depressed, and I'm a junkie, so they are afraid that I might get loose and shoot up and overdose and kill myself."

"Very well. Why is the cot chained to the CQ desk?"

"So's they won't drag me off and rape me – again."

"Thanks for clearing that up," said the Lieutenant, as he turned to face the sound of steps on the stairs behind him.

The Spec 4 appeared in the stairwell and said, "Go up to Battalion, Lieutenant – Top says the colonel wants to see you."

"Where is that?"

"Up the road to your right, sir."

"Thanks, trooper," said the Lieutenant who turned on the heels of his shined jump boots and walked out into the overcast grey morning. The first of many. The insignia of the Seventh Army to which he was newly assigned was posted at the entry to the post. It had seven stair steps on each side – the old hands said it stood for "seven steps to hell." They were right.

He would spend the next three years there, in a tank battalion with a silver lion on its crest. Pagan was their call sign. Pagan was their creed. Pagan was their religion. Hell was their home.

SADDLING UP WITH BRAVO

Having exited the barracks, his shined Cochran jump boots clicking on the floor, he entered the soft light of an overcast midmorning and turned right, along the sidewalk until he saw a low single-story building with a tank in front and the numerical designation of his new unit - 3-68 on a metal sign in front of the building. He entered the side entrance to the building and walked down the hall to a door with "Commander" on a small yellow sign over it. Turning right into the room he saw a desk with a captain, the battalion Adjutant, behind it. He walked forward, saluted, and told the captain that he was reporting in. The captain looked up briefly, rose, and motioned for the Lieutenant to follow. He walked in and stepped forward, as close to the front of the desk as he could, as it was connected to a table with five chairs, one for each of the Company Commanders in the battalion. He stepped forward, saluted smartly, and said, "Lieutenant Manteuffel reporting, sir."

The lieutenant colonel looked up from a map on his desk and saluted quickly, then rose, shook his hand, and stood back. "Lieutenant, we've been expecting you. Are you related to the German general?"

"Yes, sir. Distantly."

"Interesting," said the lieutenant colonel. "You went to VMI?"

"Yes, sir."

"We have three line-companies, a West Point company, a VMI company, and an all-others company. Ordinarily, since you went to school there, you would go to Alpha Company, commanded by Captain Coyle. It is cross attached to the 2-13 Infantry to make Team Yankee and we get Charlie Company from that battalion to form Team Mech when we deploy. But I have a special request from the West Point Company Commander for you, so you are going to Bravo Company. The other Lieutenant coming in is from West Point and he will go to Alpha Company. Where are you in from by the way?"

"Fort Knox, sir, the Infantry Training Center, and I spent a year with the Third Armored Cavalry Regiment at Fort Bliss attached to USAOTEA testing the M-1 tank."

The lieutenant colonel then said, "Welcome to the battalion, sign in with the Adjutant, and then get down to B Company. And lieutenant? See that?" He looked over at a large stake covered in black and white plastic.

"Yes, sir."

"That is a German road marker. They are placed along every road at hundred-meter intervals. During the last field exercise, one of my tanks ran over ten klicks of them – at a hundred marks apiece. Don't let that happen. Got it?"

"Yes, sir, I won't."

With that, the lieutenant colonel turned back to the map, so the Lieutenant saluted, and went to the S-1's desk, where he handed over his field 201 personnel file and orders, and asked for directions to B Company. After finding out that it was in the horseshoe shaped barracks to the west, he moved quickly down the hall and came up short for a couple of seconds. He could swear that he recognized the voice of a specialist who had worked for him at Bliss and whom he thought was being chaptered out of the Army. Shaking his head in disbelief, he assumed he was hearing things due to jet lag and lack of sleep and stopped in to pay his respects to the Sergeant Major before heading out find Bravo Company.

The western leg of the horseshoe housed Bravo Company. He pulled open a door of the same type he had entered earlier, walked in, and took a slight left where a buck sergeant sat at the CQ desk. The CQ called downstairs and then told him that First Sergeant, or "Top," was waiting for him downstairs. He walked quickly down the stairs and turned left at the bottom. It was very dark in the basement, and he looked down the hallway toward the light from an open office. He headed into the office.

A sharp Specialist 4 seated at a table beside the door was busy typing up a disposition form. Behind the large desk at the back of the room was a Hispanic First Sergeant named Guerra. The First

Sergeant looked up and quickly scanned the new officer. The Lieutenant seemed to pass inspection, "Good morning, sir," the First Sergeant said, followed by "where are you in from?"

"Fort Knox, 4th Training Brigade and a year with the Third ACR."

The first sergeant was sharp; a Vietnam veteran with service in the 25th Tropic Lightning Division. He said, "Sir, the first thing you need to do is go up to battalion and see the S-3 NCO, Sergeant Peeler. You need to read the Battalion SOP and alert procedures because we will be going out on a field exercise early in the morning. You will be going to First Platoon. The platoon sergeant is coming down from the motor pool at noon. He'll meet you out in front of the company and take you over to Coleman Barracks to get your TA-50 field gear. The old man is in the field on recon for the upcoming Reforger."

With that, the First Sergeant looked back down at his desk and returned to filling out the duty roster. In essence, the Lieutenant was dismissed, so he looked around the room briefly, then retraced his steps back out to the front of the barracks. The sun had finally burned through the overcast, and the day was pleasant, almost like a fall day back in Virginia. He looked north, across the road to the maintenance bays and a large motor pool as he made his way back to the battalion to look for Sergeant Peeler. Walking into the west end of the building again, he looked down the hall and saw a small yellow sign that read "S-3." At the threshold, he peered in to see several desks, with a Sergeant First Class and some specialists who were busy typing.

Addressing the clerk nearest the door, he introduced himself and asked if Sergeant Peeler was in. Hearing his name, Sergeant Peeler turned around from the maps he was studying and asked what he wanted. The Lieutenant told him that the First Sergeant of B Company had told him to come up to read the Battalion SOP. Sergeant Peeler nodded, retrieved a large black notebook from the locked cabinet beside him, and motioned for the Lieutenant to follow him into the empty S-3 Air's office and sit down at the chair behind the desk. Then he handed the notebook to him and stepped out, remarking, "Give it back to me when you are done; by the way I'll be leaving at noon for chow." The Lieutenant read the SOP in about an hour and returned the notebook to Sergeant Peeler well before noon. Then he walked back to B Company to meet his platoon sergeant.

A PIECE OF LUCK - THE BACKBONE OF THE ARMY

There are some people who you always remember meeting for the first time. Sergeant Robb was one of those people, but only after the fact. He was tall, lanky, and a natural athlete in an angular way, like Abraham Lincoln. He was soft-spoken, unassuming, thoughtful, and wise far beyond his years. He walked up and saluted his new Lieutenant smartly, then introduced himself. "Sergeant First Class Robb, sir, First Platoon Bravo." The Lieutenant returned the salute and extended his hand. Robb had a firm, but not aggressive grip and a ready shy smile.

"Sir, if you like, I suggest that you grab a bite to eat. There is a snack bar next to the PX, and I'll pick you up there at 1300. That way I can take you over to get your field gear, drop a couple of items off at the tailor shop, and get back here for formation at the end of the day. We need to get you ready, because we will have an alert early tomorrow and spend a night in the field. We'll come back into garrison the day after tomorrow, clean up, and get prepared to go to the field for about three weeks."

"Alright," said the Lieutenant, "I just have just one question. Where the hell is the PX? I'll figure out the rest."

"Sir, it's just out the front gate and to the left, behind the Batchelor Officers' Quarters. If you get to the Esso station, you've gone too far."

With that, the Lieutenant saluted his new sergeant, thanked him, turned on his heel, and headed back toward the transient billets. He had decided to find out how long he could stay there before he had to move out to permanent quarters. Since he had to spend a night in the woods shortly, he had to know what the plan was for his living arrangements when he returned. He walked into the building and stopped at the desk. "How long before I have quarters?"

"We don't know right now; it will probably be a month or so before we can move you into the BOQ."

"Where do I stay while I am waiting?"

"Wherever you like. That's your problem, but you have to move out of here in three days."

"What? I'm going out on alert early tomorrow and will be in the field for two days. So, I'll get back just in time to spend one night and then get evicted?"

"That's correct."

"Well, thanks for letting me know," he said, turned, and walked back out to the street.

He crossed the street and walked past the BOQ to the Post Exchange. He could see the sign for the snack bar and cut across a parking lot toward it. He slowed as he saw a busted beer stein in the grass beside the lot and a broken chair crashed down beside him. He looked up and made out the shadow of a man screaming a string of expletives before disappearing back into the room on the third floor above him. Without breaking stride, he moved quickly out of the engagement area, shaking his head as he went. While he had been in the Army for a while, and had seen some strange things, this was definitely something new. He just didn't have time to think about it as he stalked into the snack bar. He ordered a cheeseburger, fries, and a Coke, and sat down at a small table facing the door and the parking lot. The burger was thin and pretty tasteless, the fries were excellent, and the Coke tasted oddly bitter.

He finished eating and took a deep breath. Beside him was a black master sergeant visiting with a black buck sergeant who was complaining about the lack of spare parts and fuel for training. He was also appalled by the lack of discipline. Having just signed in, he was, frankly, taken aback. The old sergeant just looked at him and whistled slowly. Then he said, "Forget all that CONUS bullshit, you in USAREUR now, boy, it's a whole 'nother world."

With that, the Lieutenant looked out the window and saw a large van pull up and Sergeant Robb get out. The Lieutenant got to his feet and walked out to meet him. The sergeant snapped a sharp salute, and the Lieutenant just as smartly returned it, got into the front passenger seat, and buckled in. By that time, the van had already started moving and headed for the road where the sergeant stopped briefly before turning left and heading for the Autobahn.

As the van accelerated, Sergeant Robb turned his head slightly to the right and asked, "Where are you in from, sir?"

"Fourth Training Brigade at Fort Knox, about five months pushing infantry BT. Before that, I spent nearly a year chopped to the Third ACR at Bliss testing M-1 tanks in the desert and six months before that in a different battalion in the Fourth Brigade."

"Have you crewed an M-60?"

"Not since officer basic in 1977. I've spent most of my time in a PC and on the ground. But I did get two years on M-48 A-5s at VMI before I was commissioned."

"You went to VMI, sir?"

"Yes."

"If I might ask, why aren't you going to Alpha Company?"

"The colonel said I was specially requested by the Bravo Company Commander."

"You say you were in the Cav for a year?"

"Yes, at Bliss."

"Maybe that's it then, sir. We are getting ready to go to the field for Reforger. Due to fuel and training cutbacks, we are the only company going from the brigade, so maybe he wants a Lieutenant with some field time as there isn't much time to train."

"When do we go?"

"We'll have a company alert at four am tomorrow."

"Saturday morning?"

"Yes. We've been waiting for you, so you could train with the company. We'll load up and spend a night and two days in the local training area, shake everything down, and get it all fixed during the next week in garrison. A week from Monday we go an alert again and road march over to the railhead at Coleman Barracks, where we're going now."

Sergeant Robb looked in the mirrors to clear the rear and exited to a two-lane road leading into a large cantonment area on the banks of a river. "You say this is brigade?"

"Yes, sir. The headquarters is here as well as the division cavalry squadron, aviation, and the infantry battalion. Part of DISCOM (division support command) is here along with several signal units, and Colman Disciplinary Barracks, the confinement facility for USAREUR."

They pulled up and parked in front of one of many beige single-story buildings, just inside the main gate, and next to the railroad tracks. They entered and Sergeant Robb talked to the master sergeant in charge of the facility for a while, as the Lieutenant filled out the endless forms for the supply officer, so that he could pick up his field gear, rubber duck, parka, overshoes, sleeping bag, canteens, and waterproof bags necessary to "enjoy" life in the field. As he started through the inventory line, Sergeant Robb turned to the sergeant and said, "Make sure he gets good stuff; he's going to need it. He's my lieutenant and we're going to be spending a lot of time in the field."

The Lieutenant stepped down the line from station to station, picking up the usual field gear he was used to and a set of "Mickey Mouse" cold weather boots that were heavy as lead and awkward as hell. Sergeant Robb looked over at him and shook his head, "Probably won't use those, sir; you got any duffle bags?"

"Yes, Sergeant, two of them."

"Good, sir, pack yourself an A bag and a B bag. Put those in the bottom of the B bag."

After signing for his new equipment, Sergeant Robb and the Lieutenant each picked up a waterproof bag full of gear, walked to the van, and placed them in the back. Then they got into the front seats and the sergeant cranked up the engine. "Sir, you staying at the VOQ?"

"Yes."

"Good, I'll drop you off in front of the building so you can put your gear in your room before you come back to the company. You got any name-tags?"

"Yes, with my luggage."

"Got another field jacket?"

"Yes."

"Good. Take out the new field jacket and one of the two field shirts you just got issued and meet me back at the orderly room. I'll get you over to the tailor shop so you can get the stuff sewn onto one set while we're in the field. Keep the others and the "rubber duck" handy for the woods."

"Thanks. How long have you been in the battalion?"

"A little over a year."

"Where were you before that?"

"Fort Carson, sir."

As the van pulled over, the Lieutenant opened the door and then opened the sliding door, so that he could pick up both bags and sling them over his shoulders. He entered the VOQ and took his bags inside where he threw them on the bed. He untied the necks of the bags and took out the field jacket and one of the field shirts. Then he went out the door, locked it behind him, descended the steps, and walked back to B Company, where he saw Sergeant Robb standing under a tree at the edge of the company area.

Sergeant Robb saluted as the Lieutenant came up and they walked together toward the motor pool. As they passed the maintenance bays for B and A Companies, the sergeant said, "You have six tanks in the platoon. 1-1 is your personal tank and I have 1-4. Sergeant Webb on 1-2 is outstanding and heads up the heavy section, 1-3 and 1-5 tank commanders need some training. We also have 6-6, the old man's tank, commanded by Sergeant Fields. He's young, but steady and squared away." Sergeant Robb then walked purposefully to a small building with a metal sign in front of it declaring that the structure housed a laundry. They went inside and he handed over the two clothing items to a Korean tailor who informed him that it would take two to three days before the name-tags would be sewn on both and a unit patch on the jacket. Pocketing the claim tickets for both, the Lieutenant then followed Sergeant Robb to what was a large parking lot.

They walked in silence down the long row of tanks and other armored vehicles that belonged to the battalion he was now part of. "Sir, our tanks are in the middle of the hardstand. Second and Third Platoon are beside us. Third on the left and Second on the right. Second Platoon has the XO's blade tank in addition to the five that make up the platoon, so on alert they go out the gate after Alpha Company. That way the blade can clear vehicles out of the way if necessary."

"Where are the men?"

"They have been on detail to Battalion all day and are headed up to the barracks for equipment inspection. All the tanks are up but expect some problems when we get them cranked up. We don't get to exercise them much because of fuel constraints; you have to get a DF from the Battalion XO to move a tank to check the track."

"How often do you get to train?"

"Gunnery three or four times a year, one major exercise, and a few days a year in the local training area, out the back gate." The Lieutenant looked up at the fifty-two-ton vehicle and smiled; 1-1 was a good-looking tank and he felt the bond that any cavalryman feels with the charger that he may have to take into battle. The Lieutenant looked over his tanks and did not see any obvious problems; no rust or leaking fluids, and the tarps were tied down well over the turret. He had a good sergeant. "What's with the palm trees on C Company's tanks?"

"That's their former Company Commander's doing; painted them like the Africa Corps. He likes to think he is Rommel."

"He's the S-1 now. The new CO, Captain Butler, liked them, as he was the old XO." Sergeant Robb looked at his watch and announced that it was seventeen hundred and time for the company formation. The two turned and walked back toward the barracks. They reached the company area just as the troops were forming up: First, Second, and Third Platoons, plus Headquarters Platoon, which included Company Maintenance and Supply. Sergeant Robb took his place on the far right of the First Sergeant and the Lieutenant took a position in the rear of the formation behind his new platoon.

THE RAINBOW COALITION

The striking thing about the company was that the line platoons were segregated by race. Third Platoon was all white, with the exception of one Hispanic and Specialist 5 Edgar, who the Lieutenant knew was a Filipino as he had served with him at Fort Bliss for almost year. The platoon leader was an old hand E-7 and his assistant was an E-6 who was also the company's Master Gunner. The Second Platoon was all black, with a black E-7 serving as the platoon leader. First Platoon was a mixture of white, black, Hispanic, and possibly Asian soldiers - a veritable rainbow coalition. The Hispanic First Sergeant was a good NCO, and a no bullshit individual. He quickly set out the schedule for the next morning: an 0400 mock alert, stand to with weapons uploaded by 0500, and crew members would rotate through the mess hall for breakfast in pairs, with the company to be prepared to roll to the local training area by 0700.

There were no commissioned officers other than the Lieutenant, so he remained behind First Platoon as the First Sergeant called the NCO platoon leaders forward for a brief conversation concerning the fact that the "old man" and the XO were in the field on recon, would be getting in late, and consequently in a rare mood, so they needed to mind their business and be ready in the morning. The men standing in ranks swayed gently back and forth a little but maintained good order as they stood at ease. After a couple of minutes, the First

Sergeant called the company to attention and released the platoons to their sergeants for further instructions. Sergeant Robb informed the platoon that he would be in at 0330 and that he was pleased to introduce their new platoon leader. With that he turned and stated "Sir, the platoon is formed." The Lieutenant then marched around to the front of the platoon formation. The sergeant saluted and took his place at the rear of the formation.

The Lieutenant ordered his men to stand at ease, introduced himself, and let the men know that he was honored to be their new platoon leader. He also told them that he was newly arrived from Fort Knox, where he had served as a company training officer and executive officer in a basic training brigade. He shared that he had also spent a year at Fort Bliss assigned to the Third ACR, where he had tested the first platoon of the new M-1 tanks. He also informed them that he had spent four years in the ranks at VMI, that he intended to do his best, and that he had high expectations for them, as he had never served in any unit that was less than the very best at what it was called to do. He also asked each of them to keep an eye on him and assist him in doing his job as he was determined to learn from them. He then called the platoon to attention and directed Sergeant Robb to form the platoon for inspection.

Sergeant Robb ordered the men to open ranks and then accompanied the Lieutenant as he moved to the first man on the left and stopped in front of him. The Lieutenant came smartly to a stop in front of the man, looked him over, and extended his right hand to shake his hand, introducing himself again, personally. The man was a short and stocky African American with a broad head and grim visage, slightly smoldering and menacing, for all the world like a bull buffalo. He was a newly promoted E-5 who was known for his proficiency as a gunner, which explained why he was in charge of 6-6, the CO's tank.

The Lieutenant moved down the ranks and spoke with each man in turn, asking about previous duty stations and hometowns. At the end of the third rank, the Lieutenant turned to Sergeant Robb and said, "Sergeant, the platoon is yours." The sergeant closed ranks and walked to the front of the formation. He dismissed the men and spoke briefly with the tank commanders.

When he was done, he walked up to the Lieutenant queried, "Sergeant, it looks like the company has a black platoon and a white platoon, what have we got?

"Sir, we have the whites and Puerto Ricans that the white platoon didn't want, the blacks that the black platoon didn't want, the Mexicans and two 'carped' Guamanians. Why the segregation? The race problems in the battalion are bad, and the CO decided to separate the men by race to try and contain them."

"Very well, what is a 'carped Guamanian?'"

"We have two NCOs, one an E-6 tank commander on 1-3 and his gunner, an E-5 sergeant. They just re-enlisted under the combat arms re-enlistment program, 'CARP,' to go from non-combat specialties to armor. Sergeant Aguinaldo was in the signal corps most recently and the artillery before that, and Sergeant Fejeran was an MP. Both had service in Nam."

"Have they had armor training?"

"No, sir, the program relies on OJT."

"Really?"

"Yes, sir."

"Hmm. I know why I'm here. I went to VMI. But you seem pretty squared away - what did you do to get sentenced to this?" Sergeant Robb looked at him and realized that he was kidding, to a certain degree, smiled and lowered his head in a self-deprecating way.

"Sir, you'll figure it out. Let's go, I have to pick up my girlfriend from work. I'll take you on a tour of the area while I'm at it."

THE CHILD OF THE DUST

The Lieutenant followed the sergeant over to the van and hopped into the passenger seat. The sun had finally come out, and it was a glorious late fall afternoon. Sergeant Robb climbed in, drove out the gate past the guards, and down the street to the first left, into a small enclave of low buildings. He slowly drove by them and pointed them out. These included the officer's club, the education center, the post library, and the morale and recreation center. He then circled around the buildings and pulled up in front of the library. The Lieutenant looked over and asked, "Where are you from, sergeant?"

"Tennessee, sir."

"What part?'

"The middle of the state, sir."

"Really? I'm from Kansas myself, but my mother is from Tennessee, up by Fort Campbell." The library door swung open, and the Lieutenant jumped out of his seat.

"Stay where you are, sir, if you don't mind. That way it'll be easier to drop you off, after we get around the traffic circle, and she won't mind."

"She" was walking toward them. The Lieutenant had his answer. While she was not beautiful by western standards, she was handsome, intelligent, and exotic. She was a "child of the dust" apparently. A walking genetic history of the endless wars that had ravaged Vietnam over the centuries. She was Vietnamese, but taller and less broad of face; more in the line of Chinese. She had a slender, shapely body, long limbs, and stately walk; French, with curly black hair, honey colored skin, and freckles; African, possibly Senegalese. She was marvelous, and the Lieutenant turned and exchanged smiles with her as she opened the rear sliding door and climbed in with her purse and a couple of books.

As she seated herself and buckled her restraints, the Lieutenant introduced himself. She said that her name was Dawn and asked if he was the new platoon leader.

"Yes," he said, "what subjects do you teach?"

"Modern languages: French and German here. I used to teach Vietnamese, too, but there is not much demand for that now."

"How long have you been here in Mannheim?" the Lieutenant asked.

"A little over four years. I was in Saigon working for the DOD and Foreign Service until 1975." As she spoke, she leaned forward to place her face in between Sergeant Robb and him. She had an oriental view of personal space and liked to get close to those she was speaking to. She was also exuberant and had laughing, intelligent eyes. She asked where he was from in the United States. He explained he was from Kansas, Virginia, and some other southern states. She said that she had been in Virginia and Washington, D.C. before taking her current position in Germany. The van pulled up to the traffic circle and came to a stop near the gate to the post. The Lieutenant said, "Nice to meet you, ma'am," as he opened the door to get out.

She was already unstrapped and moving into the seat. The Lieutenant got out and closed the door, then he looked over his shoulder and said, "See you tomorrow morning, sergeant." As they pulled away, he turned toward the VOQ building. He walked up to his room and sat down on the bed to take a deep breath and rest for a minute – it had been a hell of a day, and it wasn't even over yet.

Chapter 2
First Alert

Since he didn't have any Deutschmarks and didn't feel like walking all the way back to the officers club, he went back to the Post Exchange snack bar and considered the menu of GI junk food that was available. He didn't feel like another burger, so he decided on what appeared to be a German bratwurst, a roll, and a small salad. Finally, having some time to think, he reflected a little about his personal situation.

This wasn't in any way where he had expected to be two years ago. At that moment, by his calculation, he should have already submitted his application to the JAG Corps and be waiting on staff at Fort Knox for orders to attend law school, preferably at the University of Virginia. Instead, having left his fiancée back in Kansas, he was sitting in a post near the banks of the Rhine. Not only that, he had been barred to reassignment, and involuntarily extended for the duration of a three-year tour. And he was about to roll out to the field starting at four in the morning while his body still thought it was noon.

Oh well, he bought a package of red licorice and a Snickers bar for breakfast and strolled back to his room. He had a duffle bag and a footlocker, in addition to the duffle bag of TA-50 he had been issued that afternoon. Taking a deep breath, he laid out the contents of both duffle bags and waterproof bags on the bed and packed the duffle bag that had straps like a backpack, his A-bag, to take to the field. He selected out a set of field dress and fatigues for the field as well. He undressed down to his underwear, brushed his teeth, and set his travel alarm clock for 3am. Then he cleared the bed, laid down on top of it, and covered himself with a poncho liner that he had with him since Fort Bliss. It was one of the best things that came out of the Vietnam war. He said a prayer, thought of his girl far away, sighed, and went to sleep; the single thing he had learned to do almost anywhere during the year he had spent in the New Mexico desert.

The dull, maddening, mechanical clatter of his travel alarm clock shattered the night. The Lieutenant came instantly awake, rolled toward the nightstand, felt for the small clock, and pushed in the pin to turn it off. He was still dead tired but swung out of the bed and headed to the bathroom where he shaved and rinsed off quickly in the shower. He then brushed his teeth and rinsed out his mouth before taking a drink of water. He immediately spit it out because it was simply awful, with a sour burning taste and an almost sulphureous smell, very similar to the way the air smelled, as it rolled up along the Rhine from Mannheim and the paper mills there. Well, he thought, that explained the German fondness for strong coffee, tea, wine, and beer.

He checked his watch and realized that it was 3:30am; 8:30pm by his biological clock and shook his head - it was going to be a very long day. He finished packing his things into the A-bag, putting in his shaving gear and a map case as the final items so he could get to them quickly. He then dressed in fatigues and boots, with field pants and a heavy wool field shirt over them, the prescribed uniform. Finally, he slipped the uniform baseball cap on his head, walked through the door, turned, and locked it behind him before shouldering the A-bag and walking quickly down the stairs to the street. He turned left and headed for the Kaserne and his first field duty, less than twenty-four hours after landing in country. As the veterans in the Cav said: it really was a moving train.

Arriving at the gate, he set down the bag and opened the left pocket of his field jacket to fish out his green ID card and present it to the guards at the gate. They checked it against his appearance in the dim light coming from the street lamps, then saluted. He re-shouldered the bag, saluted, and jogged to the company area using the slow shuffle he had picked up in Airborne School at Fort Benning about five years earlier, when he was still a VMI cadet.

Entering B Company and heading downstairs, he could hear the sound of the First Sergeant's voice and smell the best smell in the world - hot, black, strong coffee. Just as he prepared to go in and get a cup, he heard "Herr Lieutenant" from the adjoining doorway. Turning, he saw his Company Commander, Captain Batson, for the first time. Pressed fatigues, a Ranger tab, jump wings, and spit-shined tanker's boots made an immediate impression. The loud, "Come in here!" also had an effect. The Lieutenant dropped his A-bag, walked in, and came to the position of attention the requisite three paces from the low brown desk behind which Captain Batson sat and perused his newest officer. In a chair in the corner sat the Executive Officer, a small rat-faced man who seemed perpetually ill at ease. The Lieutenant came to attention, clicked his heels together smartly and saluted.

The captain let him hold it for a few seconds, then returned it and said, "We've been expecting you. Are you related to the German general?"

"Distantly, sir."

"Hmmm - have you been to the field before?"

"Yes, sir."

"Good. Stand to is at 0600. Here's a map of the training area. Be ready."

"Yes, sir." The Lieutenant saluted smartly. When it was returned sharply, he turned on his heel and left the office. Just outside the door, he re-shouldered his A-bag and headed down the hall to the arms room. He saw Sergeant Robb and walked up to him. "Good morning, Sergeant, what's the plan?"

"As usual, Third Platoon is loading up first. Second Platoon is behind them and then we will go. Headquarters will finish loading out after we've picked up our machine guns and signed for our sidearms. Sir, I need you to pick up the M-240 for 1-1 and follow me to the parking lot."

The company was efficient, and the supply sergeant and armorer quickly signed out arms to the NCOs who were in front of him. The Lieutenant stepped up to the window, printed and signed his name, and presented his ID card to the armorer as he was new to the company and was not visually recognized. The Lieutenant was then handed an M-4 submachine or "grease" gun, a .45 caliber pistol with ammo, a shoulder holster, and the much heavier M-240, .30 caliber coaxial machine gun that would be installed inside the turret alongside the main gun. After putting on the shoulder holster, he cleared the pistol and slipped it in. Then he cleared the primitive grease gun and slung its green strap over his neck, so it hung on his left side. Then he crouched down and slipped the larger straps for the A-bag over both shoulders and cleared and picked up the machine gun at port arms. Finally, about forty pounds heavier, he walked back down the hall and started humping up the stairs.

By the time he reached the top of the stairs and hit the doors to step out into the darkness he was breathing pretty hard and sweating from head to toe. Then he turned left and trudged down to the motor pool across the parking lot to the hardstand where his platoon of tanks waited. At last, he caught up to Sergeant Robb and walked past him to 1-1 where he slung down his A-bag and looked up to see a Specialist 5. "Who are you?" the Lieutenant asked.

"I'm Specialist Blaylock, sir. Your gunner."

"I don't remember seeing you yesterday at the end of the day formation."

"I was in the field, driving the old man until about 2000, sir. His driver was out on leave until yesterday, so I got the duty. We were out in the maneuver area for Reforger for the last two days. Good to meet you, sir."

"Good to meet you, too," said the Lieutenant, who swung his A-bag up to the soldier and then climbed aboard the tank on the left front. As soon as he got to the turret, he heard Sergeant Robb behind him, saying, "Blaylock, stow the Lieutenant's bag in the bustle rack and secure the grease gun. Sir, come with me and we'll get the 85s." The day had hardly started, and the Lieutenant was already getting tired, but he swung in behind the sergeant, and they walked in silence back to the arms room. The company was now a loud and bustling area that looked like a circus coming into town. The armorer was issuing weapons to the troops, and they were busy wrestling their gear down the stairs.

WELCOME TO REALITY

The Lieutenant followed Sergeant Robb, who cut to the head of the line and asked the Supply Sergeant for the M-85 machine guns for 1-1 and 1-4. He signed for his and the Lieutenant signed for the other. These were heavy, fifty cals, so they put a barrel in each one and slung them over their shoulders to carry them up the stairs. Humping that much steel rapidly took its toll on them and when they finally reached the edge of the motor pool, the Lieutenant had to ask, "By my count, we have over twenty men in the platoon - so why are you and I humping these machine guns by ourselves?"

The sergeant simply turned slightly and asked, "Sir, have you read the battalion SOP?"

"Certainly, I read it in the S-2's office yesterday."

"No sir, not that one, the real one - in the S-3's safe."

"No. There is a second one?"

"Yes, sir. And it says that upon alert, the battalion's officers and senior non-commissioned officers will draw the heavy weapons, install them, charge them, turn them on the troops, and force them to mount the vehicles. We have to be prepared to do that."

The Lieutenant looked at him incredulously and asked, "How do you know this?"

"Sir, I wrote it. This is the third time I've been stationed here. I was the assistant S-3 NCO on my last tour. I worked there for my first couple of months this tour and it hasn't been changed."

"So, you mean to tell me, in order to stop the Russians, if we have to, we have to drive through millions of Germans, and maybe fight our own soldiers to even get there?"

"That's about it, sir. Now, if you'll lay your 85 up on the sponson box, Specialist Blaylock will help you get it installed. Then I'll be back to check it."

The Lieutenant climbed aboard, manhandled the machine gun up to the tank commander's position, called down to his gunner, who came up through the hatch, and opened the gunport. Then the two of them manipulated the gun into position and inserted the pins to lock it in place.

By now, the remaining troops of the platoon, drivers, and loaders, had mounted the vehicles and stowed their gear. The Lieutenant reintroduced himself to his loader and driver. The loader was a small, wiry, African American, while the driver was thoughtful and quiet, older, and white. When the troops were done loading up, Sergeant Robb formed the platoon and released half of the troops and two of the tank commanders to the mess hall to eat breakfast.

Once they were on their way, Sergeant Robb swung up onto the front of 1-1 and climbed up beside the Lieutenant. After function checking the machine gun and making sure that the cupola worked as well, he turned and said, "Okay, sir, when you get in, crank the 85 barrel up to forty-five or fifty degrees, and when we move keep your head down behind the gun-mount."

"Why?"

"We still have terrorists here and they string piano wire in the trees, tank-commander high to garrote you if you're not careful. The machine gun barrel will catch the wire and break it. In the open, we can move pretty quick but be careful in the woods, especially at night."

"Lovely. Thanks, Sergeant."

"Sir, you better get up to the mess hall by HHC and get chow. I'll go when you get back. We probably won't eat again 'til nightfall if I know the old man. Oh, and sir, you'll need to get your map in a map case; it will probably rain before long." With that newfound knowledge, the Lieutenant realized that this was going to be a little more of a challenging assignment than he first thought.

He made his way back to the company area where the First Sergeant was supervising the loading of the supply truck and the arms room and walked over to the other horseshoe where Combat Support Company, Headquarters and Headquarters Company, and the mess hall were located. He walked in and paid the going rate plus surcharge and headed through the chow line. He picked up a tray and silverware and moved down the line, ordering eggs, hash browns, creamed beef, bacon, and toast. The troops appeared pretty sleepy, and the cooks who, as it turned out, were all black, were watchful and pretty sullen.

After picking up his food, he poured a mug of black coffee and looked for a place to sit. He was in a hurry, so he sat in the corner across from the entrance and ate quickly. He did notice that one white Spec 5 cook was working quickly, if not enthusiastically, loading rations and utensils into a truck that was hooked up to a kitchen trailer behind the mess hall. In about ten minutes, the Lieutenant had eaten and drank a second cup of coffee. Then he took the empty tray to the window, dropped it off, and hurried back to the motor pool and his tank.

Sergeant Robb saw him coming and met him at the edge of the company area where he handed him a small book with a lanyard attached to it. It was the CEOI, or code book, for his unit. Sergeant Robb informed him that the radios were set and he walked away toward the mess hall as the Lieutenant slipped the lanyard around his neck, put the book down inside his field shirt, and climbed up to the turret. Then he found his A-bag in the bustle rack and pulled out the map case and placed the folded map into it. The map was a little odd in that it was not a standard 1:50,000 or 250,000 map sheet, but some sort of blowup of the training and maneuver area. He studied it and located the position of his company on it as well as the location of the back gate that he assumed they would road march out.

He also pulled out the CEOI and refreshed his knowledge of how to use it. Then he reviewed the call signs and frequencies for use that day and checked the settings on the two radios on his tank: one set to the platoon frequency and the other set on the company frequency. When he was done, he checked his watch and looked up to see the other members of the platoon walking back to the motor pool. It was about twenty minutes to the pull time and the CO was already on the company freq, advising all of the platoons to stand to, crank up, and provide a situation report "sit-rep" to report their readiness to move out. A few seconds later, seventeen tank engines roared to life.

OFFICERS ARE MAGIC

Looking across the parking lot toward Battalion, the Lieutenant saw Sergeant Robb walking quickly toward the platoon. "Sir, we have a problem."

"What's that, Sergeant?"

"The Battalion Training NCO won't release any Hoffman charges to us to use in the training area without a DF signed by the old man."

"You must be kidding."

"No, sir. I sent a couple of privates down to get them, and they came to get me. I just talked to him myself. Staff Sergeant Sherman is a real pain in the ass."

"Really?"

"Yes, sir."

"Damn. Well, I'll just have to go get them myself." Setting his

CVC beside the M-85 mount, he swung down off the turret, walked to the front of the tank, slid down the glacis plate, and landed on the ground. It was getting late and the noise from the vehicles was deafening. He was sweating in his field shirt and pants as he ran to the headquarters, slipped in the back door, headed down the steps, and entered Staff Sergeant Sherman's office and supply room. Having heard the pounding of boots on the stairs, the sergeant was standing by his desk and watching the doorway when the Lieutenant came through the door. The Lieutenant came to a stop in front of the desk and said, "Sergeant, I need a box of Hoffman charges."

The staff sergeant looked him over and said, "Are you new to the battalion?"

"Yes, Sergeant, I am. Arrived from Frankfurt yesterday."

Then the sergeant shouted at the Lieutenant, "Are you familiar with the position of attention?"

The Lieutenant was at first taken aback, but decided to see what this was about, so he said "Yes, sergeant, I am."

The sergeant screamed, "Then assume it!"

The Lieutenant did so, staring at a spot on the wall behind the corner of the desk. The sergeant came up and took a position right in front of the officer and demanded gruffly, "Why isn't your name tag and unit insignia sewn on that field shirt?"

"As I said, I arrived yesterday and was issued TA-50 yesterday afternoon. I was given two sets: one is in the tailor shop for sewing on the name-tag and insignia. I was told that I would be going to the field at daybreak, so I kept one for the field."

"Do you have an ID card?" demanded the sergeant.

"Yes, I do."

"Where is it?"

"Per the Battalion SOP, it is in the left breast pocket of my fatigues."

"I need to see it."

"Very well, sergeant," and the Lieutenant began moving his right hand toward the pocket to retrieve the green identification card carried by all active-duty soldiers.

As soon as he began to move, the sergeant screamed, "I did not tell you to move! I'll get it." He moved his right hand deftly in behind the field shirt and unbuttoned the starched pocket of the underlying fatigue shirt. The sergeant then moved his right hand into the pocket and put it inside, slipping past the card, and gently squeezing the left breast of the officer. Immediately, the Lieutenant knew what the game was: he was blessed with good genes and high cheek bones. Even though he was twenty-four, he could still pass for eighteen or nineteen, and he looked like a fresh young private to the sergeant, ripe for the plunder.

Having slowly removed his hand, the sergeant gazed down and looked at what was the reverse of the card that did not provide any real information other than the blood type of the officer, O-Neg, for type "O" Negative, the universal donor. The sergeant smiled to himself again because the card was a new one which reinforced his first impression that he was dealing with a raw private E-2, naive and fresh from Advanced Individual Training, easily bullied. He sighed and almost licked his lips as he said, "Hmmm."

Then he casually flipped the card over to learn for the first time the name of his latest victim and his mouth came open in a silent scream as his eyes widened in abject horror. The Lieutenant calmly said, "I believe that you are familiar with the position of attention, since I just demonstrated it for you."

The sergeant weakly croaked, "Yes, sir."

The Lieutenant then commanded, in his best parade ground voice, "Assume it."

The sergeant, fat, dumpy, and slow, came to what passed for the front leaning rest position, a quivering pile of Jell-O.

"Have you had a chance to read my ID card?"

Again, the sergeant weakly gurgled, "Yes, sir."

"What am I?" asked the Lieutenant.

"An officer."

"No, Sergeant, you are wrong. I am your worst nightmare: a commissioned officer, a lieutenant, and worse than that, a first lieutenant. Worse than that, a first lieutenant commissioned into the Regular Army after four long years at VMI. Worse than that, a first lieutenant commissioned into the Regular Army, after four long years at VMI, who was a paratroop at nineteen; worse than that, a first lieutenant commissioned into the Regular Army, after four long hears at VMI, who was a paratroop at nineteen, and spent a year in the field at Fort Bliss with the Third ACR. Worse than that, a first lieutenant commissioned into the Regular Army, after four long hears at VMI, who was a paratroop at nineteen, and spent a year in the field at Fort Bliss with the Third ACR, who also spent a year in the training center as a training officer. Twelve drill instructors worked for me, and they, by God, shook when I walked. Now, if I might ask, are you a Christian?"

"Y-Y-Yes, s-sir," squeaked the sergeant.

"Very well, then I will give you a piece of scripture from the New Testament to remember this day by. 'Whatsoever you do to the least of my people, that you do unto me. So says the Lord our God unto us, and I say it unto you.'"

"Now, if any soldier under my command ever asks you for anything and you do not give it to him forthwith, I will immediately appear before you, drag your fat ass up those stairs, and horsewhip you through the battalion. Do you understand?"

"Yes, sir!"

"Very well, at ease. Now, be a good chap and hand me a couple of boxes of Hoffman charges."

As the NCO turned to go into the storeroom, the Lieutenant continued, "Sergeant?"

"Yes, sir?"

"I'm not the sex police, but in a few words from that greatly unappreciated American film, ***Drive-in***; 'Hey man, it's nothing personal. I'm just having a bad day and every time I look up, it's you.'" The sergeant came back and handed him the two boxes of charges. The

Lieutenant took them, said "Adios," turned to the right, and pounded up the stairs to get back to his platoon.

When he got to his tank, he gave the boxes of Hoffman charges to Sergeant Robb, who asked if he had any problems with Sergeant Sherman. "No Sergeant, and we never will again. Ever."

Sergeant Robb asked "Really?"

"Yes, Sergeant, haven't you heard? Officers are magic."

THE CHALLENGE AND PASSWORD OF THE EIGHTH ID

Sergeant Robb distributed the charges to the tank commanders and the gunner on the Lieutenant's tank for installation in the Hoffman firing devices, which simulated smoke from the main gun when it was fired. All of the tanks had been fitted with MILES gear: laser activated sensors wired to lights and sirens that activated when they were "hit." While the men were tending to that, he turned and asked Sergeant Robb, "Are we going to get machine gun ammunition to charge the heavy weapons?"

Sergeant Robb replied, "No, the ammunition is on board the supply truck which is being loaded on the west side of the motor pool."

The Lieutenant looked over in that direction and saw a large African American soldier putting ammunition in the back of a deuce and a half truck. He asked another soldier to help him load additional ammunition from a five-ton truck into the deuce and a half. That soldier looked at him and said, "You gotta be shitting me, motherfucker." The Lieutenant, hearing this, turned to Sergeant Robb and asked, "What was that?"

Sergeant Robb looked at him, smiled, and said, "Sir, essentially that's the challenge and password of the 8th ID. You're going to hear it a lot, so you might as well just go ahead and get used to it."

Hearing the radios come to life, the Lieutenant climbed up beside the commander's hatch of his tank and reaching into the tank commander's cupola, went to the platoon frequency and called for a fast commo check. Tanks 1-2, 1-3, 1-4, and 1-5, responded right away; 6-6 chimed in a couple of seconds later. He answered with, "Roger out," and prepared to move. Oddly enough, this was pretty easy for him, since back at Fort Bliss he had four radios on his track to keep up with, and here he only had two. Once the radios were up and running, he heard a call from the Company Commander for the platoons to prepare to march out the gate into the training area.

Captain Batson was in rare form, standing beside his Jeep and screaming into the radio. The Lieutenant called him and announced that First Platoon was ready. The other platoon leaders followed suit. The order to march came and the platoons left the motor pool, with Third Platoon in the lead, Second Platoon, and then First Platoon, and drove out into the training area. Then the Company Commander told them they were released for individual platoon training, and they could laager up in the training area and proceed to train on the skills that they needed at the platoon level.

Sergeant Robb went on the platoon net and ordered the tank commanders to begin coiling back from 1-4 so that all six tanks formed a circle. Then he told the NCO tank commanders to come to the center of the circle, and the Lieutenant came as well. There Sergeant Robb laid out the training plan: they were going to practice movement and movement to contact first, then laagering up, and then defensive positions. If they had time in the afternoon, they would take individual tanks and move them tactically so they could get used to operating in the maneuver area.

The Lieutenant had a visit with each of the tank commanders about what they were planning to do that day and the tank commanders told all of their crew members to come to the center as well. When they were there, the Lieutenant briefed the entire platoon on the order in which they would be accomplishing their training tasks and then added an additional comment, "I have trained several types of soldiers over the years, particularly with regard to drill, and in my view, the best way to train people is to train them until they accomplish the task correctly, then practice it one more time to make sure that everybody understands, and then move on to the next task. If we continue to repeat the same task over and over again, invariably, someone will start fooling around and if you have to go into combat, then the way you're going to execute the task is the way you did it last, which will most probably be wrong. Also, once we have learned a task well enough for today, we can move on to new tasks, accomplish a greater set of tasks and be well trained and ready. Any questions? Very well."

With that, all the crews mounted their vehicles, the tank commanders took command of their tanks and the platoon set out, in traveling formation, with tanks 1-1, 1-2, and 1-3 in the lead and 1-4, 1-5, and 6-6 following. When they came to a wooded area, the Lieutenant remembered to keep his head down behind the M-85 mount. He was glad he did as the machine gun barrel snagged and broke a wire, strung tank-commander high between two large trees at the beginning of the trail he had to use to get to the next clearing in the maneuver area. He also noted "Yankee go home" spray painted on what appeared to be old bunkers from the war that littered some of the small clearings they drove through. He was in Indian country.

The platoon, and then the individual tanks, proceeded through their training under the auspices of Sergeant Robb. He was awesome. At noon, they laagered for lunch and ate what they had with them. The Lieutenant shared his red licorice with the crews. He explained that it was his habit to carry it with him after serving in the desert with the Cav: it was the perfect field food. It would not melt, water ran off it, and it still tasted good even when covered with sand and dust.

By mid-afternoon, Captain Batson called all the platoon leaders and told them to form up the company at a grid coordinate that he sent in code, based upon a clock dial with twelve set at north: Third Platoon would cover positions one through four; Second Platoon would cover five through eight; and First Platoon would cover nine through twelve. Once the company was formed, the CO called all platoon leaders to his Jeep so he could lay out instructions for the rest of the day. The Lieutenant put on his helmet and headed for the CO's Jeep. When he arrived, the Lieutenant acknowledged Sergeant Wilson from Second Platoon, saying, "Hello." Sergeant Ramsey from Third Platoon, simply nodded his head and glowered, and the captain welcomed him with, "What the hell took you so long, Lieutenant?" Apparently, the plan was to road march in the afternoon, break for chow, do a night road march to a new location, laager up, set out security, sleep, eat, and prepare to stand to at 0700.

Captain Batson dismissed the platoon leaders to go back to their platoons and relay the orders. When the Lieutenant got back to his platoon, he called Sergeant Robb over and observed, "It seems like Sergeant Ramsey doesn't particularly care for me."

Sergeant Robb answered, "He doesn't care for anybody, because he's bucking to be First Sergeant when First Sergeant Guerra goes back to CONUS in ninety days. He's got Wilson beat out based on seniority and he's a better tank platoon leader than him."

The Lieutenant interjected, "But he's not as good as you are, right?"

"That's true, but if he can make you look bad, it'll improve his chances, because he'll make the platoon look bad and me look bad. Sergeant Ramsey is Captain Batson's fair-haired boy. Whenever he needs a tank platoon to put on a demonstration or be cross attached to the infantry to make a good impression, he sends Ramsey. He has the master gunner for the company on 3-4 and first pick of all incoming tank crewman, so he's in a good spot."

The Lieutenant interjected again, "And you, on the other hand, have leftovers like me."

Sergeant Robb stated, "Well, sir, I wouldn't put it that way."

The Lieutenant interrupted again, "Well then, all right, I know how it is and I'll just have to make sure that we look good for you so that you can become Top before I rotate out of here." He then climbed up on his tank, 1-1, and called the tank commanders over to him so we could brief them on the rest of the day.

After he did that, everyone mounted up and they waited for Captain Batson to give the road march order. He did that in about ten minutes, and they drove in single file for several minutes before breaking out into platoon traveling formations. They did that for about an hour and a half; back and forth, across the maneuver area and then formed up for chow, with the platoons taking their positions, based on the clock dial, as before. The First Sergeant made the rounds as the sun was setting and delivered dinner to everyone. It appeared to be what passed for Swiss steak, mashed potatoes, green beans, and peach pie. When the First Sergeant pulled up, the Lieutenant told him to go ahead and serve the crewman first and the officer would eat last when everyone had plenty.

After they were served, the Lieutenant took his plate, sat down next to a tree to brace his back, and proceeded to eat. Once everyone was fed and the sun had dipped down below the horizon, the captain called the platoon leaders on the company radio frequency and gave them the road march order after which they marched under the moonlight for about another hour and a half. At 2000 hours, they got the order to assemble for the night, so again they set up a circular formation, covering the same numbers from the dial, and settled in. The Lieutenant got his tank commanders together and went over the plan for the next day with them, setting up a watch schedule with their crews so that someone would be manning the radios on the all the tanks throughout the night.

Chapter 3
A Rude Awakening

The Lieutenant took first watch on 1-1, up until midnight, then he changed the frequencies on the radios based on the CEOI and told the gunner to take over. He climbed down off the tank and curled up in his poncho liner between three trees to try to get some sleep. He didn't want to be next to a track when he didn't have a tree protecting him from the tanks in case they took off. He finally fell asleep and the next thing he knew, he heard the radio crackle to life. The last person on guard had been the loader and he'd fallen asleep, so he missed the stand-to instruction on the radio.

The Lieutenant hastily climbed up to the tank commander's cupola, grabbed the radio, came up on the net, and informed the Company Commander he was up, whereupon the commander, on the company frequency, yelled at him and asked him where the hell he'd been. The Lieutenant told him he had been asleep, to which the commander snarled, "I want you to give me your grid coordinate right now," and when the Lieutenant didn't respond fast enough, the commander answered his own demand: "I'll give you a hint, Lieutenant, the sun rises in the east!"

After that humiliation, the Lieutenant got the grid coordinates, and sent them back in code to the commander, who then gave the orders for the day: after the First Sergeant issued C-rations and coffee to the troops, the platoons were going to break up into platoon-on-platoon attack and defense exercises as quickly as possible. Top made the rounds, throwing a box of C-rations to each soldier; and then the platoons immediately took up their positions. Second and Third Platoons would go first, and First Platoon would be observing; then First Platoon would defend against Third Platoon, and finally, First Platoon would attack against Third Platoon.

After those exercises, they would go back into the motor pool and wash the tanks. Apparently, the entire field exercise was set up to show that Sergeant Ransey could annihilate everybody else and be in a good position for the upcoming board for First Sergeant of B Company. Like Captain Batson, he was a political animal, so this was his environment. Hearing that, the Lieutenant walked over to Sergeant Robb and said, "We're just going to have to go out of our way to win this engagement, so we can take care of you." Sergeant Robb just gave a wry grin and said, "Alright, sir."

They worked on individual tactics and hid their tanks while the Hoffman fire and sirens from Third and Second Platoons echoed through the training area. After Second Platoon was done, First Platoon went into defensive position and Third Platoon attacked. Sergeant Robb was very good at hiding his tanks and picked off several of the Third Platoon tanks before they made it into the objective area that First Platoon was defending.

The Lieutenant's tank was knocked out at the beginning of that engagement, so he was stuck in the turret next to the flashing lights and wailing siren that indicated his demise, until the exercise was concluded. After Ramsey's platoon won the engagement, Batson told the platoon leaders to get with their platoons and prepare for a new engagement with Third Platoon on an objective about three kilometers away, with First Platoon to attack and try to take that objective. Third platoon would also be trying to attack to take First Platoon's position again.

THE PLAN

The Lieutenant called the tank commanders and Sergeant Robb over, and asked him, "What would you do, since you've been here for years?" Sergeant Robb said "Well, what he is expecting you to do is to put five tanks down here and try to pick him up as he comes through, and then he'll go around one side or the other of you and annihilate you. What I would do is," he paused, "there's a little road, a spider trail as we call them, that's right on the edge of the engagement area. Take the heavy section and run up there as fast as you can. I'll hold down the fort here along with 1-5, moving from hill to hill, hiding, getting shot at, making him believe that the entire platoon is here, while you go around and after getting up to the objective turn your tanks around and shoot all of his tanks from behind, in the grill doors."

The Lieutenant gave that a moment's thought, looked over to Sergeant Webb, the tank commander for 1-2, and his best lead tank, and asked, "Did you understand that?"

Sergeant Webb answered, "Yes, sir. We'll execute."

"Good. Sergeant Aguinaldo, keep 1-3 close behind me and guard the rear as we run for it," the Lieutenant directed.

"Yes, sir," the Carped Guamanian staff sergeant tank commander replied.

As the light section 1-4 and 1-5 moved over and took up their positions to await the attack, the Lieutenant yelled over to Sergeant Robb, "Take care of yourself. I'm starting to like you."

With that, the heavy section's three tanks went hell for leather as fast as they could up the side of the engagement area to get back behind Sergeant Ramsey. Once there, the Lieutenant called the Company Commander and happily reported, "First Platoon is on the objective; we have it secured."

Captain Batson questioned him immediately, "Are you sure? We are only five minutes into the exercise."

The Lieutenant replied, "Correct, we have it secured."

Hearing that, Sergeant Ramsey tried to turn his tanks around, but the three tanks of the heavy section shot three of them, and Sergeant Robb brought his two tanks up through them and cleaned up what was left. With that, the exercise ended earlier than anticipated, but since it looked like it was getting ready to rain, the Company Commander decided to let everybody go in early, take care of maintenance on the vehicles, and put them down for the night. As usual, Third Platoon, with Second Platoon following, took the lead. First Platoon brought up the rear with 1-1 in the middle, but unfortunately, with all of the exercise in the sand and the lack of maintenance, it had earlier thrown a track. Since his tank was broken down, the Lieutenant told Sergeant Robb to take the platoon back to the motor pool and that he'd get the vehicle up and back as soon as he possibly could.

As the rest of his platoon drove past him, the Lieutenant climbed off his tank to join his gunner who had started the process of breaking the track apart to lay it out so the vehicle could be maneuvered back up onto it and then get the track reconnected and tightened. The loader had opened the sponson storage boxes on the side of the tank and was getting out the tools necessary to do this. The Lieutenant walked over to the gunner and asked him what he needed, so he could find it and bring it to him. The gunner looked at him suspiciously and said, "Sir, I've never had an officer do that."

"Say again?" asked the Lieutenant.

"I've never had an officer work with the enlisted on a vehicle -- ever."

The Lieutenant looked at him and said, "Well, I know that you're better than I am at breaking track, and officers normally are going to meetings and drafting plans and such, but I haven't got a meeting to go to and I haven't got anything here to plan, so the highest and best use of my time right now is to get this tank up and running, off deadline, and prepared to go to war if it has to. In addition to that, I really hate getting wet and it looks like it's going to rain. So, whatever you need, let me know. I'll hand it to you or help you and we'll get this thing back and running together."

After about thirty minutes or so, they had the track prepared so that the driver could move the tank forward and then backward to get the road wheels up over the center guides. They then tightened up the track and drove the tank forward and backward to make sure it was secure. After that, the Lieutenant climbed back up to the tank commander's hatch and they drove the tank as quickly as they could back to the motor pool area to join the rest of the platoon, get the vehicle washed, checked for any problems, and cleaned out.

Once they were back in the platoon area of the motor pool, and his tank was parked, the Lieutenant, Sergeant Robb, and the tank commanders began the process of uninstalling the heavy weapons, taking them back to the arms room, and turning them into the armorer and supply sergeant. While this was going on, the crews cleaned off the tanks with hoses, as best they could, policed all of the trash out of the turrets, and locked them up. When that was done, Sergeant Robb assembled the platoon and told them that they were to go down to the platoon room in the basement of the barracks for a short meeting. The Lieutenant had Blaylock toss him his A-bag. He strapped it on and followed Sergeant Robb into the room which was about ten feet by twenty feet with a small locker, a table, a desk, and a couple of chairs. This gave him the opportunity to shadow Sergeant Robb as he explained what they would be doing for the next few days, getting the vehicles maintained, thoroughly cleaned, inside and out, and their field gear ready to go out on the Reforger exercise in about a week. Sergeant Robb then released the platoon to barracks to clean their gear.

NO ROOM AT THE INN

From talking to the First Sergeant earlier, Sergeant Robb knew that the Lieutenant was going to get a couple of night's sleep and would be leaving for the border on a reconnaissance after a day and a half.

Once the soldiers had left the room, he asked if the Lieutenant had anything that he needed handled while he was away. The Lieutenant replied, "Nothing I can think of right now, other than finding a place to stay while I am waiting for permanent quarters. I am going to be evicted from the visiting officers' quarters tomorrow. Why can't I just stay down here in the platoon room? I've been in the field for almost a year. I can lay my poncho liner and sleeping bag across this table, keep my stuff locked up in the platoon locker, and go to Battalion Headquarters to get a shower after PT. Maybe, by the time I get back from Reforger, I'll have a place to live."

Sergeant Robb looked at him very sharply, motioned him to come closer and made it plain, "Sir, you can't stay in here."

When the Lieutenant asked, "Why not?" Sergeant Robb replied, "They'll kill you."

"Who'll kill me?" asked the Lieutenant.

"The soldiers in this barracks will kill you if you stay in here overnight. Commissioned officers are not allowed in the barracks during hours of darkness. Commissioned officers are not allowed to know what goes on in here after hours," explained Sergeant Robb.

In response, the Lieutenant looked at him and said, "You've got to be kidding me."

"No sir, I'm not kidding you. It's better than it was the last tour I had here, when commissioned officers weren't allowed in the barracks at all and the drug dealers kept vicious dogs, Germans shepherds and Alsatians, to make sure that they couldn't come in."

The Lieutenant looked at him again quizzically and asked, "So you mean to tell me that I'm in a battalion where officers cannot go into the billets of their enlisted men in hours of darkness under any circumstance?"

"That's right, sir," said Sergeant Rob.

The Lieutenant then pressed further, "What do the duty officers do?"

"Sir, they come in early in the evening to make the first check, and they wait until dawn to make their second check of the arms rooms, so there is one before midnight and one after midnight. Either that or they simply falsify the checks, and they back date them into hours of darkness and come in the morning and do them before they go off duty."

The Lieutenant looked at him closely and then said, "I'll cross that bridge when I come to it."

Sergeant Robb thought a second and replied, "Well, sir, at least officers are allowed to be armed now. In the tour before this, officers were not allowed to come in at any time armed – period. Now the duty officer can carry a forty-five."

"Really?"

"Yes, sir. A few months before you got here a new lieutenant from the 2-13 infantry tried to break up a drug deal in the 3-8 Cav. He was the Brigade duty officer that night. The Cav troops did not recognize him, grabbed him, stuffed him in a wall locker, took it up to the rec room on the fourth floor of the billets, turned it upside down and dropped it. The lieutenant's neck was broken; he was killed immediately upon contact with the ground."

The Lieutenant looked at him in shock and demanded, "What happened, was anybody prosecuted?"

"No sir, there were too many fingerprints on the wall locker to determine who, if anybody, was responsible," said Robb. "And the condolence letter that went back to his parents explained that their son was regrettably killed during a training exercise involving an armored vehicle."

Again, the Lieutenant looked at him incredulously and said, "I suppose a wall locker made of steel is an armored vehicle, correct?"

"Yes, sir, under those circumstances, that's what it was deemed."

"So," continued the Lieutenant, "there was no prosecution. It was covered up and his body was shipped back to Dover as soon as possible."

"That's what it is, sir," said Robb.

"Damn," responded the Lieutenant. He was still in shock, but took out a piece of red licorice, ate it to give himself some energy, and proceeded to hump his A-bag out of the training room. He asked Sergeant Robb, who was walking beside him, when morning formation would be for PT and was told it would be at 0600.

Shaking his head, the Lieutenant left the barracks and walked over to the tailor and dry cleaners where he retrieved his field jacket, field shirt and pants, and turned in his dirty field jacket, shirt, and pants to have them cleaned and the unit insignia and rank sewn on. He carried his A-Bag and clean clothes back to his VOQ room to sort out his TA-50 field gear. He then went down to the desk and said to the clerk, "As I understand it, I'm going to be evicted from these quarters tomorrow morning, is that right?"

The clerk quickly answered, "Yes, sir, at 0800."

"All right, do you have any suggestions on where I could spend the next couple of nights, because after that, I'm going to the border and then I'll be going to Reforger."

"No, sir, that is your problem."

"Any idea about where I might store my stuff while I'm in the field?"

The clerk looked at him and said, "Well, there is storage on the fourth floor of the BOQ buildings for luggage and personal items. There are cages up there so you could put your stuff up there, lock it up, and maybe it'll be okay." The Lieutenant turned and went across the street to the BOQ, walked up four flights of stairs, and found the storage areas: multiple chain linked sections, each with a door and hasp for a lock.

Since the bank was closed and he still didn't have any marks, he decided that he would get a shower in his room and walk to the officers club for a real meal. After dinner, he came back to his room and set the alarm for 0530. Then he organized his things so that he could move his footlocker, field gear, and suitcase up to storage the next morning before he was evicted by his new friend at the front desk. Finally, he stretched out on a poncho liner draped over the bed to take stock of his situation. Oddly enough, in many ways it was much like his first night in barracks at VMI – which was strangely comforting.

Before he went to sleep, he stopped and thought about his girl. She was the one ray of hope on his horizon. To him, she was like the great lady described in the *Wild Geese.* She had more fire, generosity and compassion, and simply, more woman packed into her small frame than in any three women he'd ever met, and he loved her dearly. She was singularly self-sufficient, like him, and fiercely independent. In retrospect, years later, the movie character she most resembled when they were dating was Lana, from the movie, *Risky Business*: a five-foot, inch and a quarter, slip of a girl, with long, thick chestnut red hair, bright jade-green eyes, a comforting alto voice, and a wicked sense of humor. She was determined to live a full life and enjoy every minute of it. Like him, she was audacious, courageous, and sometimes even a little reckless. A cavalryman's woman, she could be depended upon to ride to the sound of the guns.

The next morning, as soon as the alarm clattered, he got out of bed, shaved, took a quick shower, loaded up all of his gear, and turned the key into the front desk. He then humped everything across the street, up four flights of stairs, locked it into the storage room with the padlock off his footlocker, changed into his PT gear, and went to the company. After calisthenics, and a short run led by the master gunner, he went up to Battalion to wash his face and shave with soap and a razor that he had in the pocket of his PT jacket. Then he went back down for breakfast. When he was through, he changed into his duty uniform and headed back to the Battalion. As he walked across the company area, the First Sergeant called out to him, "The old man wants to see you, sir."

So, he ran downstairs to see Batson, who had him take a seat. He told him he was happily surprised at his abilities in the field and that he needed to prepare to go with him and his driver to their border position the next morning. The Lieutenant had read in the SOP about their position, which was on the West German border near Sondheim, in the

Thüringen Wald. He said, "Very well, sir." Batson nodded and told him to go to the S-3 shop and pick up a 1 to 250 map sheet to put it in his case, so he could follow along as they drove up to the border. All the officers of the Battalion were going to make this trip, as they had to do it at least once quarterly. This would be Captain Batson's last trip as a Company Commander; he was looking forward to it and to briefing the Lieutenant on his responsibilities in case of combat. Then Batson released him for training with his platoon and to review the route and the Battalion's positions with the S-3 if he could, so that he would be ready at 0600 the next morning to go to their first battle position in sight of the Iron Curtain.

ADDITIONAL DUTY

The Lieutenant went up to Battalion, picked up his map sheet from the S-3 NCO, and walked down to the motor pool where the platoon was inspecting their tanks and repairing loose and broken items as they found them. Unlike the rest of the Battalion, they could move their vehicles and order parts without a written authorization from the Battalion executive officer. He observed his soldiers, many of them had a dull, vacant look, like a small flame was burning in the back of their eyes. He had seen this before in the desert at Fort Bliss, so he thought for a moment, then walked up to Sergeant Robb, away from the men and said, "It looks like most of the troops here are on heroin or hash or both."

"That's right, sir," Sergeant Robb replied.

"How do you manage to function?"

"Sir, I keep my platoon fed, clothed, and trained as best I can, and equipped to go to war if they have to."

"All right," said the Lieutenant, "and so we do the best we can?"

"Yes, sir, but I, and some of the other NCOs that have been here multiple times, have an additional duty."

"What's that?"

"We keep the good officers alive."

"I take it, at least so far, I'm a good officer?"

"So far, sir."

"Well, thank you, sergeant, I'll see you tomorrow at morning before I take off for the border with the old man. I will be out the rest of the day. I have to see the S-3, get some marks, and get my gear ready for the trip."

Sergeant Robb said, "Don't bother with the S-3, sir, he and the colonel are already up on the border doing a recon with the Brigade staff."

"All right – thanks for the head's up."

The Lieutenant turned and walked out of the motor pool to the tailor shop where he retrieved the uniforms that he had left there earlier. Then he carried them up to the fourth floor of the BOQ and locked them in the storage room with the rest of his gear. When he got back down to ground level, he decided to walk over to the American Express bank. There he waited almost an hour in line so that he could write himself a check for $100 and exchange those dollars for German marks. By that time, it was almost mid-afternoon, so he went to the PX and bought a couple of sandwiches for lunch and dinner, and another bag of red licorice for the road.

W that was done, he went back to the BOQ, ate a sandwich, and separated out his gear so that he had an A-bag to take with him to the border. As usual, he kept his VMI class ring, a faceted blue sapphire in ten carat gold, in the right upper breast pocket of his uniform. That way it was secure but wouldn't get caught in the machinery in the turret of the tank or other equipment that he was around. He looked closely at it – a souvenir of happier times and smiled to himself because the sparkling blue stone reminded him of the Virginia flag that could be seen through the old panes of glass from the stage in the cadet chapel. Then he felt saddened because it also remined him of his last concert with the Glee Club on that stage and the families of the five cadets who had died in the weeks before it, who had sat on the front row.

When his A-bag was packed, he laid out his poncho on the concrete floor, with his sleeping bag on top, and a poncho liner covering it. He used his field jacket liner wrapped in another poncho liner for a pillow. He neatly folded the 1:250 map of West Germany and put it in the small map case that he had used earlier in the field exercise. He then took a grease pencil and traced the route of march he would be traveling the next day with the Company Commander. That being done, he headed down to the basement of the BOQ and bought a German Fanta from the vending machine, walked back up, and had the second sandwich for supper. He set his travel alarm clock for 0500 and went to sleep exhausted.

Chapter 4
A Borderland Holiday

At 0530 the next day, the Lieutenant passed the gate guards and walked into the Bravo Company area carrying his A-bag. He proceeded downstairs to get a cup of coffee in the orderly room and ate a couple of pieces of licorice that he had in his pocket. He visited briefly with Sergeant Robb, who was preparing to lead the platoon in PT. When Robb was gone, he sat in the corner and waited for the Company Commander. Captain Batson left his office five minutes later, stopped at the entrance to the orderly room, and told the Lieutenant to follow him.

They met up with Batson's Jeep driver, Specialist Arnold. He was a big country boy, who was also an excellent Jeep mechanic, and this was what assured him his job as Jeep driver for the old man. They climbed into the Jeep with the Lieutenant in the back with his A-bag and the other two A-bags, and his 1:250 map sheet of West Germany from the Rhine to the border. As they drove out of the company area, the captain gave him a running commentary on the route; what to look for and where they would be going, as well as what his tank platoon would need to be doing as they passed the various checkpoints.

Captain Batson also had a 1:50 map of the entire route that was folded up so that he could follow along. As they moved through the villages, he would point out buildings to the Lieutenant to show him that the exact building that they saw in front of them was in fact set out on that map sheet in excruciating detail. Batson was very knowledgeable about the Battalion's positions and mission. Unlike the Lieutenant, who had already been assigned to two infantry battalions, an armored cavalry regiment, and two tank battalions since he had completed advanced camp at Bragg three years earlier, Batson had spent almost five years in the same battalion, as a tank platoon leader, scout platoon leader, S-3 Air, tank company XO, and now a commander. He was definitely on the fast track. As they approached the border, the country became more open with small villages sprinkled across the countryside. They stopped around noon for lunch at a meat market with a bakery next door where they got some cold salami, fresh bread, cheese, and mustard in a tube, like toothpaste. The Lieutenant sat on the hood of the Jeep with Specialist Arnold and ate quickly. Batson ate in the Jeep, resting a little, engrossed in his own thoughts, and planning his next assignment.

After lunch, they proceeded to B Company's battle position, on the high ground overlooking the village of Sondheim, eight kilometers from the border. Once Batson showed him where First Platoon was to take up their positions and the sectors that they were to fire in, they drove down into the valley, up to the border, and the Lieutenant saw the scar for the first time. The "Iron Curtain" was, in fact, made of iron. There were three separate fences with minefields in between, guard towers with machine guns, and mortars to cover the dead space. The fences were fascinating in that if someone tried to climb them, they were designed to collapse around their hands and feet so that they'd be captured by them before being machine gunned to death. There were also dogs and border guards who walked specific paths through the minefields, past the gates, and along the fences.

The Lieutenant had never seen anything so chilling, and it cut him to the quick to know that he had family on the other side, who were enslaved by the Russians and East Germans even as he stood there. Once they had taken a look at the company's position from the viewpoint of the enemy, they proceeded to the company's first fall-back position. His platoon, or what was left of it, would be dug in on the side of a hill overlooking the road. The Lieutenant walked his position with the captain as Batson continued pointing out the places on the road where, "you need to be killing them here and here and here." They finally fell back to the second fallback position, and he was shown on the map his other target zone. There were no more planned fallback positions; it was assumed that it would simply be a come as you are war after that.

The other officers from the Battalion, the scout platoon leader, support platoon leader, S-3 Air, all lieutenants, and the enlisted platoon leaders, were also reconnoitering their positions. At 1630, they proceeded as a group to Bad Kissingen and Manteuffel Kaserne, named for a Prussian Field Marshal in the Franco-Prussian War, who was a distant relative of the Lieutenant. There the 2-11 Cav, the famous Black Horse in which Lieutenant Colonel Shea, known as, "Wild Bill," and his old boss, "Uncle Chuck," who had served in Vietnam, had its headquarters. The S-2 and Assistant S-3 briefed them, and they met the officers who were available to get to know them, at least a little. When he introduced himself, each officer came up short when he saw his name tag but thankfully did not ask the almost inevitable question about his German family.

By 1830 hours, the sun started to descend back behind the mountains to the west and Batson told Arnold to drive to the guesthouse Maria with which he was familiar. This was a small hotel operated by Karl-Heinz and his wife where the battalion officers had an account and stayed when they were on border reconnaissance. It was a very nice, three-story building, with a garage on the first floor, restaurant on the second, and guest rooms, with flower boxes filled with bright red and pink geraniums in the windows, on the third.

When they entered the restaurant, the kitchen smelled wonderful. The Lieutenant and Batson went up to the room that they shared and laid out their shaving gear, toiletries, and uniforms for the next day. Then they went down to the dining room, where the battalion officers were assembled across from each other at one large table, which was created by putting all the smaller tables together. The drivers sat together at their own table in the corner nearest the kitchen. Everyone ordered dinner. Based upon Batson's advice, the Lieutenant ordered a quarter of a goose, French fries (which were delicious), cold salad, and a beer, Kulmbacher Mönchshof, that was as smooth as milk.

THE BOOT, THE BEER, AND THE PANZER LIED

Once they were finished eating dinner, the battalion commander and staff, and the Company Commanders, decided that they would in engage in a game called salute the boot, which was a German drinking game, and asked Karl-Heinz to bring out the boot. He came out from behind the bar with what looked like a cavalry boot made of pure crystal. The game involved a specific series of stomps, slaps, and a pounding of the table used to salute the boot. Then every man had to drink some from the boot and pass it to the next man. The key was to drink from the boot without having the boot wash over you, because it formed a bubble in the toe, and if you were new to the battalion you might try to drink it straight up. When the bubble burst, you would be washed with beer. If that happened, you had to buy the boot of beer. Otherwise, the boot continued on and on around the table to whomever finished the boot, then the person that had the boot before him had to buy the whole boot full of beer.

The Lieutenant stayed for the full three rounds of saluting the boot and never had to buy one. Then all of the battalion officers began singing the *Panzer Lied*, the tank song from the Second World War, in German:

"Ob's stürmt oder schneit,
Ob die Sonne uns lacht,

Der Tag glühend heiß
Oder eiskalt die Nacht,
Verstaubt sind die Gesichter,
Doch froh ist unser Sinn,
Ja unser Sinn.
Es braust unser Panzer
Im Sturmwind dahin."

To the Lieutenant, the irony was palpable. His distant relative, a German general, commanded the 5th Panzer Army at Bastogne during the Battle of the Bulge, and his soldiers sang the song there and in North Africa at Kaserine Pass. Even stranger, both of those times that relative faced General Patton, who also went to VMI. *It's a small world*, the Lieutenant thought, *and some things never change.*

Once they were through singing, and the echoes of more than thirty sets of combat boots stomping the wooden floor had died down, the officers and their drivers, who had participated in their own "boot" game, made their way across the room to the door. Then they slowly climbed the stairs and walked carefully to their rooms where they prepared to go to sleep. Looking down the hall, the Lieutenant noticed that Arnold, who had been drinking pretty heavily and had narrowly missed having to buy a boot on more than one occasion, looked a little green as he went to the room that he shared with two other drivers.

THE BOOT CLAIMS A VICTIM

The next morning, Arnold was obviously feeling very ill. The Lieutenant asked him, "You feeling okay?"

Arnold answered, slowly and painfully, "Not really, sir. I'm pretty sick from drinking last night."

"And why is that?" asked the Lieutenant.

"I drank that boot every chance I could; I didn't want to quit. I didn't have any money with me to buy a new one, so I had to finish the whole thing."

"Very well," replied the Lieutenant, "You've learned a valuable lesson and at least you've recovered enough to drive today, haven't you?"

"Yes, sir, I have."

"Great," said the Lieutenant, "maybe some coffee and juice will clear your head and settle your stomach."

THE WOODCUTTER

They all assembled back in the dining room and ate a quick German breakfast: poached eggs, cold cuts, cheese, orange juice, fresh rolls, and strong black coffee. After that, they loaded up in their Jeeps and prepared to drive back to Mannheim. Along the way, Batson had Arnold pull over at a small house dug into the west side of a hill. He explained to the Lieutenant that "This is where battalion will be initially. We'll go in and see Herr Eyring, the woodcutter. He's a friend of mine. I like to stop by and pick up a trinket here occasionally because he does marvelous wood carvings. Since I'm going back to the States, I thought I'd pick up a couple of things for my kids before I make my last trip back to the battalion."

They opened an ornate wooden door and walked into a beautiful workshop. Batson continued, "The owner goes out into the woodlands, cuts his own timber, and then carves animals and human figures by hand. He is amazing." Batson introduced the Lieutenant and they visited with Herr Eyring for a little bit. Batson purchased a couple of small wooden carvings of dogs for his children, then they said goodbye, and saddled up in the Jeep. Arnold, who was feeling better, put it in gear and began driving back.

Along the way, every ten or fifteen minutes, Batson would point out a tree, a bush, or a building. He would hand the map to the Lieutenant and ask him to find it on the map sheet. The Lieutenant was fascinated because everything was in fact on that map. This was very unlike the 1:50 maps you would see in the United States that had generic trees and forests; these maps were detailed right down to each individual tree or bush when you were in any kind of open terrain. They decided to skip lunch and Batson finally laid back to rest when they hit the main blacktop highway. The Lieutenant took the opportunity to rest a little as well, playing Pablo Cruise's, *Worlds Away*, in his head as the road markers flashed by.

HOME AGAIN

A little after 1700, the three of them pulled into the company area. The Lieutenant grabbed his A-bag and dismounted. Batson told him to be sure and come in early the following morning because he would need to go up to the S-3 shop and draw a complete set of maps for the upcoming Reforger exercise. Batson also directed the Lieutenant to connect with Sergeant Robb and make sure he had a map case large enough to handle those maps, go over the needs of the platoon, and go about the business of putting the map sheets together.

The Lieutenant saw Sergeant Robb and went over to him. First, he asked about where he would get a map case large enough for Reforger maps. Sergeant Robb told him he could find something workable in the PX, and that he would need to get over there pretty quickly to buy one. The Lieutenant then told Sergeant Robb he would see him at morning PT and after breakfast, they would sit down and go over the tasks necessary to get the platoon ready to go to the field.

Chapter 5
Brother Rat to the Rescue

Having coordinated for the next day, the Lieutenant strapped on his A-bag, walked out the gate, and then past the traffic circle to the BOQ where his other items were stored. He climbed up the four flights to the storage area and was disgusted. Someone had broken into the storage area, destroyed the lock, rifled through his belongings, and smashed the footlocker that he had since he was a Rat at VMI. Nothing was taken since his valuables were with him in his A-bag. His B-bag was still there with some uniforms, and his suitcase was opened but nothing had been stolen.

He decided that he would get something to eat and maybe buy a new lock and map case before attempting to get another night's sleep on the hard floor. He headed across the parking lot to the PX and purchased the map case and a stronger lock with a key. As he was coming back around the BOQ, he saw an individual who looked familiar to him from the rear. The Lieutenant ran up behind him and said, "Hey, Love Boat, is that you?"

The soldier turned around and he knew immediately he was looking at Petrovich, who was from the same class at VMI, and therefore his Brother Rat. "Love Boat" was his VMI nickname; the

Lieutenant's was "Devil Man," a play on his last name. Petrovich recognized him immediately and asked him when he had arrived in country. The Lieutenant answered, "About three days ago," and asked if Petrovich would let him stay on his couch for a few weeks while he was on Reforger, as he was supposed to get quarters when he got back.

Petrovich happily declared, "Sure, come with me," and they headed into his quarters. It was a one-bedroom apartment on the first floor. Petrovich gave him an extra key, so the Lieutenant went back to the storage area, packed everything in his A and B bags, and carried them, with his suitcase, back to his new "room," leaving the busted footlocker behind. He set the bags down next to the couch and they left to go to the officer's club to have roast beef, vegetables, and pie, as it was "Beef and Burgundy" night at the club. Both avoided the Burgundy, since they had to get up early, without headaches.

After dinner, they went back to the BOQ room and talked about all of their assignments after graduation from VMI. Petrovich had gone to basic course at Fort Gordon, Georgia, and then was shipped to Germany, arriving on station about a year and a half before the Lieutenant. Petrovich was a signal officer and was leaving the next morning for about two weeks to set up communication equipment and maintain communication sites for the upcoming Reforger. The Lieutenant told Petrovich that he expected to depart Monday at 0400 and would get back about three weeks later, that way he should only have to impose on his Brother Rat for two or three days at most, before the Lieutenant returned and was assigned his own room.

The two of them went to sleep early since they had to get up well before dawn the next morning. The Lieutenant set his alarm clock for 0500. When it went off, he realized that his classmate had already left. The Lieutenant dressed in gym gear and tennis shoes to go to morning PT and upon arrival, formed up behind his platoon. After a two-mile run and some calisthenics, he coordinated with Sergeant Robb to meet at 0800 to plan the day. He also gave Sergeant Robb his BOQ room number and told him to pass it on to the First Sergeant so the company could find him if he was needed. Then the Lieutenant went back to his room, shaved, showered, and had some licorice for breakfast.

GETTING READY TO ROLL

Upon his arrival back in the company area, he saw Sergeant Robb, who told the Lieutenant that he needed to get up to Sergeant Peeler in the "3" shop to get the full map set and that he would probably have to spend most of that day, and maybe part of Saturday, trying to put all the maps together. Sergeant Robb had scheduled an inspection of the vehicles, equipment, and soldiers for 1000, and after the inspection was complete, he planned to release the men to do final preparations before the alert on Monday and have some weekend time with their families.

With that, the Lieutenant went up to headquarters and saw the operations sergeant, who gave him a huge stack of 1 to 50,000 map sheets. He asked if the colonel and S-3 were in and was told that they would be at brigade in meetings to coordinate their control/umpire duties with other officers from the Battalion who would be observers. He took the maps back to the BOQ room and started laying them out. He soon realized that they would take up the entire living room, and he was also going to need scissors and a lot of tape. He looked around the room and found scissors and tape in the drawer of a small table in the corner. He said a quick prayer, thanking his Brother Rat for leaving them with him, and laid out the first couple of maps.

He got back to the company area about 0930. He saw Sergeant Robb in the orderly room, and the two of them walked over to the motor pool to check the tanks and their crews. They inspected all the vehicles and the men to make sure that everything was ready to roll on Monday morning. When the inspection was done, he asked Sergeant Robb to call the tank commanders and crewmen together. He let them know that they were released for personal business and final preparations, but that they would need to be ready at 0400 on Monday, two days away.

It was Friday and he had nothing pressing, so he went down to the orderly room and asked the First Sergeant if the Company Commander was around or had any instructions for him.

"None, other than to be ready at 0400 Monday, sir," said the First Sergeant.

The Lieutenant breathed a sigh of relief and realized that he finally had some time to get ready to go to the field again. First, he went to the PX and bought some tape to replace the tape he used putting his maps together, then he went back to his room, laid out his maps, and took inventory of his clothes, to see what he had that was clean and what was dirty. He then took the dirty clothes over to the laundry and made arrangements to pick them up on Saturday, the following morning. He headed back to his room and then began the laborious process of laying map sheet to map sheet putting them together like a giant jigsaw puzzle and then trying to tape them together so that the lines on the maps flowed continuously from one to the other.

After a couple of hours, and a chicken sandwich and a Coke from the PX, he came back to the maps, took a deep breath, and decided he would take a break from assembling the maps to write his girl. He took a piece of stationery and envelope out of the attaché case that he had been carrying in his A-bag, and penned her a short note to let her know that he'd arrived in Germany, been in the field, and was getting ready to go to the field again, but he was safe and sound and had a great platoon sergeant. He gave her the APO Box number to send him mail, and the designation of his unit so any letter that she sent him would come to him. He also told her that he would send her information on a wedding date as soon as he was able to get with the Battalion commander and operations officer to check the training calendar.

By about seven in the evening, he had two thirds of the maps put together and figured that he would get the rest together by mid-morning the following day if he started early. Then he went over to the PX and purchased a few TV dinners, a couple of frozen pizzas, some canned soup and Carnation breakfast bars, as well as several bags of red licorice. Finally, he watched a little TV on the Armed Forces Network and turned in to get a good night's sleep.

The next morning, he got up, made instant Maxwell House black coffee and had it with a breakfast bar. By about 1030, he had his maps together and folded them all up to try to get them into what was essentially a huge plastic bag, almost as big as his tank turret, that was his Reforger map case. Fortunately, it looked like it would double as a waterproof blanket. He had no idea where they were railroading into, so he folded up the maps as best he could, put them in the map case, and laid out his uniforms for the field. Then he remembered that he needed to pick up his laundry, so he went out the door and down about a block to get it and the uniforms that he had left to have his name tags sewn on.

He brought everything back to the room and began the process of packing out. By 1300, he was pretty much finished and decided that he might as well explore the area a little. There were small streetcars called *strasse*, trains that ran into downtown Mannheim and another set that went out to the suburbs. He decided to take some marks, buy a train ticket, and try to go to downtown Mannheim, as there were supposed to be some beautiful parks and German stores where he could buy some local food items. The trains came together at the station right outside of his unit. Seeing a train coming in from the suburbs headed toward downtown Mannheim, he made the assumption that he could hop on that train and would go to the city. Unfortunately, as it turned out, that train actually traveled to Käfertal, where he was stationed, from the suburbs and would return to them. Consequently, he had a ticket to go the wrong direction.

A CHILD OF THE BLOOD

The electric train pulled almost silently out of the station at a fast clip. After two hundred yards, a tall, slender, older German came through the car of the train asking for tickets. The Lieutenant showed him his ticket and the German became upset. Then he started to yell at him to show him his identification, apparently because it looked like he was trying to steal a ride by buying a ticket for the wrong direction. The Lieutenant calmly reached into his wallet, pulled out his identification card, holding it upside down and gave it to the German, who flipped it over and saw his last name. Upon seeing the name, he immediately came to the rigid position of attention described in Frederick Forsythe's novel, *The Odessa File*, that was not like a normal German soldier, but the SS position where the fingers of the hands are pointed directly down, and the individual is ramrod straight. The Lieutenant was shocked at this and asked him if everything was all right. He just kept saying "Entschuldigung, entschuldigung" or "excuse me, excuse me."

Undoubtedly, he recognized the last name of the Lieutenant's relative, who was a four-star general of the SS and well known in Germany as a national hero. The Lieutenant asked for his identification card back, and the German gave it to him immediately. The former soldier then asked him if he was a child of the blood. The Lieutenant said, "Yes," as he was of German descent, he was a child of the blood, and he was a "von Manteuffel." The old soldier was amazed and then the Lieutenant added, "I am an officer."

The German asked, "An officer?"

"Yes, a First Lieutenant of the Panzer and cavalry." At this, the German practically burst into tears. The Lieutenant gave him all of his tickets and told him he certainly didn't mean to cheat the German people or the railway, so he would buy new tickets when he got to the end of the line. The soldier tried not to take the tickets, but the Lieutenant insisted and when the train came to a stop, he got off at what would soon become one of his favorite places: Viernheim.

IN THE COMPANY OF A FELLOW TRAVELER

Viernheim was a small village northeast of Mannheim. It had a shopping center with a large department store and several local eating establishments. The Lieutenant walked to the department store and decided that he would buy a German long sleeve brown shirt, like students would wear and some nondescript brown, plastic soled suede shoes that he noticed Germans of his age wore as well. Once he had those items in a bag, he walked back into downtown Viernheim to the train station and asked the ticket agent if there was a place to eat nearby. He was directed to what would become his favorite restaurant in the world, The Threshing Floor, or Nick's, operated by a man of that name, from Greece. The Lieutenant would discover that Germany was populated by numerous Greeks and many Italians who had fled their home countries to make a better living in Germany. After saving their money for several years, they then planned to retire back in either Greece or Italy or wherever else they came from, after they had enough money accumulated to purchase a house.

Nick was fantastic. He made Greek and German food that he fired on a grill with Greek spices and served it with French fries and cold salad. It was wonderful. He was also a tremendous host and wanted to know everything about where the Lieutenant had been and how he was getting used to the area. Having enjoyed a great meal, the Lieutenant told him that he was going to get married, bring his wife back, and when he brought her back, the first place he was going to bring her to eat would be Nick's. Nick was beaming as he was newly married himself and hoped to have a child soon. Saying goodbye to Nick, the Lieutenant went to the train station and purchased new tickets to head back to his unit area. After he made it back to his room, he took a shower and promptly slept for half a day.

He had already decided that he would be taking Sunday off, other than packing out and preparing for the 0300 alarm that would send him on his way to Reforger, which he assumed was going to last from two and a half to four weeks. He had no idea about any of the particulars since he had no orders or overlays for movement. When he got up, he ate a breakfast bar and assessed his situation. It looked like he had enough clean fatigues to change clothes about every four days in the field, two sets of field pants and field shirts, a field jacket with liner, a rubber duck, a poncho liner, and sleeping bag, which would make up the bulk of his kit.

He put the sleeping bag, the rubber duck, field jacket, field jacket liner, rubber duck, poncho liner and his extra food, red licorice, and breakfast bars in a water proof bag in his A-bag and he put extra changes of clothes, his overshoes and any other items that he did not expect he would need right away in the second water proof bag in his B-bag. After that, he made a TV dinner for his evening meal, watched a little Armed Forces Network television, and went to sleep at 1930 as he knew 0300 was going to roll around really fast.

Chapter 6
The Load Out

The next morning at 0300, the alarm clattered, the Lieutenant woke up, shaved, took a shower, checked and finished packing his A and B bags with shaving gear and an extra poncho liner. Then he shoved a breakfast bar into his mouth, drank a couple of cups of instant black coffee, and walked to the company area. Once he was there, he walked to the motor pool and gave his gunner his A and B bags to put in the bustle rack of 1-1. He headed to the arms room and met Sergeant Robb there around 0400 with the company in full alert mode. They signed for the small machine guns, grease guns, and .45 caliber pistols, and took them up to the tanks and then went back down for the heavier M-85 machine guns.

Once the machine guns were installed in the vehicles, the crews and tank commanders began rotating to the mess hall for breakfast. As before, the Lieutenant and Sergeant Robb alternated for breakfast as well, and once everyone was fed, they checked the vehicles to make sure that all the systems were operating. At 0500, Captain Batson came through in his Jeep and issued CEOIs with the radio frequencies and call signs for the day and every other day that they expected to be in the field. He also asked the Lieutenant if he had his maps put together in his case. The Lieutenant answered, "Yes, sir!" Satisfied, the captain drove away. Once all of the tanks had the correct radio frequencies and call signs for the day, the Lieutenant did a quick commo check with all of his tank commanders, who answered in turn, plus Sergeant Fields on 6-6, the commander's tank, that would be following 1-5.

By 0600, all the vehicles were ready, but as they prepared to pull out, the Company Commander pulled up and gave the company the order over the radio to stand fast. He had just been issued fire resistant NOMEX flight suits by the brigade and all armor crewmen were required to wear them for the entire exercise. Consequently, the Lieutenant and all of the other tank crewmen had to strip down to their underwear on top of their tanks, across the street from the housing area, and then get dressed again. The Lieutenant took his wallet, note pad, pen, and VMI ring out of the front pockets of his fatigues and transferred them to the pockets of the flight suit as best he could, crammed the fatigues into his B-bag and hopped back in the tank commander's seat. The Supply Sergeant had loaded up the extra small arms and machine gun ammunition, and other items to be taken to the field. The mess trailer and truck were sent over from Headquarters Company, with a driver and a cook, to accompany B Company throughout the exercise and prepare hot meals when they could.

After fifteen minutes, the company began road marching to Coleman Barracks. Third platoon was in the lead, followed by Second Platoon, with First Platoon bringing up the rear. The headquarters' vehicles, executive officer's Jeep, First Sergeant's Jeep, supply and mess trucks, and maintenance were sandwiched in between Second Platoon and First Platoon. Once the vehicles arrived at Coleman, they pulled up by the railhead, and the lead tanks began the process of sliding up a ramp onto railcars so that they could be fastened down with cables.

Once his platoon had pulled up behind the last tank of Second Platoon, Sergeant Robb came over to the Lieutenant and asked him to step around behind the tank. The Lieutenant dismounted and came around to the right rear. There, Sergeant Robb informed him that Sergeant Ames, the tank commander for 1-5, had come to him and demanded that he be left in the rear, as his wife was pregnant and expecting to deliver a baby while they were in the field. Sergeant Ames was the youngest tank commander and a very handsome young African American. Sergeant Robb told him what he really wanted to do was stay back and screw everybody else's wives, while they were in the field and cause a lot of trouble. He just didn't like to go to the woods. The Lieutenant looked hard at Sergeant Robb and told him, "That's not going to be a problem. I have a fix for that."

THE INOCULATION

As the lead vehicles were being loaded up and the connecting cables attached to them, Sergeant Ames dismounted and came up to the Lieutenant and told him, "Sir, I have to stay back because my wife is pregnant. She's going to deliver while I'm in the field and I can't leave her." The Lieutenant just began thinking back to the inoculation he'd been given by Senior Drill Instructor, and acting First Sergeant, Craver when he was in his first company and picked up his first cycle of trainees at Fort Knox. The inoculation was a short master class on how to deal with "slicky boys," as Craver called them, and Ames was one for sure.

The Lieutenant said, "Well, let's sit down and talk about this," so they sat down Indian style in the "kumbaya" manner that was currently in vogue in rap sessions between management/leaders and soldiers. The Lieutenant asked him what was troubling him and Sergeant Ames told him that he was uncomfortable going to the field with his wife expecting. The Lieutenant told him they would certainly get him back if she went into labor, but the young sergeant really didn't want to hear it.

After trying to explain to him that he was going to be taken care of and not to worry about it one more time, the Lieutenant looked at him, and said, "I sense that what we have going on here, in the words of that great American film, '*Cool Hand Luke,*' is a failure to communicate. But be of good cheer. The Army thinks of everything. It's the responsibility of the higher ranks to establish communications with lower ranks, so it's incumbent upon me to give you a message that you can understand. I picked this up from a couple of drill instructors in the training center at Fort Knox, when I worked there pushing basic training, so here goes:

Get down mofo and beat-a-yo feet,
In the rhythm of the boogie to be,
Don't give me no jive, don't give me no lip,
Don't give me no grief, brother - get a grip."

The sergeant's eyes got wider and wider, and the Lieutenant added, "Or perhaps you'd like the shorter version." Getting no response, the Lieutenant continued, "Very well, you shall have it. Get your ass on that tank."

He immediately slipped his hand to his left breast, pulled out his .45 caliber pistol from the shoulder holster, racked and jacked it to put a round in the chamber, took it off safe, and held it up, canted down and to the left. "What you have here before you is the single most dangerous thing in the United States Army, a .45 caliber pistol, round in the chamber, off safe, in the hands of a lieutenant. The damn things have a tendency to go off, so I expect we'll have a training accident. You see, I don't even want to be here. I was involuntarily shipped here, and I've been involuntarily extended for three years. I should be on my way to law school, but instead I'm in a tank battalion with you. The easiest way for me to get to law school is to have something terrible happen, get investigated and sent back to the States, put my paperwork in, and go to the University of Virginia."

The sergeant's eyes just kept getting wider and wider, so the Lieutenant continued, "And if this thing goes off, the first thing it's going to do is blow your right nut off. Now, you're going to lurch forward so the second round, because all semi-automatic weapons in the West ride up and to the right, is going to catch you through the heart. You'll be killed instantly. But don't worry, I'll say only good things about you at the memorial service, your wife will cash in the Serviceman's Group Life Insurance, and I'll lead the assembled congregation in a rousing rendition of Amazing Grace. I spent nine years in a Methodist Church choir - I know all the verses."

The sergeant continued to look at him with his mouth open, so finally the Lieutenant added, "Echoing a line from that great American film, '*Network,*' am I getting through to you, Howard?"

The young sergeant shook his head "yes," turned around, and climbed back on his tank. The Lieutenant went over to Sergeant Robb and told him, "I don't think we're going to have a problem with Sergeant Ames anymore."

Robb looked at him quizzically and said, "Really?"

"Certainly, didn't I tell you? Officers are magic. Let me know if he tries another end around on you."

Sergeant Robb said, "I don't think that is happening, sir," and they both went back to their tanks to get them up the ramps to the rail cars.

THE CAT WITH NINE LIVES TAKES A TRAIN RIDE

When all of his tanks were loaded on flat cars and tied down with cables, the Lieutenant went over to Sergeant Robb, who had assembled the platoon in formation, slightly away from the tracks. The Lieutenant went over the upcoming schedule: they would be riding the train until sometime in the late evening, after which, they would offload the tanks and then road march into the forest. He did not have the location that they would be marching to, but he was sure the Company Commander would lead the company in a Jeep, and they would lagger up in the woods once they'd reached that location. Their mission was to simulate the Corps reserve, which would normally be a full brigade. They were a company which was one ninth the size so in the engagements, each tank would have to be killed with MILES gear nine times before it would finally be dead.

In a normal exercise, a vehicle would be knocked out and it might sit for a day or two before the unit would be reconstituted. In their case, they could expect to road march day and night, and fight engagement after engagement. Sergeant Robb then told the men to prepare to board the trains. They had previously taken their A and B bags and locked them up in the turrets of the tanks so they would be available when they got to their destination. The Lieutenant's gunner had the lockset for his tank so once he ensured that all the other vehicles were locked up, he proceeded, with Sergeant Robb, to see where the men were billeted on the train. The troop cars were dingy grey affairs that had bunks that would pull down or fold out to sleep approximately ten men in a car. Everyone was pretty much covered with diesel, grease, and dust at this point, so the cleanliness of the cars was not really an issue. Everyone was tired as well, so the men climbed on board, made themselves comfortable, and promptly went to sleep.

The train moved out in the early afternoon, and from time to time, would pull onto a siding and sit motionless for twenty or thirty minutes to allow other trains to pass by. Troop convoys had the lowest priority of anything in the German rail system. Around 2200, the train pulled into a small station. The Lieutenant had no idea where they were, other than it would be closer to the East German border.

He peered out the window and saw the sign that read, "Bad Hersfeld." He knew that it was one of the villages where a squadron of the 11th Armored Cavalry Regiment was stationed. As the train pulled into the station, light rain began to fall. The Lieutenant shuddered because he hated being wet, and when tanks were wet, there was oil and grease on them that made them almost impossible to get on and off, without slipping and oftentimes falling.

As soon as the train pulled to a stop, the exit doors of the troop cars opened up. Sergeant Robb called the members of First Platoon together and formed them up in two ranks. The Lieutenant came over and took command of the platoon. He told them that they were in Bad Hersfeld and that as their vehicles came up to the loading dock, they would mount the tanks, unlock them, get their equipment out and put it in the bustle racks, put tarpaulins over the equipment so it didn't get wetter than it had to get, then disconnect the cables that attached the tanks to the flat cars and drive the tanks out onto the road by the station. Once again, they would be following the headquarters maintenance vehicles, a cherry picker crane, and an M880 tracked maintenance vehicle, which could tow and recover tanks and carried most of the company's mechanics. and an armored personnel carrier. As the wheeled vehicles, the mess truck, supply truck, and the Jeeps of the Executive Officer, Company Commander, and First Sergeant, were already at the station, having made the trip by highway.

When the tanks and other armored vehicles had moved off the rail cars and lined up on the road beside the train station, the Company Commander and headquarters vehicles moved to the head of the line and the company began to move again, with Third Platoon in the lead, Second Platoon in the middle and First Platoon in the rear. While the vehicles were lining up, the light rain fell heavier and increased in intensity as the company road marched through the darkness, up into the mountains and forests surrounding Bad Hersfeld.

The Lieutenant was having a pretty difficult time trying to keep the gun of his tank from hitting trees by slewing the turret as he moved through the woods and hold on to the map sheet, which was inside the waterproof cover, but still completely covered in rain. At least he was able to put on his rubber duck raincoat at the train station, so he wasn't soaked, but the rain was still driving into his eyes and he had no real idea where they were located as they crawled along the small spider trail dirt roads up into the forest to a small clearing that served as an assembly area.

Shortly before midnight, they reached the assembly area and the company laagered, with First Platoon taking the sector from 5:00 o'clock to 8:00 o'clock this time. Once they were in position, he and Sergeant Robb walked from tank to tank, to site the firing sectors for the guns and make sure that the tank commanders had the radio frequencies switched over for the next day, as it was now past midnight. Then each tank crew set up a guard roster. Again, the Lieutenant took first watch. After two hours, he woke Blaylock to relieve him and crawled into his sleeping bag along the gun on the mantle, with a poncho liner over him and his combat vehicle crewman helmet near him, as it had the radio handset built into it.

FUN AND GAMES

Once the sun began to rise over the mountains, Captain Batson walked around the laagered vehicles and checked with the platoon leaders. Upon seeing the Lieutenant, he asked him, "Where are you, Lieutenant?" The Lieutenant answered, "Somewhere in central Europe. I don't have a grid coordinate."

This enraged Captain Batson, who told the Lieutenant to come to the side of B 1-1 and look out through a clearing in the woods that overlooked a small village. He asked him how far it was to the intersection of two small dirt roads at the edge of a field near the village. The Lieutenant looked at the intersection and told him, "One-two-fifty meters." The captain then took out a plastic protractor from his field jacket pocket and placed it on his 1:50,000 map, checked it and said, "That's correct," and then asked the Lieutenant to look a little farther away to the next intersection.

The Lieutenant promptly said, "One-five-hundred meters." The captain checked again, and that was also correct, so he then asked him to give him the distance to a small copse of trees farther away. The Lieutenant looked and told the captain, "One-seven-fifty meters."

Batson placed his protractor back on his map and said, "Hah, I've got you. It's eighteen hundred meters based upon the map."

The Lieutenant looked at the captain and asked, "What's the resolution of your map sheet?"

Batson replied, "One to fifty thousand."

Then the Lieutenant asked, "What's the accuracy?"

Batson said, "To the nearest fifty meters."

The Lieutenant then queried, "So, one-seven-fifty could be correct?"

Batson, slightly peeved, said, "That's true," and asked the Lieutenant to look at his map. He showed the Lieutenant on the map sheet where the company was located and then pointed out the grid coordinates to the location they would be road marching to, in order to pick up an armored vehicle launch bridge from an engineer company as an attachment to their unit.

Batson then showed him where they would laager for the next evening, and the Lieutenant circled those grid coordinates with a grease pencil on his map case. As soon as Batson was gone, the Lieutenant called the tank commanders to him and relayed the orders that he had been given. Once that was done, the First Sergeant came around with breakfast and a C-ration for each crew member. As the First Sergeant pulled away, Sergeant Robb came over and asked, "Sir, I overheard you and the CO earlier. Where did you learn to estimate distances?"

"I spent almost six months in the desert at Fort Bliss building tank gunnery ranges with a hand-held laser range finder. After a month or so, you get pretty good at estimating range. My best buddy and I also worked there as down range controllers for nine thousand rounds of main gun firing over all range bands from five hundred to three thousand meters."

Sergeant Robb thought a minute and then said, "Better get ready to roll, the old man is headed for his Jeep."

THE REMOUNT

Batson gave the order to prepare to move out and road march to the liaison point to pick up the engineer vehicle, the Lieutenant ordered his driver to back up and a huge plume of black smoke poured out of 1-1's engine compartment. Sergeant Robb immediately looked over, and on the radio, told him that he had blown his engine. With the engine blown, he ordered 6-6 to come up. He took the tank commander's position on 6-6 and left that tank commander with 1-1, to be dragged by the maintenance vehicle and repaired while they were in the exercise. The crew of 1-1 also exchanged positions with the crew members of 6-6. The Lieutenant's gunner grabbed his A-bag, which included a sleeping bag, poncho liner, and field jacket, and threw it over to him; he stowed it in the bustle rack behind him on 6-6. He had his shaving equipment in the A-bag as well, but all his extra clothes were in the B-bag which he left on 1-1.

CODEINE AND CROSS ATTACHMENT

The Lieutenant called the tank commanders to him by radio, sat on the side of his new tank, and quickly relayed the orders to them. He showed them the route of march, the intermediate stopping point, and the final destination for the day. Once that was done, they waited for the headquarters vehicles, and Third and Second Platoons, to march out of the forest to the main blacktop road. Then, First Platoon, with the Lieutenant in the lead, followed them. It was almost pleasant. The day was bright and sunny as they road marched, at about ten miles an hour, until midafternoon. They pulled over to the side of the road and parked the tanks in a quick herringbone defensive position, with the lead tank tube being directed forward, the rear tank's tube directed to the rear and all the rest alternating left and right. When the tanks were in position, in the distance, they could hear a large vehicle approaching. This vehicle was essentially the chassis of a tank, with an extending hydraulic bridge located on top of it, that could span sixty feet.

What was unusual in this circumstance was that the driver and the commander of the bridge had been drinking codeine cough syrup all day and had erroneously switched the intercom and radio connections on their vehicle, so they were talking about it as they road marched up to the company. In short, they were high as kites. The Company Commander, hearing this, made sure they stayed in the rear, more than sixty feet behind the last tank, 1-5. They continued for a couple of hours to the final assembly area to lagger for the night. Once they were in position, just inside a stand of trees, the Lieutenant went over to Sergeant Robb and told him there was no way any of his people were going to be sleeping on the ground with those cough syrup guys with a bridge running around in the trees. They had a C-ration for supper and set up guard duty again for the night, preparing to march to a major assembly area the next day.

The next morning, the loader, who had the last watch, woke the Lieutenant up. He was sleeping in his sleeping bag which was tied to the main gun and the mantlet with a poncho liner over him. He climbed out of the bag, rolled it up, and stored everything in the A-bag. Then he climbed off of his tank to check the other tanks in the platoon. While he was doing that, the First Sergeant brought them breakfast, which consisted of powdered eggs, bacon, toast, and what appeared to be creamed beef. The First Sergeant also brought along cartons of Vollmilch, which was radiated milk that did not have to be refrigerated. The Lieutenant had never seen this before, but being thirsty for something besides water he took a carton and sat on the sponson box on his side of the tank beside the turret and ate breakfast. He was having a hard time figuring out how to get the milk out, so he pulled his Swiss knife out of his field jacket and took the awl and punched a hole in the little silver circle on top of the box. Then he squeezed the milk into his mouth like he was milking a cow. Since he was raised on a dairy farm as a small child, this was nothing new.

However, Sergeant Robb looked at him in wonderment and said, "You know, sir, we have straws that come with these."

The Lieutenant said, "Really? In my ten months in the desert at Bliss, I never saw a straw, so I assumed we didn't have any. But thanks for the head's up."

The remainder of the morning was restful. He and the crews inspected the tanks and did preventive maintenance. They also reviewed the map sheet sets to make sure that they knew where they were currently located. Everyone got a little bit of sleep as well. Early in the afternoon, Batson came on the radio and called all platoon leaders to his location. There he laid out his map sheet on the hood of his Jeep and explained the route of march to the assembly area where they would be going before sunset. Since his company represented the entire corps reserve, for the initial few days, they would be following the main body which consisted mostly of the 4th Infantry Division and 1st Infantry Division units that came from the United States and the 4th Infantry Division units that we're in Germany, as well as tank battalions from the Third Armored Division. The opposition force, at least in their area, would consist mostly of German army units from the divisions in the Central Army Group.

When Batson was through briefing the platoon leaders, they went back to their platoons, laid out their maps, and relayed the route of march and objective area that they would be moving to. The crews then function checked all of their weapons, both sidearms and the machine guns, and inspected the main guns as well. Since they were in USAREUR, they had a complete upload of live ammunition including a round in the main gun itself. The only thing that kept the tanks on safe was that their firing pins were taken out and secured with the senior NCOs and officers.

Around 1400, Batson gave the order to march and the AVLB backed out of the trees and lined up with the headquarters vehicles. The tank platoons, in their normal order, proceeded to road march almost three hours to an assembly area for what would be an attack position. When they got to the new position, the company formed up on a small hill near a tree line two hundred meters from the other units. The First Sergeant drove up and gave every crew member a C-ration for the evening meal and to tide them over when they were given the road march order, and he said that they would be moving early in the night.

IN DEEP SHIT

The Lieutenant decided that he would seek out a little peace and quiet to review the map sheets and eat part of the C-ration, so he went over to the tree line and sat down with his back against a tree. Immediately, when he sat down, he realized, to his horror, that the ground was wet and hot. He jumped up and swept back some leaves and realized that he was sitting in a large pool of human waste, because some trooper that had previously been in this position had had a gastrointestinal attack of some sort. The Lieutenant was consequently covered in shit and his clean clothes were in his B-bag on 1-1, which was being dragged several kilometers to the rear. To compound matters further, the radios came to life, and he was ordered to proceed to the Company Commander's position to be briefed on the upcoming attack. It was also starting to rain. The Lieutenant hurried back to his tank, climbed up to the tank commander's cupola, and pulled on his rubber duck waterproof jacket that was folded up in the storage compartment behind the tank commander's seat. He listened to the radio as the Company Commander advised that all team leaders were to proceed immediately to his Jeep for further instructions.

Since the area around the tanks was fast becoming a sea of mud, the Lieutenant got into the bustle rack behind him, grabbed a pair of overshoes, and pulled them on over his combat boots. Once those were on, he moved slowly around the turret to get to the edge of the tank, to creep down to the front of the vehicle and dismount. Unfortunately, the entire turret was covered with dust that was becoming mud and oil, so in his overshoes, he slipped and fell about four feet to the ground, cracking his left hip as he hit the sponson box before leaving the tank. Shaken but undaunted, he asked the gunner to throw him his steel helmet which the gunner did. The Lieutenant placed the "pot" on his head, grabbed his map sheet off the ground, and took off toward the Company Commander's Jeep.

It was there that Batson laid out the general scheme of maneuver for the platoon leaders, but everything was subject to change, since they were the corps reserve, and no one knew where they were going to be committed. The Lieutenant went back to the platoon and relayed that information to the other tank commanders. He then carefully climbed back up to the tank commander seat on 6-6, just in time to hear a radio call from Batson ordering him and the other platoon leaders back to his location for additional briefing.

It was raining much harder now, and the turret of the tank was extremely slippery, so he did the best he could to hang on to the side of the turret as he attempted to move toward the front of the tank and dismount. But again, his overshoes slipped on the oily surface of the tank, and he cracked the same hip on his left side as crashed to the ground with his map sheet and steel pot.

Putting himself back together again, he went to the Company Commander's Jeep and was briefed on additional contingency plans. He then went back to his platoon and relayed that information to his tank commanders. Once again, he climbed back up to the turret of his tank and attempted to organize his map sheet and flashlight and other items he might need for the night road march. No sooner had he completed that task than he received another order on the radio to come back to the Company Commander's Jeep. Again, he got around to the side of the turret where it was extremely slick, and he slipped and ended up cracking the same area on the same left side for the third time that night. The only upside was that the rain and had washed off most of the crap.

He made his way to Batson's Jeep and was told that they would be moving out in approximately an hour. They would road march to an attack position where his company would attempt to set up on a side road to ambush an enemy formation. With this new information, the Lieutenant again went back to brief his tank commanders beside 6-6 and then, yet again, climbed up to the turret of his tank. He was beginning to wonder how anybody was going to ambush anyone in this organization since every vehicle had a rotating yellow light about ten inches in diameter on top of it, and everyone on either side could tell it was coming, literally, two miles away.

In any event, the Lieutenant grabbed a couple of bites of red licorice and prepared to lead his tanks on the road march that night. After an hour or so, the order came to move out of the assembly area. The platoons left the area in their normal order, a little after 2000. The only difference was the rain was coming down in sheets by now. This made it almost impossible to see the road well enough to keep the gun from hitting trees, while he was trying to read his map sheet through the plastic case with a flashlight with a red filter on it, in the inky darkness. Another problem was that there were so many vehicles on the road that massive traffic jams ensued, so the Lieutenant's platoon ended up sitting still on the road for hours waiting for other units to move through. By daybreak, around 0530 the next morning, his company was able to fight through the press of other vehicles and get to an assembly area in an apple orchard on a farm about fifteen kilometers from where they had originally been assembled.

APPLES – THE FORBIDDEN FRUIT

The platoons moved into the orchard and laagered in separate circles throughout it. Once the tanks were in position among the trees, the crews scrambled through the tanks to find what food they could. The Lieutenant ate some more licorice and shared some with his crew. He hoped that the First Sergeant would arrive with breakfast by midmorning but was not optimistic. He noticed that some of the soldiers had dismounted their tanks and were climbing up into the trees to get apples.

Sergeant Robb immediately dismounted, ran up to the soldiers, and told them to stop. This was for two reasons: first, you couldn't take the fruit that belonged to the villagers, but second, if you paid attention, you could see a large wagon full of yellow fluid that the farmer was spraying on the trees. This was urine from the livestock he kept in his barns, and he used it to fertilize the trees and the surrounding ground with the urine as it contained ammonia and nitrogen. If you ate the apples without washing them thoroughly, you would become violently ill.

Once that was under control, the Lieutenant ordered the crews to do preventive maintenance checks and fix anything that was broken, while they waited for breakfast. It finally arrived in the back of the First Sergeant's Jeep and consisted of cold cream beef, powdered eggs, and diced ham along with some bread. The First Sergeant also distributed C-rations to all the crewmen. About that time, the radios crackled, and Captain Batson ordered the platoon leaders to his Jeep. There he informed them that they should expect to be in the orchard for about two days, as the company was being held in reserve and would not be committed, until after about forty-eight more hours of fighting. Over the next two days, the Lieutenant got as much sleep as he could, rested the crews, and waited for the promised orders to move so he and his platoon could begin engaging opposition forces.

A NIGHT ON THE SET OF YOUNG FRANKENSTEIN

On the evening of the second day, Captain Batson came back from Corps headquarters, called the platoon leaders to his Jeep, laid the map of the surrounding area on the hood of the Jeep, and briefed them on the route of march for their first assault. They were to travel approximately twenty kilometers and engage German opposition forces in a small village. The German forces had Leopard tanks and were dug in around it.

The Lieutenant went back and met Sergeant Robb, who had the tank commanders assembled near his vehicle. The Lieutenant briefed them on where they were going, showed them on the map sheet where the engagement would occur, and then ordered them to saddle up and prepare to march. The Lieutenant turned to climb back on his tank, but then looked over his shoulder and asked Sergeant Robb, "By the way, where is 1-1? This NOMEX uniform is covered in human waste, dust, diesel, smoke, oil, and hydraulic fluid but all of my clean clothes are in my B-bag on that tank. Any idea when we might see it?"

"I don't know," said Sergeant Robb, "last I heard, it was being dragged by an M88 on loan from the 4th Infantry Division Forward, the unit that were supporting."

With that unhappy news, the Lieutenant clambered back up on his tank called all the tank commanders in his platoon by radio for a commo check, and after they had answered, prepared to march. They left the orchard at twilight, with Third Platoon in the lead, Second Platoon in the middle, and First Platoon in the rear. A little after midnight they moved into the village that they were supposed to assault. On the edge of the village were several fields and the Lieutenant ordered his platoon into one of them because it appeared that it did not have anything in it.

The platoon laagered peacefully, but after half an hour, a crowd of villagers surrounded the tanks with large wooden pitchforks, flashlights, and actual torches; then they began screaming and beating on his tanks. The Lieutenant shouted over the sounds of the mob and asked Sergeant Robb, "What the hell is going on?"

Sergeant Robb dismounted, inspected the ground, and shouted back, "We are in the middle of a field of giant sugar beets, sir. They are underground for the most part and are the largest cash crop in West Germany. We squashed a bunch of their beets, and the villagers pissed."

The Lieutenant got down on the ground, faced the crowd, and told Sergeant Robb to have one of the German-speaking tank crewmen explain to them that they could apply for compensation for the lost beets. While they were doing that, a platoon of German

Leopards came into position on the other side of the field and shot his tanks. They then disappeared into the village as his MILES gear shrieked and the lights began flashing. The Lieutenant was frustrated by his inability to counter the German force. He was also frustrated because light rain had begun to fall and they had to get out of the field, which was now surrounded by more Germans with pitchforks.

In the midst of that mayhem, Batson called him on the radio, so he climbed back up onto the tank. The captain sent him a grid coordinate in code that was the location they were to move to in order to prepare for the next attack. Both of the other platoons had also been engaged by opposing forces, taking heavy losses, so the company pulled out of the village and road marched about twenty kilometers, before taking up a position on the side of a dirt road in the herringbone formation. Here they were told they would have about twelve hours to do maintenance, eat what they could, and rest before the next movement.

Late in the afternoon, Captain Batson drove down the line of tanks in his Jeep. He had been to a German bakery and brought several dozen small loaves of fresh German bread and threw them up to the crews, who enjoyed them thoroughly, as they hadn't had any fresh food in a long time. As luck would have it, they ended up spending the night along the side of the road and eating what they had left of their C-rations, as they had outrun the mess facility and would not have a hot meal that day.

THE PAST IS NOT OVER

Early the next morning, they made their way through several small German villages. As usual; it was foggy, gray, and drizzling with intermittent light rain. The Lieutenant felt like he had been born in his rubber duck. He had been wearing it almost constantly over the last ten days. As they moved through the second village, he looked to his right and saw that there was a German civilian on a balcony, peering down onto the road that his tanks had to travel to get through the village. He had a large basket of green apples that he was throwing at the tank crewmen, trying to hit them in the face or the side of the head, as they drove past him.

Seeing this, the Lieutenant went on the platoon frequency, called Sergeant Robb and informed him that they were taking enemy fire from what looked like a German civilian, veteran of the Second World War, throwing apples at the troops.

Sergeant Robb always had a plan. Like any good tank platoon sergeant, he kept spare parts, and in particular wedge nuts, solid pieces of steel, in the side compartments behind him, inside the tank commander's cupola. His tank pulled up behind the Lieutenant's tank, and when the civilian began to throw apples, Sergeant Robb leaned out of the tank commander's cupola of his tank and threw a wedge nut sidearm up, over the top of the balcony, and right into the mouth of the former soldier who was showering them with apples. He pitched over his chair backward into his room and was not seen again by the platoon as the tanks moved quickly through the village. The Lieutenant thought of Faulkner whom he read on the bus trips between his home in Kansas and VMI: the past isn't over; it isn't even past.

GREASE GUNS AND GUMMI BEARS IN THE CHECKOUT LINE

The company moved on, for about nine or ten kilometers, into a clearing on the side of the highway they had been traveling on, and they laagered up quickly. This clearing was on the edge of a village that had a small supermarket close by. Captain Batson drove up and announced to the company that if they had marks, they would be in this position for at least a couple of hours so they could dismount, take their weapons, and go shopping for food if they wanted to. By this time, the Lieutenant was beyond hungry, and he was almost out of licorice, so he dismounted his tank and had the gunner throw him a grease gun. He had his .45 pistol in his shoulder holster and his gas mask around his waist. He then walked over to the supermarket, selected a shopping cart, and began to look for things that might withstand the weather and be decent to eat.

He picked up several packages of gummy bears, fresh bread, German mustard in tubes, sliced salami, and cheese spread. When he was finished, he went to the checkout counter, which was a surreal experience since he was heavily armed trying to negotiate a gas mask and a grease gun through the aisle to pay for his groceries. He realized when he got to the check out that although it was an odd experience for him, the Germans had been doing this for years. They didn't bat an eye. He put his purchases in some plastic bags and walked back to the tanks. He assembled the platoon and distributed food to the crews. The company spent the afternoon and early evening hours in that clearing.

Close to midnight, Batson received a movement order and relayed it to the platoon leaders along with the grid coordinates of a new assembly area several kilometers to their West. As usual, the rain began again, and the company rolled through a torrential downpour to this new assembly area, which was a large mountain, with several small clearings around the base of it, and literally hundreds of vehicles. Yellow lights were flashing and it looked like the entire mountain was on fire. The Lieutenant understood from Captain Batson's order that all of these vehicles were the remnants of friendly forces that had been engaged for the last week and were reconstituting in this area. His company, which was simulating the corps reserve of one armored brigade, was to assemble there and prepare for cross attachment to different units, to strengthen them when they went into the counterattack.

THE LONELY LIEUTENANT

The company spent the next night there and almost all the vehicles shut down for the crews to sleep and turned off their yellow lights. Throughout the entire night, a new infantry second lieutenant with a thick Puerto Rican accent kept coming up on the company net. He was looking for Captain Batson to cross attach his platoon to the armor company to simulate a task force. Unfortunately, the new lieutenant was unable to locate the position of the of the company and spent the entire night on the radio plaintively saying, "Delta niner-niner, this is Whiskey three-six, over. Where are you?"

This went on for hours. Finally, Batson said, "I'm going to turn on the yellow light on my Jeep. That's where I am, so you need to come to me by navigating to the light." The lieutenant said he would, but unfortunately, at that same time, the vehicles in the assembly area from the division his company was supporting got the order to move and all of their lights came on at the same time.

At that point, in frustration, Captain Batson asked the Lieutenant to put his noncommissioned officer on the radio. He told him to locate the grid he'd been given, get the platoon there, and be prepared to march in ten minutes, and if he had to - simulate that a sniper had killed his lieutenant. By this time, it was dawn, and all the vehicles were cranking up, preparing to move out. The Lieutenant was heartened by the sight of the First Sergeant in his Jeep coming up with hot breakfast for the first time in several days.

As usual, the Lieutenant made sure that all of his tank crewmen ate. Then the First Sergeant came back and handed up a Styrofoam plate with cold fried eggs, sausage, hash browns, and toast on it. But just as the Lieutenant got the plate in his hands, his platoon got the order to move. His driver heard the order, and immediately backed the tank up quickly, so the plate fell out of the Lieutenant's hands, so the last time he saw his breakfast it was it was being run over by his own tank treads as the vehicle reversed and then shot forward to line up for the march to the next assembly area.

After five hours, the company moved into a large, wooded area that contained several units from the 4th Infantry Division. Upon arrival, Captain Batson directed the platoons to the units that they would support. The Lieutenant's platoon was directed to the Second Battalion of the 28th Infantry. After he laagered his tanks, he went with Sergeant Robb over to the Infantry Battalion Commander in his M-577 Tactical Operations Center track.

When they arrived, they were told to wait outside until they were asked to come in for a briefing. Thirty minutes later, they both went in to see the lieutenant colonel who was commanding the 2-28 and his S-3 Operations Officer. He welcomed them and showed them on the map sheet both where they were and where he anticipated they would be committed in the middle of the night or early the next morning. This was, of course, subject to change. With that done, they checked their codebooks for command frequencies and call signs for the next several days, coordinated them with the S-3, and then the Lieutenant asked if there was hot food available for their platoon. They looked at the colonel who said that there would be, later in the afternoon, and he would make sure that his S-4 coordinated with the Lieutenant to ensure that his troops were fed.

They spent the rest of the day in that clearing, checking their tanks, and sleeping as best they could, since they knew they wouldn't be moving for the balance of the evening. By 1800, the S-4 came by and told them that he had tasked the support platoon leader to make sure that food was brought to the tank crews. When it arrived, they ate and prepared the vehicles for the night march.

At 0200, with light rain falling again, the Lieutenant got orders to march with the infantry battalion. The battalion moved out of the wood line with the tanks following behind it. They followed the infantry battalion's tracks for about fifteen kilometers and halted in a new assembly area where they prepared to assault with the opposition forces from the Third Armored Division and a German division.

Having arrived at that assembly area, the Lieutenant ordered his tanks to laager and await further instructions. Shortly after noon, the Lieutenant was called to the battalion commander's track, where he and Sergeant Robb were briefed on their mission. They were to take the platoon a little less than three kilometers north and meet an infantry lieutenant who would lead them through the woods on a small trail to try and sneak up on the enemy.

LUMBERJACKING

The infantry lieutenant led them deeper and deeper into the woods and the trail became narrower and narrower until the Lieutenant ordered his driver to stop. He dismounted and ran forward to the infantry officer and asked him where he planned for the tanks to go. The Lieutenant didn't believe his tanks could go any farther in the woods. The infantry officer signaled the Lieutenant forward to look at the trail which had sets of fresh tracks from armored vehicles on it. The Lieutenant looked over his shoulder and asked Sergeant Robb to come forward.

Sergeant Robb joined them and told both officers that the tracks they were looking at were Marder tracks. The Marder was a German infantry vehicle which was significantly narrower and lighter than his tanks. In addition, the trail was going closer and closer to a small stream, and the ground was getting marshy so besides the possibility of being stuck in the trees, if they did get forward, the tanks would be in hull depth mire and likely would wind up stuck in the mud. To make matters worse, the tank platoon was so far into the woods that it was impossible to turn around.

The Lieutenant told the infantry lieutenant that there was no way to go forward, and he would need new orders once his tanks were out of their predicament. Then the Lieutenant turned to confer with Sergeant Robb, who told him that the only way he could see them getting out was to do something that he hadn't had to do since Vietnam. The Lieutenant asked him what he meant and Robb told him that he needed all of the tanks to completely button up with the crews inside for protection. Then he would go back to the last tank, which was 1-5, and command that tank with his driver to back up and knock over enough trees to make a clearing sufficient for them to turn their tanks around with their gun tubes at max ord.

Sergeant Robb proceeded to do exactly that and the Lieutenant slipped out to the side to watch him in action as he backed the tank into one tree then moved forward into another tree and around in a semi-circle to another tree to knock them over until he had a clearing on dry ground large enough for the tanks to turn and get back out of the woods to the asphalt road. By the time they did that, it was midafternoon and the unit they were supposed to cross the swamp with to ambush the enemy had already left the area.

THE AMBUSH

By mid-afternoon, the platoon received orders over the radio, and moved to a new assembly area, where they prepared to assault with the infantry battalion against a couple of German Panzer battalions. Having arrived at that assembly area, the Lieutenant ordered his tanks to laager and await further instructions. They stayed in that position for the balance of the day and into the night. The unit was spread all over, so they ate their Cs and tried to sleep. Around 0230 the next morning, the Lieutenant was ordered to the infantry battalion commander's track.

When he arrived there, he was informed that his platoon was going to be cross attached to another infantry battalion and the colonel gave him a grid coordinate for the commander of that battalion. He and Sergeant Robb went back and briefed the tank commanders as to where they were going and what they could expect. They road marched five kilometers to a new location in the woods and came to a halt on a small dirt road behind the armored personnel carrier command track for the Second Battalion of the 22nd Infantry nicknamed, "Triple Deuce."

The Lieutenant and Sergeant Robb climbed into the back of the S-3 track where the battalion commander was waiting for them. He showed them on the map board where they were and where he wanted the platoon to go, to ambush a battalion of tanks that he anticipated would be moving into their area along a small road skirting the wood line across a small valley from another road. The Lieutenant's platoon was to take up a battle position along that road, disperse the tanks, and then fire on the lead and trail vehicles as soon as they came within range. They would have to march approximately twenty klicks and would need to leave within the hour.

The Lieutenant checked to make sure all his radio frequencies were correct since he was assigned to yet another new unit. The lieutenant colonel told him to stand by on the radio and wait to move out until he gave the order. The Lieutenant and Sergeant Robb went back to their platoon and briefed the tank commanders on the plan. They then lined up the platoon vehicles with the heavy section 1-1, 1-2, 1-3 in the lead, followed by 1-4 and 1-5.

Before they left, Sergeant Robb again briefed the tank commanders reminding them to maintain good discipline with regard to alternating gun tubes with their machine guns elevated against air attack and that this applied particularly to the lead vehicle, 1-2, which he ordered to the front and the trail vehicle to the rear, with the loader of that tank acting as an air guard for the platoon.

At 0600, the infantry First Sergeant from Triple Deuce company moved through the platoon giving the crews some warm coffee and C-rations, two each for the day. The crews took time for a light breakfast and awaited the order. After an hour, the infantry battalion commander came over the command net and ordered the tank platoon forward. The Lieutenant relayed that order to the tank commanders, and they began their road march to the objective. By about 1030, the ambush position came into view, and the Lieutenant ordered his tanks to deploy with each tank having a primary firing position and two alternate positions. They would be online in the ditch near the roadway, hidden from view, and waiting for the tank battalion that they would ambush.

As the trail tank, 1-5 came out of the trees into the open area where the rest of the platoon was located. A helicopter came down out of the sky and observed the tank, then moved up and observed the other tanks before pulling back to the wood line. After about ten minutes, the lead tank of the battalion they were going to ambush came into view and the Lieutenant ordered his tank commanders to stand fast until that tank had moved to a position on his left flank, near where 1-5 was located, and the last vehicle in the opposing tank battalion was within view of 1-2.

He then ordered them to fire in their sectors, working from the front to the center and they began doing so. Each tank fired one round from its initial battle position and then moved to an alternate fired a round and then moved to another alternate, fired a round, then went back to its original position, so that it fired a round from each of the three positions and then started the system again. They took the tank battalion by surprise and decimated it.

After about five minutes, it was completely destroyed. Yellow flashing lights and MILES beacons were flashing everywhere. When the engagement concluded, the Lieutenant relayed the status to the infantry battalion commander who ordered the tank platoon to move into a wooded area approximately one klick west of where they were currently located and await further instructions. The commander also informed them that they were going to be detached from his battalion and attached back to their company.

The Lieutenant immediately ordered his driver to back up and wheel his tank to the left. He then ordered the remainder of the heavy section, followed by the light section, to road march behind him to the wood line and take up concealed positions where they could rest, eat, and do preventative maintenance on the tanks. Once they were under the cover of the trees, the crews began eating C-rations. When they were through with lunch, they slept, in shifts, and checked the tracks of their vehicles to make sure that they were tight enough and inspected underneath the vehicles for any leaks. Around 1500, Captain Batson came on the company net. and sent the Lieutenant the grid coordinates for an assembly area five klicks north of their current location, in a wood line surrounding a large open field.

The Lieutenant gathered his tank commanders together to show them where they were going. When they reached the large clearing, they were the first to arrive, so they took a position farthest away from the entrance. After about an hour, Captain Batson and the headquarters section arrived. They moved their vehicles over to the left of the road and the Lieutenant was overjoyed to see that the M88 tank retriever was dragging his tank, 1-1. He dismounted and ran over to Captain Batson to get an update. Batson told him that Third Platoon should be with him before evening, but that Second Platoon was attached to another infantry battalion and would be linking up with them the following morning. The exercise was coming close to an end, and all the units were trying to reassemble for eventual transport back to Mannheim.

Batson ordered the Lieutenant back to his platoon to stand by for further instructions, set out guards, and prepare to sleep there that evening. The Lieutenant saluted and ran over to the M88 and his tank. There was no one on the tank, so he told the M88 driver that he would be scrambling up to get his B-bag as he needed clean clothes.

The Lieutenant mounted the vehicle, which was attached to the M88 by two tow-bars. He grabbed his B-bag and climbed back down off the vehicle. He asked where the crew for the tank was and the M88 driver told him that they were in an armored personnel carrier with other members of headquarters having a late lunch.

He went over and saw the crew of 6-6. Sergeant Fields, the 6-6 tank commander, told him that it looked like 1-1 would be operational the next morning. The Lieutenant told him that if that was the case, they would swap tanks at the assembly area where Second Platoon was supposed to meet them tomorrow. The Lieutenant then walked back over to First Platoon with his B-bag and told Sergeant Robb what would be happening. Then he ordered Sergeant Robb to set up security, firing sectors, and put them in rotations to get some rest. He then pulled out a clean field jacket, his second set of NOMEX, a t-shirt, and underwear and went into the woods to change.

After he was dressed in clean clothes, he went back around to the tank and threw the B-bag up on the right-side sponson boxes, before he went to the front of the tank and climbed up. Happily, there was still some licorice left in the back of machine gun ammo tray, so he ate a piece. The Lieutenant hadn't slept in for about a week by now, so he told Sergeant Robb that he was going to lay a sleeping bag and poncho liner up on the gun and mantlet of the tank and try and get a nap before evening.

A VISIT FROM THE CINC

No sooner had the Lieutenant nestled down into his sleeping bag and pulled the poncho liner over his body and face than he heard the whoop-whoop-whoop of a helicopter above him. He threw off the poncho liner, slipped out of the sleeping bag, climbed up to the tank commander's cupola, and looked into the open field. The helicopter descended gently into the middle of the field and a tall officer dismounted from the seat behind the pilot. The tall officer began walking toward Captain Batson, who came to the edge of the clearing and saluted. He then engaged in conversation with the officer for a period of time and pointed over toward the Lieutenant's platoon. Seeing this, the Lieutenant grabbed his helmet and dismounted his tank. He went over to Sergeant Robb and told him, "It looks like a general officer is going to come over to see us. I can see the silver stars on his headgear. In my experience, it is never good news when a general comes out of the sky to talk to you. I'll go out and meet him and see what he wants. If it's an ass chewing, it's mine."

The Lieutenant squared away his uniform as best he could, given that he had not shaved in a several days and still smelled to high heaven and marched out toward the center of the clearing. As the tall officer approached, he could see that there were four stars on his ball cap. When the general was about three paces away, the Lieutenant announced, "Lieutenant Manteuffel, First Platoon Bravo Company, 3-68 Armor, sir." At the same time, he saluted.

The general returned the salute and held out his hand. The Lieutenant shook it and as he pulled his hand back, the general asked, "Are you related to the German general?"

The Lieutenant answered, "Distantly, sir."

"Where are you from?" asked the general.

"I'm from Kansas and the Virginia Military Institute," said the Lieutenant.

"Ah, VMI."

"Yes, sir."

"How long have you been in this country?" asked the general.

The Lieutenant replied, "Me, sir? Or my family?"

"Both."

"Sir, members of my family have served as soldiers in this country for almost seven hundred years. As for me, Herr General, ich habe drei wochen im Deutschland verloren."

The general looked amazed and asked, "Did you say three weeks?"

"Yes, sir, about three weeks."

"Where were you before?"

"The Fourth Training Brigade, at Fort Knox, and the Third Armored Cavalry Regiment, at Fort Bliss, sir."

"Ah, the Cav," said the general.

"Yes, sir, the Cav."

The general then asked the Lieutenant, "Do you know who I am?"

"Sir," said the Lieutenant, "I believe you to be General Crosson, Commander of United States Army Europe and the Seventh Army, and the commander of the Central Army Group."

"That's correct," said the general, "I'm the commander of USAREUR, the Seventh Army, and CENTAG. Do you know why I came down to talk to you?"

"No, sir," replied the Lieutenant.

"I've been the commander of USAREUR and Seventh Army, and COMCENTAG for over a year, and I have been on many exercises. I've been up and down the battle line for the entire Reforger exercise that you've been part of. I came down to tell you that your platoon exhibited the best road marching, movement to contact, and fighting of a tank platoon that I've seen."

The Lieutenant immediately interjected, "It's not my fault, sir."

The general looked shocked and said, "I'm not finding fault, lieutenant; I'm complimenting you."

"I know," replied the Lieutenant, "but it's not my fault. I've only been here three weeks. I'm blessed with excellent soldiers and the single best tank platoon sergeant in the United States Army, Sergeant Robb. I know this to be true, because when I was with USAOTAE, at Fort Bliss, I was assigned with the Master Gunner of the United States Army, who was supposed to be the best tank platoon sergeant in the Army. He's not as good as Sergeant Robb. So, I appreciate the compliment, sir, and I'll pass it on to my sergeant and my soldiers right away."

The general saluted him and said, "Thank you, lieutenant."

The Lieutenant returned the salute. The general about faced and went back to his helicopter. The Lieutenant did a quick about face and went to see Sergeant Robb.

When he reached his platoon area, he told Sergeant Robb to form the platoon in two ranks. When the platoon was formed, the Lieutenant went to the front of the formation and Sergeant Robb went to the rear. The Lieutenant had the troops go to parade rest and then ordered them to stand at ease. He said, "I suppose you were wondering why a four-star general came down out of the sky today to talk to me. He told me that I had the best tank platoon in the field that he had seen since he's been here as the commanding general of USAREUR for a year. I told you three weeks ago when I took over this platoon, we would have the best platoon in the United States Army. As you know, the best tankers in the United States Army are in Germany and this platoon is the best platoon in Germany. That makes you the best tank platoon in the Army. You don't have to take my word for it; you have the word of the commanding general of the United States Army Europe and the Central Army Group."

"I want to thank Sergeant Robb for his work in making this happen and thank each and every one of you as well. Now, we need to try to get some rest. If I know Captain Batson, he'll wake us all up at about four in the morning and we'll road march somewhere. I believe the next part of the exercise is going to be an attack on a defended position. To do that, we'll have to assemble Second and Third Platoons with us. As soon as I have a grid coordinate for our assembly area for the attack, I'll pass it on to the tank commanders." The Lieutenant then brought the platoon to attention, ordered Sergeant Robb to post himself at the front of the platoon, and turned it over to his NCO.

As he was finishing, he could see Captain Batson coming across the field toward him. He walked out to meet him and was immediately confronted with the question, "What in the hell did the CINC USAER want with you?"

"Sir, he came down to tell me that I have the best tank platoon in Europe."

"He what?" demanded Captain Batson.

"Yes, sir, he watched us road march in support of Triple Deuce, set up an ambush position, and then move into this wood line after the attack was concluded. He watched the entire engagement, and he came down to tell me that I have the best tank platoon in the United States Army in Europe."

Captain Batson shook his head in amazement, turned around, and went back to his cot. As the Lieutenant walked over to his Jeep, Third Platoon came into the clearing, and took up a position across the field from First Platoon. The First Sergeant also appeared with C-rations and distributed them to Third Platoon and then First Platoon. The Lieutenant ate part of a C-ration ration and stored the rest for later. He then tried to get some more rest, but the radio cackled to life and Batson ordered him to his Jeep for further briefing.

The Lieutenant grabbed his Reforger map, slid down the front slope of the tank with his steel pot on, and ran over to Captain Batson's Jeep. Sergeant Ramsey was already there when he arrived. Batson laid his map out on the hood of the Jeep and pointed to a position on the edge of a clearing at the base of a hill ten klicks north of where they were located. He then said that at 0500 he would order them to stand to, and eat something quickly, after which, they would road march to that assembly point and attack the hill at dawn. A German Panzer Grenadier unit would defend the hill.

The Lieutenant went back to his platoon and relayed the information to his tank commanders. He then told Sergeant Robb that he was going to sleep for about four hours. He told Blaylock to take the first watch and to get him up at midnight, so he could change the radio frequencies, and he would stay up the rest of the night to wait for stand to and the movement to the assault.

Chapter 7
A Well-Deserved Rest

The Lieutenant climbed back into his sleeping bag, with his boots still on, sticking out of the end, and his poncho liner laid over him. His combat vehicle crewman helmet with its cord was by his side, so he could monitor the radio. Finally, he closed his eyes and tried to get what passed for sleep for the next few hours. As he did so, he thought about his girl again. She was probably getting ready to go to work as a quality control checker in a factory that made IV bags.

He thought back to August of 1977, two years earlier, when he suffered massive heat stroke at the end of a five-mile run in extreme heat on his first evening at Ranger School. The heat stroke made him go in and out of consciousness several times, and he was evacuated by ambulance to the Fort Benning hospital when he passed out for the last time. Around midnight that night, he woke up in intensive care, with two one-liter bags of Ringers Lactate on full drip, one in each arm to hydrate and bring him back.

When he looked up, he saw the logos on the IV bags and recognized them as being made in his hometown in Kansas. He thought, "You know, I know the girl who is the quality control checker on the line that makes those bags. I know she does a good job. These bags are going to work and I'm going to make it." By 0800, he was discharged and transported back to the Ranger Training Battalion where the Ranger cadre and students gave him a standing ovation because it was the first time in the history of the school, according to the chief Ranger instructor, that a Ranger student ever continued to do pull-ups while unconscious.

Unfortunately for the Lieutenant, the next day his left knee was cracked in half in the hand-to-hand pit, and although he went ahead through orienteering and the patrol phase, he could not jump into the mountains without, in all probability, losing a knee. Consequently, he had gone back to Fort Knox and an assignment there that led him to Fort Bliss and back home for Christmas break where he began dating his girl.

At midnight, back in the present day, Blaylock climbed over and shook his right shoulder to make sure he was awake. He got up, rolled his poncho liner and the sleeping bag, and put them in his A-bag in the bustle rack next to his B bag, pulled on the CVC, and settled into the tank commander's seat. He then checked the radio frequencies against his codebook and waited for Captain Batson's orders. At 0500, the radio cracked to life in his helmet, and he received the order to stand to and be prepared to road march to the objective by 0600. The Lieutenant went on the platoon frequency and alerted all of the tank commanders and crews. He ordered them to eat a quick breakfast of C-rations since he expected Captain Batson to have them moving out of their position within the next thirty minutes.

As anticipated, that's what happened. Batson came on the company frequency and issued the order to prepare to march. The Lieutenant went on the platoon frequency and ordered tanks 1-2 and 1-3 to follow him down the road out of the clearing with Sergeant Robb in 1-4, and 1-5 trailing to follow the Third Platoon, Captain Batson, and the headquarters vehicles to the assembly area for the attack.

CROSS OF IRON

With his tank spun around onto the edge of the clearing, the Lieutenant waited to see the yellow flashing lights of the Third Platoon and the headquarters vehicles. After about five minutes, he saw Captain Batson's Jeep pull out, followed by Sergeant Ramsey and his platoon, then the First Sergeant and the headquarters vehicles. The Lieutenant ordered his driver to pull in behind the M-88 pulling 1-1 and they proceeded for an hour to another large open field where they took up a position on the east side. Captain Batson then ordered First and Third Platoons to take up positions online facing the hill to their west.

After a little less than thirty minutes, one lone tank from Second Platoon pulled up and took a position between Third Platoon and First Platoon. The field was extremely foggy; visibility was down to about twenty feet. As the sun came up, he barely made out a little of the crest of the hill on the other side of the field. When the mist burned away a little more, Captain Batson gave the order to attack that position and all the tanks attacked online. The MILES equipment began shrieking as the tanks were hit by nearly invisible gray Leopards on the backside of the hill. A few of the German tanks began blinking and sounding as well. After a few minutes, the tanks on the hill withdrew, and the tanks of Bravo Company that were still alive moved up to the base of the hill. Looking out across the open field, there were dead tanks scattered here and there.

The one lone tank from Second Platoon was 2-1, Sergeant Wilson's. That tank had been hit and came to rest near 1-1. The Lieutenant sat on his cupola and observed as Wilson jumped down off of it when he saw the Company Commander approaching. Batson walked up to him and yelled, "Sergeant Wilson, where is the rest of your platoon?"

The sergeant turned, looked at Captain Batson with tired eyes, and told him, "Sir, I am the rest of my platoon. I've got four tanks broken down along the front or stuck in the mud with the infantry battalion we were supporting. They're scattered all over this map sheet."

He and Batson then laid the map out on the captain's Jeep. Sergeant Wilson showed him where his tanks were when he last had contact with him. Batson then told Sergeant Wilson to get in the back of his Jeep and leave his tank under the care of his gunner so they could go from position to position and round up Second Platoon tanks to bring them back so he could reassemble the company for a simulated assault in an urban area. *Swell,* thought the Lieutenant, – *what could go wrong?*

Captain Batson jumped in his Jeep, with Sergeant Wilson in the back, drove over to First Platoon and shouted up to the Lieutenant to assemble his platoon in the wood line and do preventive maintenance and to prepare to march at 1030. On the way out of the clearing, he stopped by and gave the same instructions to Sergeant Ramsey in Third Platoon. Finally, Captain Batson went over to the M88 crew and ordered them to disconnect 1-1 and follow him to retrieve any tanks from Second Platoon that were stuck in the mud. 2-1's gunner moved to a position just to the right of the First Platoon's tanks and the crews began preventive maintenance, ate some more breakfast, and rested as best they could, while awaiting the arrival of Second Platoon. Two hours later, Captain Batson and Sergeant Wilson came back into the clearing followed by 5-5 and the other four missing tanks from Second Platoon.

DOWNTOWN

Once he was back in the clearing, Captain Batson ordered the platoon leaders to assemble at his vehicle. The Lieutenant dismounted from his tank, with his helmet on and map sheet in hand, and jogged over to the Jeep. Second and Third Platoon leaders came over as well. Once they arrived, Batson explained that they would road march to the city of Kassel and once they were there, the company would go online and attack through the town at noon. This surprised the Lieutenant because most Germans go home for lunch over the noon hour. The traffic in a major city would be horrific, making it extremely difficult to try to drag seventeen tanks through. He asked Batson to make sure that he understood what they were going to do.

Captain Batson looked at him scornfully, and said, "Yes, Lieutenant, that is what we're going to do." Then he ordered the platoon leaders to prepare their platoons to road march onto Kassel with Third Platoon in the lead, Second Platoon in the center with Headquarters Platoon, and First Platoon in the rear. Third Platoon would be following Batson's Jeep.

The Lieutenant looked over to the M-88 and it looked like 1-1 was disconnected and moving on its own. He ran over to the company maintenance sergeant and asked if 1-1 was up. The sergeant said, "Yes, it is." So, the Lieutenant ordered the crew of 6-6 to bring 1-1 over to his platoon area. Once it was there, the crews changed places on the vehicles, and everyone made sure they had the appropriate A and B bags and their other gear.

The Lieutenant was happy to see that he still had at least half a bag of licorice, and a couple of breakfast bars stashed in the back of the storage compartment behind the Autobahn seat for emergencies and he transferred a couple of cans of his favorite Cs there to eat later if he got the chance.

The First Sergeant drove up at that point and threw a box of C-rations to each man. The Lieutenant caught his box in his right hand and was overjoyed to see that it included diced fruit – most excellent; he thanked his lucky stars. Then he took the licorice and put some in the ammo tray while he downed a breakfast bar. Shortly thereafter, his radio came up on the company net and the platoons were ordered to road march to Kassel. The Lieutenant stored his C-ration behind him in the tank commander's cupola, and First Platoon followed Second Platoon as it exited the clearing.

The company cruised for about an hour on a good blacktop highway traveling to the southwest through a large park. As they exited the park, Captain Batson ordered the platoons to spread out online on the edge of the woods looking into Kassel. Once all of the tanks were online, he gave the order to attack and the platoons moved out abreast, grinding down roads and highways as best they could to fight their way into the core of the city, with cars, small trucks, and the occasional tractor hauling produce, going in all directions to avoid them. Seventeen tanks in the core of the city created massive traffic jams. Horns were honking, people were screaming, and as it progressed, this part of the exercise bore the hallmarks of something that Captain Batson was doing for the sheer hell of it, since it was his last couple of weeks in Germany.

After about an hour and a half, First Platoon had worked its way around to the southwest, to Highway 49. The Lieutenant ordered his commanders to follow his tank, with 6-6 bringing up the rear to cross the river on the far edge of town and move quickly to Highway 7. That would eventually lead them back to Bad Hersfeld, where they would load out by rail the next day.

THE GHOST

When he reached the intersection of Highways 49 and 7, the Lieutenant stopped with his platoon behind him. Captain Batson and the other two platoons had crossed through the city to the north of him until they reached Highway 7. They then turned south and as they passed his platoon, the Lieutenant ordered his tanks to follow him and close up behind him. They rode marched for about two hours and then the company laagered in farmland with each platoon circling its tanks separately, in a different field.

The Lieutenant ordered his crew to drive into what appeared to be a large fallow field. When his tanks were all in position, the Lieutenant took a deep breath, reached back into the storage compartment behind the seat in the tank commander's cupola below and pulled out the C-ration meal that he'd been saving for a while.

He set about to eat the ham slices, cheese and crackers, and John Wayne bars, and hopefully the fruit, if he had time.

He climbed out of the cupola, crouched down on the back of the tank, and ate lunch quickly – zipping the cans of ham, crackers, and cheese with the P-38 on his dog tags and drinking the juice. He ate all of it saving the fruit and candy bars, so he would have something to look forward to a little later. The sun was warm, and he still had his rubber duck on against the cold wind, so he laid back on the tank and tried to catch a short nap since he hadn't really slept in over two weeks.

No sooner had he begun to doze off than he heard this tremendous banging noise coming from the right front sponson box of 1-3 that was on the edge of the field. The Lieutenant rose up and saw that there was a German civilian with a cane and one wooden leg, standing beside the tank. He was banging on the sponson box with the cane, screaming at the Guamanians for their officer. Neither of them spoke German, so they were both speechless and puzzled, to say the least.

The Lieutenant knew in a moment the old German was calling for him, so he rose up slowly, climbed up beside the cupola, grabbed his helmet, and slid down the front slope of the tank to the ground. He then began walking toward the German civilian who continued to bang on the side of the tank. As the Lieutenant approached, he started to feel hot, so he took off the rubber duck, making his field jacket visible. The German turned, looked at him, and immediately came to the position of attention. As he approached the civilian, the Lieutenant asked Sergeant Robb if he would bring him one of his soldiers who spoke German. Robb went over and spoke with the gunner from 6-6, who came up and began to translate for him.

The Lieutenant asked the civilian what was bothering him. The German replied that there was nothing bothering him. The Lieutenant, looked at him quizzically and asked, "Then why are you beating on my tank?" The German said that the tanks were sitting on one of his potato fields, crushing the potatoes. The Lieutenant explained to him that he had laagered the tanks there in error. He thought it was fallow and did not realize there were potatoes underground. He also explained that he was of German descent and that he would not have ever done anything purposefully to harm this farmer or his country, save in time of war.

The German farmer, still in the rigid Wehrmacht position of attention, said only, "von Manteuffel." The Lieutenant then asked him if he had served in the war. He said that he did. The Lieutenant asked if he served under General Manteuffel. He said that he did, in the 7th Panzer Grenadier Division in Manteuffel's battalion, then in his brigade and when he took command of his division in Russia.

When General Manteuffel came back from North Africa, after his assignment with Romel, and commanded Gross Deutschland on the Eastern Front, the German had transferred to that division. And, when General Manteuffel commanded the Third Army in the final defense of Berlin, the German served under him there as well. That was where he lost his leg. As he recounted this, the Lieutenant could see tears forming in the man's eyes.

The Lieutenant told him if he would just fill out a claim form for the lost potatoes, he would make sure that he was compensated for them, and then turned to walk away, thinking, *My family has been fighting the same people on the same ground for seven hundred years, with no apparent progress.* When he was back to his tank, the translator called after him and said, "He's not moving, sir. He wants you to tell him something."

The Lieutenant then walked back to the old soldier, looked him in the eyes, and said, "Tell him this von Manteuffel again rides in Germany to take up arms against the Pagan Slav, comfort, comfort ye my people, someday soon you will again breathe free."

The Lieutenant turned to walk back to his tank again to finish lunch, but the translator said, "He's not leaving; he needs you to tell him more."

So, the Lieutenant turned, walked back to the civilian and said, "Very well, zehen Jahren, ten years, you have the word of von

Manteuffel, it shall be done." When the old soldier heard those words, he took his cane, put it back up in the palm of his hand, and hobbled off to his house.

Sergeant Robb, who had been observing the exchange, came up behind him and asked, "Who are you, sir?"

The Lieutenant responded, "Who am I? First Lieutenant Robert Manteuffel from Kansas and the Virginia Military Institute, late of the Third Armored Cavalry Regiment and the 4th Training Brigade. Why do you ask?"

Sergeant Robb said, "No, sir. Who are you really?"

"Who am I really?"

"Yes, sir, who are you really, that general officers come down out of the sky to talk to you, and German civilians come to the position of attention when they see your name and cry when you talk to them. Who are you, sir?"

"Sergeant," said the Lieutenant, "I am a ghost from the past, come to set my people free." With that, the Lieutenant turned and climbed back up on his tank, to eat his fruit and the John Wayne bars that he'd been saving for the better part of a week.

FUN AND GAMES

He was finally able to get an hour of rest before the quiet of the mid-afternoon was shattered when several helicopters flew over them, fast and low. Sergeant Robb jumped up on his tank, grabbed his Combat Vehicle Crewman's helmet, and ordered the tank commanders to move into the wood line a little farther. When the Lieutenant heard this, he jumped in the tank commander's seat and used the intercom to direct his driver forward between two large trees. After his tank was moved, he slid off the glacis plate and went over to Sergeant Robb to ask him what prompted him to do that. Sergeant Robb told him that he'd been on many Reforger exercises, and that on at least one exercise, a helicopter had flown down over his vehicle and tossed a smoke grenade into the loader's hatch.

The pilot thought this was pretty funny until Sergeant Robb reached down, grabbed the smoke grenade, and lobbed it back up into the cargo compartment of the helicopter, which flooded the cockpit with smoke. The helicopter was forced down. Sergeant Robb took his tank to the helicopter and took everybody prisoner.

"The problem," Sergeant Robb explained, "is that tanks in Germany have a full combat load in them; everywhere they go, so there are sixty-four rounds of tank ammunition in each and every tank. That includes seven rounds of white phosphorus, that when ignited, will burn at nine hundred degrees. It's simply not anything you joke about or take lightly."

The Lieutenant told him that he understood and thanked him for having the tanks moved, and he would remember that in the future. At this point, everyone was exhausted after three weeks in the field, so they walked to the vehicles, checked the tracks to make sure they were tight, and did what preventive maintenance they could, sorted through their clothing and food, got rid of trash, and waited for the First Sergeant to come with dinner. When it was almost sundown, Guerra rolled up in his Jeep with the trailer attached. It contained a meal that passed for meatloaf, mashed potatoes, green beans, and peaches. When the troops had eaten, the Lieutenant had dinner and climbed up onto his tank to take the first watch for his last night in the field on Reforger.

ON THE ROAD AGAIN

The next morning at sunrise, the First Sergeant began making rounds to each tank crew to serve breakfast and issue a final C-ration. When the crews had eaten, Batson called the platoon leaders to his Jeep and showed them the route of march back to Bad Hersfeld, where they would railroad out for transport back to Mannheim. The Lieutenant went back to his tank and called the tank commanders over so that he could show them on the map where they would be going when they got the order to march. Shortly thereafter, Batson came on the radio and gave the order to march with Third Platoon, as usual, being first, Second Platoon second, and First Platoon in the rear, after the headquarters element. When the company arrived at Bad Hersfeld, the tanks moved one at a time up the loading ramp onto flat cars where they were secured for the trip. After all the tanks were loaded and locked, the Lieutenant and the other platoon leaders boarded the rail cars, along with the crew members, to sleep all the way back to garrison.

The train pulled into Coleman Barracks late in the afternoon. They unloaded the tanks as quickly as possible, formed up, and prepared to road march back to their company area. Batson came up in his Jeep and gave the order to follow him, so the three tank platoons fell in behind him for the dusty trip back to the motor pool. Upon arrival, they turned in their CEOIs to the S-2 NCO and cleared and turned in their small arms to the supply sergeant and the armorer. Then they cleaned the machine guns and the officers and non-commissioned officers turned them in. While they were doing that, the crews checked the vehicles and cleaned them as best they could. It was dusk, after 1930, when they were finally done.

Guerra formed the company and turned it over to Batson, who congratulated them on the job they had done and released them until the following morning. As usual, the crews had rotated in and out for the evening meal in the mess hall earlier. The Lieutenant had a couple of pieces of licorice and looked forward to getting back to his room, for a shower and a pizza or a TV dinner. By the time they were done, and the Lieutenant had carried his A and B bags back to his temporary BOQ room, he decided he would just have a can of soup, clean up, and go to sleep.

When he walked through the door, Love Boat asked him how his exercise went. He told him it had gone just fine, but he was really glad to be back with a shower and bed again, even if it was a couch. The Lieutenant then asked his friend how his Reforger went, and Petrovich told him that he had traveled over the entire area by Jeep from west to east and then in reverse. He got to see a lot of the countryside and even ate a few meals in decent German guest houses and bakeries along the way.

While the Lieutenant said he hadn't been as lucky on the culinary front, he did gain a lot of experience, and his soldiers did a good job, so he was as happy as he could be. Since he was exhausted, the Lieutenant told his friend he just planned to take a hot shower, eat, and go to sleep as quick as he could, because the next morning after PT, he would be looking for the battalion commander and operations officer to try to see when he was going to be able to get married.

Chapter 8
Getting Back to Job One

The next morning, after a short run for physical training, the Lieutenant went down to the orderly room to see if he could visit with Batson. Guerra looked up from his coffee and said, "Certainly," and canted his head left toward the old man's office. The Lieutenant went through the door, saluted, and asked Batson what the company would be doing that day. Batson told him that the troops would be cleaning their gear, checking the vehicles, doing maintenance, and pretty much getting everything cleaned up for the next time they would have to go to the field. The Lieutenant then asked if he could have the morning to go up to the battalion and coordinate with the battalion commander and operations officer with regard to a wedding date. Captain Batson said, "Of course," and told him to take most of the day to handle any other personal business that he needed to take care of. The Lieutenant told him that he would get his uniforms in the cleaners and also check with housing to see when he could get a room in the BOQ.

His quick shower was followed by a breakfast bar and a cup of instant coffee. He went over to the transient billets to see if he was assigned quarters yet. The same surly clerk at the desk checked and told him that if he came back the next morning, he would be assigned a BOQ room and given keys to it at that time. The Lieutenant was excited to finally have a place to live and went quickly up to battalion headquarters to see the Adjutant and try to schedule a meeting with the battalion commander.

He entered the building at the west entrance and walked quickly past the Duty NCO and Sergeant Major's offices down the hall to the Adjutant's desk. He approached the Adjutant, saluted, and told him what he needed. The Adjutant checked the colonel's calendar and told him that he should come back at 1300, right after lunch to catch the colonel and the S-3, before their afternoon planning session. Since it was midmorning, the Lieutenant decided he would go back to his quarters and pick up his dirty clothes to take them to the dry cleaner's and cash a check. He also wanted to make sure that the change of beneficiary he made at Fort Knox, before he left the States, was in effect, so that his fiancée would receive the fifty thousand dollars from his Servicemen's Group Life Insurance policy in the event of his death before their marriage.

The first two errands took over an hour and a half. When he was done, he went back to Petrovich's apartment and made a TV dinner for lunch. By 1230, he was back on his way to battalion headquarters to see the colonel. He again entered the building on the west end, and when he presented himself to the Adjutant at 1300, the Adjutant escorted him into the S-3's office and a small conference room where Captain Pendergast, the S-3, and the colonel, were gathered around a calendar to schedule the battalion's training, maintenance, and gunnery for the next few months.

The Lieutenant came into the room, saluted both officers and then asked if he might speak to them. They said that he could, so he asked them if he could find out from them a good day to set for his wedding date back in the United States. His wife's parents had been married on April the sixth, and the fifth was the closest date to that day so that was her preference if it was possible.

Colonel Shea looked up and said, "You know, lieutenant, if the

Army wanted you to have a wife, you would have been issued one." Then he smiled a little before he quietly reviewed the papers he had received from brigade regarding training in the next year and told the Lieutenant that the month of April was simply not possible, since they would be going to gunnery at a place called Baumholder that month. A disappointed, but unfazed Lieutenant then asked, "Is there any way I can get married in the month of March?"

Shea and Pendergast looked at the calendar again, conferred briefly, and determined that at least the first two weeks of March appeared to be clear. The Lieutenant then told the colonel and the S3 that he would put in for leave for three weeks from the end of February, to get married on the eighth, return to Germany on the fifteenth, and have a week to get his wife settled in. Both senior officers agreed that was the best course of action, so the Lieutenant saluted, left the room, and went as quickly as he could to the telephones on the second floor of the bank building.

PILGRIM'S PROGRESS

As soon as he got to the room full of telephones, he took a five mark piece out of his pocket, put it in the slot on the phone, and dialed the number for his girl's apartment back in Kansas. It was early morning there and she worked nights, so she should be home and awake. She picked up the phone immediately and he told her what the situation was regarding a wedding date. She understood the situation but was not happy because the eighth of March was during Lent, and she would need to make arrangements with the church and her uncle, the Monsignor, to have the wedding on that date. She said she would make all the arrangements, and he told her that he loved her and missed her and was looking forward to that date. By now, after feeding in several more coins to keep talking, he had burned through the marks he had with him, so he told her that he had a place to live at last, and that the call would end soon.

They said goodbye. He sighed as he hung up the telephone and made his way to the finance building. When he got there, he asked the clerk at the desk if he could see someone concerning his SGLI. After a ten-minute wait, he met with a Specialist Four from the finance corps. He told her that he needed to check to make sure that his life insurance beneficiary was changed to his fiancée so that if anything happened to him before they were married, she would have enough money to start her life over.

She looked in the file and saw the change form, but it had not been activated since he had been enroute to Germany; she told him that she would make that change if he signed the final paperwork. Satisfied with her assistance, the Lieutenant then asked her if she knew where the transportation office was so that he could check on delivery of his car that had been shipped from New Jersey. She gave him directions and informed him that the transportation offices were also there. They handled delivery of hold baggage and household goods as well, so he could check on that as well.

He proceeded to transportation and gave the clerk his name and Social Security number. Then he asked if his hold baggage had arrived from the United States. It had, so he told the clerk that the Lieutenant expected to get quarters the next day and would come back to set up a delivery, since it was in country. He also asked about his car, which was on a ship that had not yet come into Bremerhaven but was expected within the next two weeks. The Lieutenant told him that he would check back, and once it was in port, he would make arrangements to pick it up.

By this time, it was past noon, so he made himself a sandwich in Love Boat's room and went back to the company, where he found Sergeant Robb in the motor pool with his platoon. He asked Sergeant Robb how things were going, and Sergeant Robb replied, "Just fine, sir, catching up on all the maintenance on the vehicles. We've got some broken torsion bars, missing end connectors, and some pieces of dead track that need to be replaced. But it's so hard to get parts, we'll be working on them until we go to gunnery."

The Lieutenant asked, "Do you need me for anything?"

"No, sir."

"Great, then I'll go over to see if the old man has any details for me."

GETTING SETTLED

The Lieutenant walked across the street to the company barracks and proceeded downstairs to the First Sergeant's office. He asked Guerra if Batson was in, and after Guerra nodded "yes," the Lieutenant entered the commander's office. He approached to within three steps of the desk and saluted, saying "Good afternoon, sir."

"Good afternoon, lieutenant, what can I do for you? Stand at ease."

The Lieutenant told him that he needed to tend to some personal matters in the afternoon, pick up his field gear, and get it squared away. Also, he would be signing for a BOQ room in the morning and would need time in the morning to move into his new quarters. Batson told him that it would be fine but that he should coordinate with the First Sergeant to get a Jeep and driver to take him to brigade the next day at 1300 for nuclear, biological, and chemical (NBC) officer training, as he was going to be the company's new NBC officer.

The Lieutenant said, "Very well, sir," saluted, did an about face and left the office. Then he went back to the motor pool to let Sergeant Robb know he would be picking up his field gear, laying it out, and getting it squared away in his quarters the balance of the day. Sergeant Robb told him to be sure and separate out the Nomex uniforms, as he had been told that all of those were to be turned back into the First Sergeant to be sent back to Brigade. The Lieutenant said he would.

He also had some banking to do. He jogged to the American Express bank, where he wrote a check for some American currency and some marks. Then he went to the dry cleaners, picked up all of his field gear and uniforms that had been cleaned and pressed and went back to his Brother Rat's room to lay everything out and pack it so that he could move it the next day. By late afternoon, he had cleaned up his overshoes, shined his boots, packed his A and B bags with additional uniforms, and set aside the two clean Nomex uniforms to turn back into the First Sergeant the next morning.

When he was finished, he sat in the one club chair in the room to take a short break when Love Boat came back in through the front door. The Lieutenant told him that he was happy to say he would be drawing a key from housing and moving into his own quarters the next morning. The Lieutenant asked if his old classmate would like to go to dinner with him at the officers club or anywhere else that he wanted to go to show his thanks for putting him up when he was essentially homeless. Petrovich was something of a connoisseur and knew his food, so he told him that he would really like to go to Europa Haus in Viernheim, because it served excellent Hungarian goulash and schnitzel.

So, they jumped in Love Boat's old BMW and after a fifteen-minute drive, came to the hotel. It had a large dining room, and they had a delightful evening, recounting their latest adventures in the forests of Germany and reliving some of the old days back in Virginia. His classmate especially liked the story about the Lieutenant's tank platoon being attacked by villagers in the dead of night in the rain, with wooden pitchforks and torches, as it sounded a lot like a scene from young Frankenstein. The Lieutenant said it was, and he was just happy to be back out of the woods where he could sleep on a bed and take a shower once in a while, without worrying about pissed off locals and drugged out troops.

When they got back to the room, they went to sleep immediately, as both of them were still exhausted after running around the clock for almost a month. His buddy had spent most of Reforger driving from signal site to signal site checking on the communication teams. These were usually three person teams and as his classmate recounted, usually two happy guys and one tired girl. That was Europe in the seventies.

HOME SWEET HOME

The next morning, he opened his eyes just before the alarm sounded, so he switched it off, dressed in his physical training gear, and slipped out of the room to go to the company. His Brother Rat was still sleeping since he did not have to report until later in the day. After PT, he told Sergeant Robb that he would be moving in the morning but would be back in the company area about noon to get a Jeep and a driver and go to brigade for NBC training. He ate in the mess hall, since he didn't know if he would be able to get lunch or when he would be back for dinner at the end of training period.

After breakfast he went back to his buddy's room, shaved, packed the rest of his gear, and almost ran over to housing. There he was issued a key to a room on the second floor of the building across the street from where he had been staying. It was also the building where what was left of his footlocker remained in the fourth-floor storage area. The Lieutenant thanked the clerk. With key in hand, he walked across the street and up the stairs to what would be his home for the next nine months.

As he opened the door, the first thing he noticed was that he was standing directly in front of the bathroom that consisted of a bathtub with shower, sink, and a small chest of drawers. Looking to his left, he saw a living room with a sofa and chair upholstered in black vinyl, a tall reading lamp, coffee table, end table, and a small dining table with two chairs. There was one window that looked across the street to the BOQ on the other side. Looking to his right, he saw a small bedroom with one twin bed, a nightstand, a chest of drawers and a wardrobe for hanging clothes.

From the layout, it looked like this would have been a room shared by two German junior officers during the war who had a shared bath. The two rooms had been combined for the benefit of Americans who generally expected more space. Excited, he went back to his classmate's room and picked up his A and B bags and carried them across the street to his new quarters. He secured them there so he could go back and get his suitcase and attaché case and the balance of his stuff on the second trip. Once he had all his personal belongings and field gear in his apartment, he took stock of his situation.

The first problem was that the voltage of all the outlets was wrong for any electric devices that would be coming in his hold baggage. The second problem he had was that he needed bedding. There was a small refrigerator in the bedroom, so he decided to go to the Exchange to get some sheets and pillows and some food. As he walked down the hallway, he explored the building a little.

There were washing machines, a government telephone, that was only good for calling the units, and vending machines in the basement. There was a center stairway and stairwells on each end of the building as well. Next door to his room was a communal kitchen with a stove, oven, and a dining room table that would seat six to eight people.

Given what was available at the Exchange, he decided to forego the ugly brown sheets for the twin bed, but picked up some white pillowcases, and a couple of small pillows. He would just use poncho liners for his comforter and sheet until he could get his things from Knox delivered. He also got some more packages of licorice so he would have something to eat before he went to training at brigade. Having gone back to his room, he made his bed, as best he could, set up his alarm clock on the nightstand and decided that he should go back to the company to link up with his driver so he could make it to brigade for training. When he got back to the company, he checked with Guerra, who told him that the driver would be coming up from headquarters maintenance with his Jeep in ten minutes so he went up the stairs to ground level and waited for the driver to bring the Jeep around for the drive through the training area to brigade headquarters.

NBC – THE DISMAL SCIENCE

When the Jeep arrived, the Lieutenant climbed in and asked the driver if he knew where they were going. The driver replied, "Yes, sir, I've been to brigade headquarters several times." The driver slipped the gear shift into first and drove out of the company area through the dirt road that entered the training area where the Lieutenant had been on his first exercise. They drove for half an hour through the woods and came out in the middle of Coleman Barracks, which was home to the cavalry squadron, and infantry battalion, the aviation troop, supply, support, and signal units.

Once they arrived at brigade, the Lieutenant got out to go into headquarters. The driver parked to the side of the building and prepared to take a nap until the Lieutenant was ready to go. Upon entering the building, the Lieutenant asked the Duty NCO where NBC training was and he was directed back to the operations officer's office and the brigade chemical officer. That officer, a captain, led him back into a conference room where there were several other newly arrived lieutenants and a couple of new NCO platoon leaders who had been delegated the additional duty of NBC officer for their respective units.

For the next three hours, they were trained on various nuclear, biological, and chemical weapons, and the steps necessary to counter the employment of those types of weapons. They were taught how to use the wind and other variables to calculate the distribution pattern these weapons would exhibit when the wind hit them. They were also trained on how to calculate the amount of radiation the soldiers would absorb based upon the types of weapons employed so that they would know what condition they would be in and how best to use them in further combat operations.

On the whole, it was a pretty dismal subject and a dismal afternoon. Newly trained, he received his certificate of training from the brigade chemical officer, went out to his Jeep, and went back with the driver to his company area. By the time they got back to the company, the troops were already at evening formation so the Lieutenant jumped out of the Jeep to take his place behind his platoon. Once the First Sergeant had finished his instructions and announcements for the day and the soldiers were released, he spoke briefly with Sergeant Robb to find out what was in store for the balance of the week.

PREVIEWS OF COMING ATTRACTIONS

Sergeant Robb told him that the tanks would be going into "Q-service," which was quarterly maintenance. Each tank would be placed in a maintenance bay at battalion. Its engine would be pulled out, all the connections and fittings of the engine would be checked, and the vehicle gone over by the battalion motor officer, a warrant officer who worked for the battalion executive officer and the battalion maintenance officer who was a captain or first lieutenant.

Sergeant Robb told him that the rumor was a new Company Commander was coming in from the States, a West Pointer named Borman, so in the next week or so they would have to do a full field layout of the tanks and all of their equipment so that Captain Batson could sign the equipment over to the new captain. Oddly, it occurred to the Lieutenant at that time that he had never signed for the equipment in his platoon, as he had been preparing to go to the field or in the field almost continuously since he arrived in Germany. Sergeant Robb told him not to worry, he had already signed for it and it was all accounted for. Robb also told him that when he arrived, Captain Batson would be going with Borman and the Lieutenant to bring him up to speed on the battle positions and mission.

After the tanks were through with Q-service, Robb expected that they would spend about a week in simulated gunnery training, after which they would have an early Thanksgiving dinner with the troops and their families, and road march the battalion to a place called Baumholder for almost two weeks of tank gunnery and maneuver training after which they would return right before Thanksgiving.

After that, they could look forward to Christmas and the half-day schedule.

The Lieutenant thanked him for that information and gave him the building and room number for his new quarters so that he would have it and could pass it along to the First Sergeant to update the alert roster. Sergeant Robb also told him that they were coming to the end of NCO week, which was a week each month when senior noncommissioned officers pulled battalion staff duty, since they were habitually short commissioned officers. The Lieutenant should be on the next duty roster and could expect to pull 24-hour staff duty in the next few days. As a result, he needed to go up and see the S-2 to be trained thoroughly on how to translate the coded alert messages when they came in to battalion from brigade, so he could call a full alert of the battalion if it was ordered but certainly not to if the message was only a test. The Lieutenant thanked him for that information as well and then told him that he would see him the following morning at PT.

Chapter 9
Monday - First Night - Welcome to Hotel California

The Lieutenant walked quickly back to his new quarters, looking forward to sleeping in his own bed for the first time in months. As he came up the stairway to his floor, he picked up the scent of roast beef with vegetables and potatoes. He peeked in the kitchen and saw an older officer, with a couple of junior officers, sitting around the table. As he walked in, he said, "Hi, my name's Bob. I'm moving in next door," and gestured to his right. The older officer looked up and said, "I'm Buzz Tally, the dentist assigned to the clinic here in BFV."

The older of the two junior officers, who was a huge man, looked up and said, "I'm the chief warrant assigned to the ordinance company here. I go by Bigfoot."

The last junior officer looked up and said, "I'm Carey Pavel, currently the S-3 Air, but I'll be taking over Combat Support Company next month."

The Lieutenant looked up and asked Pavel, "Is that in 3-68 Armor?"

"Yes, I'm currently a first lieutenant, but I'll make captain next spring. How about you?"

"I'm the platoon leader for First Platoon Bravo, but I've been in the field so much that I haven't had a chance to meet anybody in the battalion other than the old man and Batson."

They were all in jeans, shorts, or athletic wear, so he told them that he would step over into his quarters change and come back shortly. When they asked him if he would join them for dinner, he said, "Absolutely." Then after changing into jeans and Adidas sneakers, keeping his olive drab tee shirt on, he came back into the kitchen. Buzz looked up pointed to the refrigerator with his bottle, and said, "Get yourself a beer – the opener is on the counter somewhere."

The Lieutenant opened a bottle of Spaten, took a long drink, and leaned back against the counter. Pavel watched him a moment and asked, "Where were you stateside?" He told them that he went to school at VMI and had asked for assignment to the border Germany, but ended up assigned to Fort Knox, and then got chopped TDY to Bliss to test M-1 tanks near White Sands, before being involuntarily shipped to Germany and involuntarily extended before he could submit his application for law school and eventual transfer to the JAG.

Buzz asked when he arrived; he told him, "About a month ago. I went to the field the evening of the second day I was here. When I got back from that, I went to the border for a couple of days, then after I got back from that, I finally found an old classmate from VMI that I could move in with for less than a week to get my shit together and go to the field for Reforger."

Buzz shook his head and said, "Well, that explains why we haven't seen you around."

The Lieutenant answered, "Exactly."

Before he could continue, Pavel looked out the window to the BOQ across the street and said, "It looks like the show's about to get started."

The Lieutenant walked up and took a seat at the end of the table. He looked to his left, out the window. The sun was beginning to set; it was almost 1900. He could see a young woman stripping her clothes off in the window of what appeared to be a BOQ room almost opposite them. When she was done removing her clothes, she stood in the window fully exposing and playing with her breasts. She opened the window and slowly climbed through to sit on the sill in the corner near them. After a few seconds she spread her legs as wide as possible and looked coldly out at the world in general. The Lieutenant looked at the others, who were looking out the window, and asked, "What the hell?"

Buzz turned to him and said, "The thing is - she was dating an officer in this building and he broke up with her so she's upset and sits in her window every evening, like the German girls in the red-light district downtown on Penis Strasse, to attract attention, and perhaps a new boyfriend. And to try to make her old boyfriend jealous."

The Lieutenant asked, "Is it working?"

Buzz shook his head and answered, "Not so far."

The Lieutenant asked, "Who's the officer she had this relationship with?"

Buzz whispered, "Well, that's Bernie; he lives across the hall."

Pavel chimed in, "Bernie Franz is actually attached to CSC and he's the FSO."

The Lieutenant then asked, "So, he's the fire support officer for our battalion – he coordinates all of the artillery?"

"Yes."

Bigfoot, who was quiet until then, added, "Bernie also sells Turkish hash - that's one of the reasons they broke up. She was getting too serious and she was making demands on his time that he couldn't meet, He's busy with his side business."

The Lieutenant looked around the room in surprise and even a little dismay but then asked, "How does that work?"

"Well, every couple of weeks, Bernie goes on a USO tour bus to Amsterdam. Amsterdam is a free port and he can load up on Turkish hash, good high-grade stuff. He just stashes it in his luggage to bring it back here. USO tour buses are never searched crossing the border into Germany so he comes back here and plies his trade – selling to the officers in the battalions in this village. The Turkish gangs from downtown control the enlisted market. The stuff they sell isn't as good because they cut it with crap. Bernie is discreet and careful not to encroach on their turf, even the Red Army Faction terrorists won't cross them."

The Lieutenant then asked, "So, she thought he was getting serious, but he was just looking for a good time right, and she misses him, and the hash, too?"

"You got it," said Pavel.

"Very well," the Lieutenant said as he shrugged. Then he added, "How do the communal meals here work? I don't have hold baggage in yet and don't even have anything to cook with, if and when it gets here. I delayed my household goods shipment because I got engaged right before I got posted here. It looks like I'll be going back to CONUS in March to get married, bring my wife back, find married quarters, and get her stuff shipped in."

Buzz looked up and said, "Don't worry, we've got plastic plates and extra silverware in the cupboards. If you'll wash the dishes, we'll have dinner in about five minutes." The Lieutenant nodded back to him. It was getting dark fast and the show was over, so he caught Pavel's attention and asked if there were any other battalion officers living in the building. Pavel slipped over beside him and said, "Axel Borman, the S-2, lives downstairs right below you. John Pallas is on the first floor, too. He's living there with his fiancée, a German girl named Siggy. He'll be marrying her in the next six months or so as well. He's currently the Headquarters' Company Commander but he'll be taking Alpha Company when we get back from gunnery in a couple of months. And Axel will be taking over for him. We're due to have a rotation of commanders with me going to CSC. We just got a new Charlie Company Commander who is getting settled in and there's a new commander coming in for Bravo in the first part of November."

Meanwhile, Buzz got up and took the roast out of the oven. Pavel continued, "We have an officer from 5-68 here as well: Lieutenant

Rauch up on the third floor, on the side that faces the PX. You'll notice him at least once a month because on payday he gets dead drunk and throws all his furniture and some of his utensils out into the parking lot. When it's all busted up, he has to go to finance and pay for it, gets new furniture delivered, and the cycle starts over."

The Lieutenant looked up and said, "I've seen him. He almost hit me with some of his stuff the first day I was here."

Pavel nodded, "Be careful when you walk around that side of the building. Steer clear of him, he's got a lot of issues."

"I take it he's not happy about being here?"

Buzz interjected, "That's putting it lightly."

The Lieutenant then looked around the room and said, "So, we have this young lady over there who's placing herself buck naked in the window of her quarters every evening. How long has this been going on?

"Oh," replied Buzz, "about two weeks. She's Signal and was out in the field doing Reforger setup. When she got back, she got the news from Bernie. Since then, this has been a nightly occurrence."

The Lieutenant shook his head in disbelief and asked, "Does anyone take exception to this?"

Buzz said, "No, you're in Europe now, and anyway, she'll have to quit in a couple of weeks when the weather turns colder and we get sleet, snow, and freezing rain."

The Lieutenant thought a moment and said, "Well, I'm not a narc and I'm not the sex police, and USASEUR is sure as hell a whole 'nother world, right?"

"Yes, indeed it is," said Buzz. "Let's eat."

So, the Lieutenant ate dinner and chatted with his new found companions. He waited for them to finish, then washed the dishes, and put them back in the cupboard. Buzz made a plate to take back to his room. The Lieutenant looked at him quizzically as he left but didn't say anything.

When he was gone, Pavel told him, "Buzz is taking a meal back for his roommate to eat when she gets in from work. His wife was badly injured in an automobile accident outside Fort Hood and she is in a wheelchair back in Harker Heights raising his two daughters. His roommate is a slim, blonde, nineteen-year-old British girl who teaches English to children in the DOD schools and night school for soldiers and dependents who have difficulty with it."

The Lieutenant now recollected that Buzz was definitely not wearing a wedding band, then took a breath and repeated, "I am not the sex police."

"Fair enough, Bob. Just a head's up. So far, you've only seen the tip of the iceberg. Good night."

The Lieutenant just nodded and slipped out to his room. There he took a shower in his own shower for the first time in a long time and changed into shorts and a clean t-shirt. Then he placed a poncho liner over his entire bed and lay down to try to get some sleep, to ready himself for the next day in the insanity of his new home.

TUESDAY - READY FOR DUTY

The next day dawned clear and cool. He rolled out early and went into the company for physical training. He met Sergeant Robb at the back of the formation and asked him what was on the calendar for the day. Sergeant Robb told him that the platoon would be preparing for Q-service so he would be free unless Batson had something for him to do. When the company returned from the morning run, he went up to Batson and asked if there was anything on the calendar for him that day as he needed to make arrangements for his hold baggage to be delivered to his new room. Batson said, "No. You are released for personal business."

He walked over to the mess hall, and after eating a large breakfast, jogged back to his room for a shower so that he could begin what would be a long day of waiting. He decided that he would deal with the hold baggage first, so he went to the Transportation Office, where he took up a place in line behind about a dozen dependents who were trying to get their household goods delivered. Finally, by mid-morning, he was able to sit down with the clerk who checked the manifest to see if his hold baggage had arrived. The clerk said it had arrived and was in storage, so they made arrangements for it to be delivered to his BOQ room in the afternoon of the following day. Once that was done, he went back to the German post office on the second floor of the American Express Bank building and placed a call to his girl back in Kansas to let her know that he finally had an apartment, and that he couldn't wait to get her to Germany in March of the next year.

They hung up quickly as the overseas call was running through his marks quickly and he looked around the room. There was a line of twenty or so soldiers waiting to pay their telephone bills, get telephones installed, buy train tickets, or pay their utility bills if they lived on the economy. This was a new feature of German life as far as the Lieutenant was concerned, in that everything revolved around the post office. For example, if you fail to pay your utility bills, you couldn't get a train ticket and they shut off your phone. Centralized government at its best.

He made his way back to the company area and met with Sergeant Robb to see if there was anything else he needed to tend to. Sergeant Robb said, "No, sir, but you need to go up to battalion and check with the Adjutant to find out when you are scheduled to be the battalion duty officer. Since you're new to the battalion, just returned from the field and are single, you should expect to be scheduled for duty on a regular basis, including weekends, and probably all holidays."

"That's fine. I don't have much to do until I get my girl here anyway – and I can eat cheaper."

The Lieutenant walked briskly up to battalion headquarters, passed through the door by the Sergeant Major's office, said hello to the Duty NCO, and walked back to the Adjutant's desk. He asked to see the duty roster. The Adjutant, Captain Steves, formerly the Charlie Company Commander, who fancied the Afrika Corps fame, pulled it out and said, "Lucky you, LT, you have duty Friday night.

You report at 1500 and if all goes well, you will be relieved at 0800. You need to sit down with the Duty NCO. He'll go over the SOP with you and explain how things work here. You also have to meet with the S-2 and get trained on breaking down flash traffic, coded messages, that will be a big part of your duties. Dismissed."

The Lieutenant went back down the hallway to the stairs past his old friend, the supply sergeant, who avoided him, to the S-2's office where the NCO and safe were located. He asked the NCO if he could train him on breaking coded messages. The NCO said, "Sure, sir," and opened the safe to get out the codebooks and showed the Lieutenant how the codes change from day-to-day so that when he was on duty, he would use one set of codes before midnight and a new set of codes after midnight similar to the CEOI.

Once he was trained, he walked down the hall and back up the stairs to the Duty NCO's office where he sat down at the duty officer's desk. The NCO gave him the duty notebook to review. He read as quickly as possible since the entire office was filled with a thick haze of combined stale and fresh cigarette smoke. In his experience, after two years, this was not unusual. Almost all NCOs on duty smoked continuously to stay awake.

After he finished reading the operating procedures, he went back to his company area and waited for evening formation. He saw Sergeant Robb and explained what the Lieutenant would be doing over the next couple of days and that he was scheduled for duty on Friday night. Sergeant Robb nodded and said, "Be sure and check with me before you go on duty Friday. Also, I hear that next week, while the platoon is in Q-service, you can expect to go over to Brigade for a war game called 'Pegasus' that will take at least a day. Once all our tanks are through Q-service, we'll prep for gunnery. First, we'll spend about half a day in the training area, shaking them down for mechanical problems, and going over maneuvers. Then we'll have another half day on sand tables up in the rec room to train on distribution of fire, followed by another day with the whole company rotating tanks through a miniaturized firing range that uses M-16s attached to the gun tubes to shoot small metal targets."

"Thanks Sergeant. If it wasn't for you, I'd never know what's going on around here."

"No problem, sir, it's my job."

ANIMAL HOUSE

As soon as the First Sergeant released the company after evening formation, the Lieutenant told Robb, "I'll see you in the morning," returned his salute, and hastened back to his quarters. He took a quick shower, changed into jeans and an old VMI t-shirt before heading down to the communal kitchen. When he got there, Buzz was heating up the leftover roast and boiling some noodles to go with it. His roommate, a slender, petite blonde with striking grey eyes, had also decided to come down for dinner since she didn't have a class to teach that night. They opened bottles of beer and exchanged small talk while waiting for the others to come by.

Carey Pavel stopped by to wait for dinner and gossip. He introduced the Lieutenant to Tom Storch, a fresh young ordinance second lieutenant who popped in, and promptly sat quietly in the corner. As usual, right on time, the lovelorn lady across the street stripped naked in her window, but no one paid attention, as they talked about what was going on in their various units.

The Lieutenant asked the assembled crowd how long it usually took to ship a car through Bremerhaven. Buzz told him, "At least sixty days," so the Lieutenant figured that he would pick it up when he got back from gunnery, which would take another three weeks. While they were waiting for the roast to heat up, Bernie came down the hall and passed silently by the kitchen. He was fairly short and a handsome young officer, sandy-haired, like Robert Redford, with bright shining eyes. He was carrying a pizza back to his room. When he opened the door, you could smell the smoke from burning hash clear across the hallway. He shut the door quickly, making it plain that he was settling down for the night and didn't expect to be disturbed.

Once the meal had heated up, Buzz sliced a fresh loaf of French bread and put out a plate of butter. Everyone lined up, made a plate, and sat down around the communal table. The Lieutenant asked Buzz's roommate, who had introduced herself earlier as Elizabeth, how she met the lieutenant colonel. She told him that she had run into him in the commissary and struck up a conversation. Buzz chimed in, "I was lucky to meet her since I'm here alone."

He went on to explain that he really didn't like stateside assignments. He liked to live overseas and CONUS depressed him. As a dentist, he had served two tours in Vietnam, but that was mostly with Graves Registration, identifying the bodies of dead soldiers, many of whom had been rotting in the jungle for months, by matching their remains against what dental records they had in Saigon. Once he got back to the States, he was on the west coast for a while but still tasked to do identification of remains, in addition to being a dentist, since he had so much experience. That was one of the factors that made him want to get to Europe where he could serve as a dentist for soldiers and their dependents and forget the dead.

As they were eating, the mood improved and the volume of the conversations increased as the beer kicked in. The Lieutenant mentioned that Bigfoot was absent. Buzz responded, "I think he is entertaining this evening. I saw him earlier bringing up a couple of large pizzas, a bottle of Jägermeister, and a case of beer."

The Lieutenant looked up quizzically and added, "That seems like a lot of food for one or two people." Buzz explained, "He's probably entertaining two or three of his female troopers who were coming over to party."

The Lieutenant again thought to himself, *This Q is nothing short of "Animal House" on the Rhine, but again, I'm not the sex police. I just need to make it through the next few months to get married and move into the married quarters. Then, we'll get through the rest of the tour, and, with any luck, back stateside to law school.*

He looked over at Lieutenant Storch. Tom was assigned to the same ordnance battalion as Bigfoot. He seemed pretty sullen, almost depressed, so the Lieutenant didn't ask him any questions and waited until everyone had finished eating. When the Lieutenant had washed the dishes, and everyone except Pavel had left, the Lieutenant asked, "What's the deal with Storch?"

"Unfortunately, Tom has fallen madly in love with Elizabeth, or at least he thinks he is, so he was completely depressed by the fact that she is shacking up with Buzz." *Again,* the Lieutenant thought to himself, *There is no end to the excitement to be had in this particular building,* so he excused himself to try to relax and get some sleep.

Getting to sleep, however, was not as easy as he thought it would be. When he stretched out on his poncho liner, he could hear several muffled noises, plus banging against the wall through his headboard, and what sounded like a small dog yelping. He hadn't yet met the officer living next door on that side so he simply tried to cover his ears with a pillow and get to sleep as quickly as he could. Again, he wasn't the sex police, and glad of it. Fortunately, his tour on the Mexican border with the Third ACR had well prepared him for almost anything concerning sex, booze, drugs, and relationships. He faded off to sleep, thinking of his girl, and hoping for some semblance of a normal life, whatever that was.

WEDNESDAY – MOVING IN

The next morning, he was up before the alarm. After pulling on his jogging suit and Adidas sneakers, he ran over to the company for his first cup of coffee, the best part of every day. After PT, he talked to Sergeant Robb to see what was on tap for the day. Robb told him that the platoon was still working on quarterly service but that the crewmen who had completed it would work on a sand table on the fourth floor of the barracks during the morning. The Lieutenant told him that as soon as he got cleaned up, he would grab a bite to eat and then join Sergeant Robb and the troops for the sand table exercise, which involved moving small model tanks in a large sandbox with terrain configured from the sand to simulate movement and firing positions in preparation for the upcoming Table 9: platoon gunnery, the ultimate test for tankers in his division. He told Robb that he would need to be absent in the afternoon to await the arrival of his hold baggage in his quarters. Once that was done, he should be pretty well set, with the exception of getting his car from Bremerhaven after gunnery.

He liked training and spending time with his troops, so the morning passed quickly. At noon, he stopped by the PX to get a cheeseburger and Coke and then went up to his room to await the arrival of his hold baggage. From his recollection, this consisted of a toaster oven, a small hot pot, a radio, an alarm clock, silverware, "Texas ware," aka ugly green and yellow plastic dinnerware that he had picked up at Fort Knox, bed linens, a red VMI comforter, civilian clothes, field manuals, and his small personal library.

After 1400, and the obligatory "beer break" that was customary in Germany, he heard a knock on his door. When he answered it, there were two locals standing in the hall with two boxes each. He asked them to enter the apartment and place the boxes on the couch in the living room. When they finished, one of the Germans produced a receipt. Once he signed for his household goods, they departed, and he opened the boxes to sort through them.

He placed all the electric devices on the small dining table and looked around the room for outlets. Since the building was of German manufacture, the outlets were 220 volts so he immediately knew two things. First, he was going to need transformers to transform the voltage and second, adapters for all of his American power cords so that they would fit in the German sockets. He then arranged his civilian clothes, hanging some, and folding some, before he made his bed with real linens and his old red comforter that had seen him through for years at VMI. The sheets and pillowcases were bachelor brown; he had picked them up in the PX at Fort Knox when he was first assigned there.

Next, he took out the field manuals and placed them on the bottom shelf of the small bookcase by the doorway and then pulled out the books that made up his personal library. They were a reflection of his personality. Like all VMI cadets, and most cavalry officers, he was an incurable romantic, but tempered by realism, and a healthy dose of cynicism. They included, *A Rumor of War* by Philip Caputo, *How to do Almost Everything* by Bert Bacharach, *Zen and the Art of Motorcycle Maintenance* by Robert Pirsig, *The Lion Heads* by Josiah Bunting, *Aerobics* by Kenneth H Cooper MD, *The Great Train Robbery*, by Michael Crichton, *The Wild Geese* by Daniel Carney and his personal favorite, and *The Dogs of War* by Frederick Forsyth. In addition, he had a Webster's New World Vest pocket dictionary, a 1970 - 71 edition of the *Officer's Guide*, various field manuals, and a worn copy of *The Good News Bible*.

This last still contained notes about scripture on three by five cards and a small booklet titled, *Manna in the Morning*, by Stephen F. Alford. These last dated back to his time with the "Navigators," a Bible study group he was in during basic course. He attended meetings with one of his classmates who encouraged him, and he had a really good time with the husband-and-wife team who led them, and the other young officers. The only downside was a badly sprained ankle, the result of an accident whitewater canoeing. In any event, it brought back good memories when he reviewed those items, especially Proverbs, Chapter 19, verse 14: "A man can inherit a house and money from his parents but only the Lord can give him a sensible wife."

DEAD MAN'S JOURNEY

He thought for a minute about the truth in that scripture and remembered the night that he drove for hours on Highway 54 from Fort Bliss north across New Mexico on the Jornada del Muerto, "Dead Man's Journey," that snaked up past Socorro and Last Chance to the west and then across the vast nothingness of the Llano Estacado,

Staked Plains, that ran through Texas and Oklahoma. He did not see a soul as he drove; past the Organ Mountains, White Sands, Alamogordo, and the Mezcalero Apache Reservation, through miles and miles of desert on a moonlit night going home for Christmas leave.

He listened to cassette tapes to keep awake: Gordon Lightfoot, Steven Bishop, and then, James Taylor. He rushed along, over ninety, with his silver-grey sports car whining in overdrive. Then, a tremendous feeling of loneliness came over him and it grew and grew until he pulled the car over at a roadside stop. He looked up at the stars and the moonlight on the sand, and prayed Adam's prayer: "Lord, I am your creation. Make of me what you will, that thy will not my will might be done. I love you, but I am so lonely. I need a friend, a companion, someone to share the fullness of life with. I need a woman and I need her now."

He spent that Christmas with his family in his old hometown. That visit included singing in the choir for the candlelight Christmas Eve service at the Methodist church with his old friend, who was now in medical school in Kansas. He told him that the annual New Year's Eve party for their high school crowd would be at a friend's place, a house near his parents' house. His friend told the Lieutenant that he hoped to see him there. The Lieutenant said he would.

SOMEONE

Almost everyone else he knew had gone on with their lives by that time, so he spent the days following that Christmas in his father's study, reading and catching up on his sleep, knowing that he would be headed back to the desert again in a few days. The weather was cold and harsh in northwest Kansas in December, with driving snow and drifts frozen between the houses. In the early evening of the twenty-eighth of December, the phone rang. He picked it up and answered. It was a girl that he knew, but only a little. His best friend had moved to Colorado and she had been his girlfriend when he left.

She asked if he would like to go out for a cup of coffee since she hadn't seen him in over a year and a half when he had go on active duty. She also remarked that he had a new car and she wanted to see it.

He said, "Fine, I'll pick you up in ten minutes." He pulled up in her driveway, got out, and went to door of her garage apartment. She met him at the door and followed him to the passenger side. He opened the door for her as she came around and got in the car. She liked the silver-gray Celica a lot: it was sporty and had a tinted glass moonroof that made it interesting. She asked how he'd been and he told her that since the last time he saw her he had been to Fort Knox, Kentucky; Fort Benning, Georgia; and now was assigned on temporary duty to Fort Bliss, Texas. He explained, "I'm hoping to finish up at Fort Bliss in the next month or two and then return to Fort Knox, probably to a basic training unit where I can serve, until I have two years active service so I can submit my application for law school and transfer to the Judge Advocate General Corps as an Army lawyer."

By this time, they had arrived at the restaurant. It was a twenty-four-hour diner. They went in and sat at a booth about halfway down the aisle, almost the exact same place he had sat with her and his best friend when he had met her for the first time two and a half years earlier. A chance encounter that would change his life forever.

Chapter 10
Silver Eagles

He opened the refrigerator to make sure it was working. It was, so he went over to the PX and bought some small adapters for his appliances, frozen dinners, instant coffee, and breakfast bars. Then he went back to the company area for evening roll call. When Guerra released the formation, he went over to Sergeant Robb and asked him if he could locate a couple of transformers for him. Robb told him, "No problem. They will cost twenty-five dollars apiece and I'll bring a couple of them in the morning."

The Lieutenant thanked him and went back to his room where he changed into his usual evening attire before heading down to check in with Buzz and see what was for dinner. He gave Buzz fifty bucks for the communal groceries, before he grabbed a beer and sat down. Buzz was busy making a roast chicken, with vegetables, and German rolls. While Buzz opened the oven to check on how the meal was coming along, the Lieutenant looked out the window and remarked, "It looks like our lady is not taking up her usual place in the window tonight."

Buzz looked out and said, "Yes, it's getting dark earlier and way too cold, so it appears she's finally given up."

Buzz asked the Lieutenant how things were going. The Lieutenant replied that he finally had his hold baggage and was looking forward to getting settled in and seeing some of the country as soon as he could get his car picked up.

The Lieutenant asked Buzz how things were going with him. Buzz said, "I'm happy as can be. My number finally came up on the full colonel's list and my eagles will be pinned on the day after tomorrow. By the way, Bob, do you have a set of greens I can borrow? The only uniform I ever used is a short sleeve shirt and green pants without a hat."

The Lieutenant said, "No problem. Do you have the brass? If you do, I can fix up my greens blouse and pants, and loan them to you."

"Yes, I do. I'll get the brass to you after dinner."

Buzz then asked him, "What are you doing for the next couple of weeks?"

"I'll be here in maintenance and training for another week or so. Then we'll be road marching to Baumholder for gunnery."

Buzz looked at him questioningly, "Baumholder?"

"Yes, I hear it's a real garden spot."

Buzz shook his head and grimaced, explaining, "I've never been there personally, but by all accounts, based upon what I've heard from dentists stationed there it is, in short, worse than awful. The post is on a mountain close to the French border, where neither the Germans nor the French want to live because the weather is always terrible."

The Lieutenant just shrugged and asked, "Is dinner ready?" There wasn't anything else to say; the hits just kept coming.

When they finished dinner, since it was just the two of them, they put the leftovers in the refrigerator and did the few dishes they had together. Buzz made a plate to take to Elizabeth, who was busy grading papers in their room. The Lieutenant followed him to the doorway. He said good evening to Elizabeth and picked up the Dental Corps brass and silver oak leaves from Buzz.

Back in his room, he pulled out his copy of the Officer's Guide and used it to place the branch insignia, leaves, and other devices on the greens properly. Then he took his greens, a green long sleeve shirt, black tie, and belt back to Buzz's apartment and knocked on the door. When Buzz opened the door, he handed them to him. Buzz told him that he would be able to return them Friday afternoon when he was off duty.

The Lieutenant said, "Don't worry about it. I have battalion duty coming up Friday. I'll catch up with you over the weekend." Then he went back to his room to shower and change before draping a poncho liner over his entire bed. He read a couple of chapters of *The Dogs of War* and fell fast asleep.

THURSDAY – THE GAMBLER

After PT next morning, Sergeant Robb came over to him and told him that he had two transformers in his van and they were really heavy so he suggested, "Sir, I'll take you in my van over to the BOQ so you just have to carry them up to your room." When he got to the van, the Lieutenant looked at the transformers and picked one of them up. It must have weighed over twenty-five pounds. When they got in front of his quarters, the Lieutenant hopped out, grabbed both transformers, thanked Robb, and said, "I'll pick up some cash at American Express and pay you as soon as I can."

"That's fine, sir. Your tank is going into queue service today and should come out tomorrow."

The Lieutenant thanked him again and fought his way through the front doors of the building and up the stairs lugging the fifty pounds of dead weight. Once he got to his room, he sat both of them down, opened the door, and placed one in the living room and the other in his bedroom. He plugged his radio into the one in his bedroom and his hot pot and toaster oven into the other one on the small dining table in the living room. Then he took a shower, got dressed, and sat down to read for a few minutes, waiting for the bank to open. When he saw 0830 on his alarm several hours later, he shot down the stairs to the bank to cash a check to get fifty bucks to pay Sergeant Robb.

He went back to the company area and found Sergeant Robb at battalion maintenance where Q-service was being done on 1-1. Since it was his tank, he told Robb that he wanted stay with the tank and the crew to help do the service and watch them to make sure he understood everything that had to be done to his vehicle. Sergeant Robb told him he would be taking the balance of the platoon to the sand table for another exercise and then release them for the rest of the day to get their gear and laundry cleaned and ready to depart for Baumholder next week.

Sergeant Robb also reminded the Lieutenant that he had duty Friday night, and that the battalion would have its annual Pre-Thanksgiving Baumholder dinner for the soldiers and their families that Saturday. The Lieutenant thanked him for reminding him, and that once he got off duty, he would try to get a little sleep, but he would make sure he was back for the dinner.

He stayed with his crew, helping them as best he could, staying out of the way, and visiting with the battalion maintenance chief warrant officer, whose sister was married to Kenny Rogers, of all people. They talked about the battalion, Germany in general, and what he could expect when they were in Baumholder. The chief told him that he could expect to experience all four seasons in one day. "By that I mean it will be warm at the base of the mountain, where the troop billets for four battalions and division support command are located. As you drive your tanks, climbing up to the gunnery ranges at the top of the mountain, you will go through rain, sleet, snow, and then freezing rain. And, by the time you get there for gunnery, you'll probably be on hold because of fog. So, you will have gone through heat, rain, snow, and ice, all in one afternoon."

With evening coming on, the Lieutenant went to final formation and told Sergeant Robb that he would be back in the morning and back with this tank for Q-service through Friday. Sergeant Robb told him he would be sure and check in with him Friday afternoon, as the whole battalion would be loading up to march out on the following Thursday. The Lieutenant thanked him and went back to his room.

Once there, he went down to the basement, picked up a can of Coke from the machine before he made his way back upstairs, to have it with a frozen dinner, hot out of the toaster oven. After a quick shower, he picked up a pen and started a long letter to his girl. As he did that, he thought of the song, *Sad Eyes*, and held close the last letter he had from her – it still smelled like the *Windsong* perfume she liked to wear.

FRIDAY - ROCK 'N' ROLL FANTASY

Friday dawned, chilly and gray. He beat the alarm, jumped up, and jogged over to the company area where he saw Sergeant Robb waiting for him. Robb told him that the First Sergeant had a meeting of the NCOs that morning and the old man wanted to see the Lieutenant after PT. After a short run, he went downstairs and reported to Captain Batson in his office.

Batson returned his salute after a couple of seconds, and said, "A new Company Commander, a West Point classmate named Borman, is coming in and he will take over the company before we go to Baumholder. I need you to get with Sergeant Robb and lay out all the equipment for 6-6 and your entire platoon for inventory with the incoming Company Commander at 1000."

Then he added, "I understand that your platoon is coming out of Q-service today. As soon as all your tanks are up, you need to take them on a short road march, about an hour, in the training area, and back, to shake them out and make sure that everything's ready to road march to Baumholder next Thursday. On Monday, you will be accompanying me and Captain Borman to brigade to participate in your first 'Pegasus' war game exercise. That will take until the middle of the afternoon. When we get back, you will need to make sure that your platoon is ready to march to brigade for tank gunnery training on the miniature firing range using an M-16 attached to the tube next

Tuesday. Captain Borman and I will be on border recon Tuesday and returning late that same day. Wednesday morning there will be a short change of command ceremony and a company inspection to make sure that everything is ready for the road march the next day. Dismissed."

The Lieutenant saluted, executed an about face, and left his office. As he stepped in the hall, he thought, *And tonight's my first night on battalion staff duty. It's going to be a long day.* He went over to battalion maintenance to see how his tank was progressing in Q-service. He quickly learned that his tank was almost through and the crew was preparing to return it to the motor pool. Sergeant Robb's tank, 1-4, had been pulled into the maintenance bay already and the engine was removed. Sergeant Robb's crew and the battalion maintenance mechanics were busy trying to get everything done to complete the Q-service by the middle of the afternoon.

Trying to do a full field layout of equipment and tools for each of the platoon's tanks, plus 6-6, with two of them in maintenance and four on the ready line, was going to be a challenge. Sergeant Robb, however, had everything under control. He had his crew layout their equipment beside 1-4. Then he and the Lieutenant walked down to the ready line and had the crew members for each of the other tanks lay out all of their tools, helmets, radios, and the other equipment with the exception of the machine guns, grease guns, and pistols, that would be inventoried by Batson and Borman, in the arms room by serial number.

Around 0945, the Lieutenant and the crew members sat down and relaxed a little bit as they waited for the inspection. A few had the blue flicker in the backs of their eyes, but none of them was obviously stoned, so he beathed a sigh of relief and shot the shit with them. Blaylock, the Lieutenant's gunner, asked if he had any licorice on him and the Lieutenant answered, "Regrettably not, but I'll be loaded up for the trip to Baumholder."

Finally, a little after 1030, the two captains came up to the platoon. The Lieutenant and Robb stepped forward. The Lieutenant saluted and reported, "Lieutenant Manteuffel, First Platoon." The captains moved from tank to tank, counting all of the pieces of equipment known as BII, or Billy, as it was called. Everything was accounted for, so the Lieutenant, Sergeant Robb, and the two captains walked up to battalion maintenance, where they reviewed the equipment laid out for 1-4.

When they were through, the Lieutenant signed for the tanks and the equipment for the first time since he had arrived in country, due to his being in the field almost continuously and Sergeant Robb had signed for it earlier. When they were through, Borman thanked the Lieutenant and Sergeant Robb before he turned to depart the building and follow Batson back to the company area.

ROUND IN THE CHAMBER – OFF SAFE

Sergeant Robb turned to the Lieutenant and said, "Sir, I'm hoping to get 1-4 out by 1430 or so, but I wanted to talk to you because I know you're going on duty tonight. First, before you go on, you need to draw your service .45 and shoulder holster from the company arms room, with three full clips and an ammo pouch. Put one clip in the pistol and rack it to put a round in the chamber. Then, take that clip out and put in a fresh clip so you have an extra bullet. After that, put the first clip, and the remaining full clip, in the ammo pouch and put it on your LBE web belt. Be sure and attach the lanyard from the .45 to the holster and be sure to take your steel pot with you when you report to the S-1 at 1630."

"Once you have the instructions for the evening from the Adjutant, check in with the Duty NCO and have him call down to all of the company First Sergeants and CQs to let them know that you're going on duty, that way they'll know who you are. Then you simply need to follow the SOP, stay in contact with the Ready Reaction Force at the motor pool, and radio the detail that's on guard duty in the preposition stock point area. Make your checks of the arms rooms in the companies once before midnight and once after midnight. You know we discussed how most of the officers do this."

The Lieutenant turned to him and said, "You know, I've given that some thought and I won't falsify a government document. So, let the word go forth to friend and foe alike, I will make my checks. I'll make the first check as early as I can in the evening and the second as late as I can the following morning. I will check each and every arms room and I will initial the check forms correctly. I will have the big duty officer's ring of keys with me and I will be shaking those keys the entire time I'm walking through the barracks so everyone there will know I'm coming. They'll have no reason to run into me. I'm not a narc and I'm not the sex police. I won't go looking for trouble, but I won't run if trouble comes looking for me. I am my mother's son, a hillbilly from central Tennessee, Scotts Irish to the core. She taught me several things and this one in particular: 'You can back up boy, but you can never back down.' I'm making my checks with my .45 drawn, round in the chamber, and a full clip, off safe, so the only thing that will keep it from firing is the palm switch on the hand grip."

"Now, should any of our friends try to rush me, I'm going to shoot the first one in the right knee and, since all automatic weapons in the west ride up and to the right, the second bullet's going to catch him crotch high. I don't know much about drug dealers, but I'm hard pressed to believe that there is one anywhere who wants to do twenty to life in Leavenworth with one leg and his balls blown off. So, that's how it is. Let the NCOs know and pass it on to their troops."

Sergeant Robb answered, "Very well, sir. I'll do that. I'm going to catch a little lunch and then come back to try to get 1-4 finished up. After lunch, I'll have the rest of the platoon meet you up in the day room so you can go over individual and platoon gunnery and maneuver tactics on the sand table. Then, if you would, release them to barracks to get their equipment ready and be prepared to road march at 1500."

"Thanks," the Lieutenant replied, "I'll see you at 1430, then."

COULDN'T GET IT RIGHT

After lunch at the snack bar, the Lieutenant went into the barracks and up to the dayroom on the fourth floor, where there were card tables, a pool table, and the sand table, where he would be working with his platoon. The Lieutenant had a good time going through various scenarios with his tankers, building on their experiences in the local training area and on Reforger. Several of his soldiers had been to Baumholder. They expected, for the most part, to be on fog watch and freezing on the gunnery ranges waiting for conditions to clear long enough to shoot something. The Lieutenant took that to heart and as soon as he had run through all of the scenarios he could think of, he released them to be prepared at 1430 to go down to the motor pool and get ready for a short road march.

He hurried back down the three flights of stairs and out the front door to Battalion Maintenance. When he got there, he found Sergeant Robb more agitated than he had ever seen him before. Apparently, his tank's Q-service was done, with the exception of a couple of pieces of dead track and a filter. Robb told him that he could scrounge up the track, because most of the platoon sergeants kept a couple of pieces of track strapped to the sides of their turrets for just such an occasion. His problem was a filter, that had been ordered two days ago but had not yet arrived from division to brigade maintenance for distribution to the battalion. As a result, it didn't look like his tank would be ready to roll until late in the afternoon. The Lieutenant told him, "If we can't get out into the maneuver area by 1530, you will have to take the platoon out and do the shakedown yourself since I'm going on staff duty."

The Lieutenant then walked down to the motor pool and informed Aguinaldo, his Guamanian staff sergeant and senior NCO, who was aware of the situation and told him to have the platoon standby until he was given further instructions, and to make sure that the tanks were cleaned up and everything prepared for the upcoming gunnery. He went back to Battalion Maintenance and waited with

Sergeant Robb for any word of the missing filter. At 1600, the phone in the maintenance warrant's office rang. The chief told the Lieutenant and Robb that they were wanted in Bravo's XO's office ASAP. They ran across the motor pools and up the steps to the office where the XO, Lieutenant Marley, and Batson were waiting.

LYIN' EYES

As the Lieutenant and Robb walked through the office door, Captain Batson turned to the Lieutenant and told him, "I had Captain Borman stay in the company to finish inventorying the arms room and the supply room with the First Sergeant. And the supply room. That's because I wanted to talk to you. Why is 1-4 still on deadline? We've reached the reporting period and it's Friday, so that tank won't come off until Monday. And, as you can tell, I am royally pissed."

Before the Lieutenant could answer, Marley, who most resembled a small, cornered rat, chimed in, "The filter we need to get it off deadline is at Brigade. I let," he motioned over to Robb, "Sergeant Robb know, but he's never gone to get it."

The Lieutenant looked sideways at Sergeant Robb and could tell, immediately, that the conversation had never happened. Captain Batson then turned to Marley and asked, "Is the filter we're missing at Brigade now?

"Yes, but Brigade is shutting down and the last parts run just arrived without the filter on it. It's so late now that by the time we get a vehicle to Brigade, the parts NCO and Brigade Maintenance will be gone for the day and the weekend."

Seething, Batson scowled at the Lieutenant and said, "Very well, Herr Lieutenant: you will not be attending the 'Pegasus' exercise with me and Captain Borman on Monday. You will make arrangements to have that filter picked up. Sergeant Robb, you will have it installed first thing Monday morning, after which, you two will take your platoon to the training area and shake all the vehicles down. You will then have all of them washed and ready for the road march to

Brigade to shoot the mini-gunnery range Tuesday afternoon. And, when Captain Borman and I return from Brigade on Monday afternoon, you'll also stand for a full inspection by us. Oh, and by the way, if by chance you can get the dead track replaced and the filter installed at any time between now and Monday, let me know. This evening from 1900 to 2100 I will be at the officers club for the battalion hail and farewell. Herr Leutnant - you will be hailed in your absence since you're on duty. Other than that, I will either be in the company area or my quarters for the rest of the weekend. And Herr Leutnant, tomorrow night, Captain Coyle and his wife are having a farewell party for me, him, and the junior officers in our companies since we are departing for CONUS together. We're going to be going to the advanced course together at Knox, since we extended for command here. I understand you know Harry."

"I do," said the Lieutenant, "he's VMI '74, my dyke's class: a first class-man when I was a first-year cadet."

"Good. His address is 31C Jackson Strasse. I'll see you there."

With that, Batson shot out of the office, headed for battalion maintenance to go over the deadline report with the BMO before it was submitted to the battalion commander and the brigade XO.

The Lieutenant looked over to Sergeant Robb and quietly said, "Leave us."

Robb quickly departed the office. The Lieutenant focused his eyes directly on the executive officer. He could feel the anger welling up in him and he coldly addressed Marley. "I could tell by the look on your face, and my sergeant's face, that the conversation you related to the commander never happened. You lied. I'm going on duty tonight, so I don't have time to tend to you, but if you ever lie about me, or one of my troops, again, so help me – I'm not a butter bar fresh out of Knox. We're the same rank...I will kick your ass from here to Battalion and back. Do you understand me?"

The XO turned his gaze down toward his desk, and sheepishly just said, "Yes."

The Lieutenant hurriedly exited the office, still steaming, and went down to his platoon on the ready line. He told his Guamanian sergeant that since they were missing a filter on 1-4, they would be taking the platoon out in the maneuver area on Monday. He also told him that Sergeant Robb would be coming down to take over the platoon since he had to go draw a sidearm to go on staff duty.

The Lieutenant then set off for Battalion Maintenance to see Sergeant Robb. When he came into the maintenance bay, where 1-4 was waiting, he walked up to Robb and asked him, "What the hell's going on?"

"Well, sir, Lieutenant Marley is looking out for his OER. Since Captain Batson is leaving, he'll be rating him on the way out. Also, he was Third Platoon Leader before he became executive officer, so it helps Sergeant Ramsey and him if he makes me, and you, look bad."

The Lieutenant thought for a minute and said, "All right. Well, I'll tell you, that's not going to happen again on my watch. From now on, I'll keep a close eye on him. I already told him if he crossed me again, I'd kick his ass. It's the Cav way. I haven't been this upset since I stood on a major's desk at Bliss screaming at him that if he 'ever cut the rations behind my soldiers again, so help me God, I'd make him wish he never met me.' Oddly enough, that's not an uncommon occurrence in the Third ACR. And, as my mentor there, Major Riley, told me that day, 'Lieutenant, you're learning.' Anyway, I just told Sergeant Aguinaldo that you'd be coming down to take over the platoon, and I'd be going on duty. Looks like I'll see you Monday. If you need me for any reason, I'll be here until seven in the morning. After which, I'll be in my quarters, here for lunch, or at Captain Coyle's for about two to three hours starting from 1900 tomorrow night."

NEW KID IN TOWN

The Lieutenant was running a little behind schedule, so he jogged over to B company, went downstairs, and told the First Sergeant he would be drawing a sidearm for duty. He went to the arms room and told the armorer that he needed his .45 for staff duty with three full clips, a shoulder holster, lanyard, and an ammo pouch. The Specialist Five armorer issued him the equipment. He signed for it, put a clip in the pistol, slipped it in the holster with the lanyard attached, put the other clips and the ammo pouch in the pocket of his field jacket, and left to go to his quarters. There he buttoned a liner in his field jacket against the cold of the evening, and picked up his load bearing equipment, two poncho liners, and his helmet for duty. With all of those items in hand, he proceeded to Battalion to see Captain Steves and report in. The Adjutant told him that he did not have any special instructions other than what was in the duty notebook that was down with the Duty NCO, a Sergeant First Class, from Charlie Company.

He saluted the Adjutant and walked down to the duty office, where he set all of his equipment down at the duty officer's desk and introduced himself to the Duty NCO. He told the NCO, "If you want to get chow, go ahead now. I'll eat when you get back." The NCO left immediately. The Lieutenant sat down by the phone at the NCO's desk, pulled the duty notebook out, and reviewed all of the provisions again, especially those dealing with the Ready Reaction Force, in case the Preposition Stock Point was attacked for any reason. That being done, he leafed through a couple of magazines, *Armor and Army*, and skimmed the *Army Times* newspaper while he waited for the staff to leave for the night and the NCO to come back.

When the NCO returned from the mess hall, he sat down at his duty desk and took the .45 from his shoulder holster. He removed the ammo pouch with two magazines from his field jacket and attached it to the web belt of his load bearing equipment. The third magazine was in the .45 already so he racked and jacked it to put a round in the chamber. Then, as Robb had instructed him, he ejected the clip from the .45, took one of the fresh clips out of the pouch, and slammed it in the butt of the pistol so that it contained a full clip in addition to the extra round. Then he returned the clip he had ejected to the ammo pouch and attached the lanyard to the pistol butt and the shoulder holster. It wasn't too cold yet, so he returned the pistol to his shoulder holster and left his jacket at the desk before he walked down to the mess hall. He paid the standard rate for his meal, ate quickly, and headed back to the duty office so he wouldn't miss anything.

Back in the duty office, he took the duty officer's key ring which had the keys to the offices in the headquarters, and the file cabinets as well. It was massive. The sun was setting rapidly; it was a pleasant, clear, and cool evening. He decided to walk around behind the barracks near the parade ground. He clicked the pistol off safe, came around the east corner, and passed the personnel center. He could see there was still considerable activity in the medics' offices. He walked by quickly, saying, "Hello," and kept moving moved around the barracks and the mess hall back door so he came into Headquarters Company from the west.

He nodded a silent greeting to the CQ, who was on the phone. Then he went downstairs to the arms room, where he checked the locks and initialed the check sheet on the clipboard attached to the portal in the cage where weapons were issued. He made a notation on the standard issue green notebook in his left breast pocket and proceeded out of the building and across the courtyard to Combat Support Company. After saying hello to the CQ, a buck sergeant from the Scout Platoon, he went in the basement of that billet and made his check there in the same way. He repeated the routine for Alpha, Bravo, and Charlie Companies, visiting briefly with that CQ in each one. Wherever he went he jangled his keys so that the troops in the barracks knew he was coming. All the barracks were the same: dark and depressing walls everywhere, similar to the first day he arrived, with the exception of the noise from soldiers talking to each other in their rooms, and the mixed music from competing stereos mingled with the strong smell of smoke drifting down from the rooms above.

When he was done, he went back to Battalion and asked the Duty NCO if there was any message traffic or calls for support. The Duty NCO, who was a chain smoker, said, "No, but if things go as usual, we should expect a flash traffic message in the next hour or so and after that, maybe another one about two in the morning."

I LOVE THE NIGHTLIFE

The Lieutenant looked at the Duty NCO and asked him what duty officers would normally do when they weren't making their arms rooms checks or decoding messages. The NCO told him, "Sir, the officers usually go down to the battalion commander's office. There's a sofa there and they try to catch a nap in order to get at least some sleep during the night because they usually have to go on duty the next day." The Lieutenant told him that he understood. He estimated he would be getting off duty about 0800 the next morning and he would be back at around 1130 for the battalion's Thanksgiving dinner. Then he would try to get some sleep before going over to Captain Coyle's quarters for that farewell event, so it looked like he would need to grab as much sleep as he could during the evening hours. He thought to himself, *It would also be good to get away from the chain smoke in the duty office.*

He walked down the dark hallway to the colonel's office and found the couch in a corner next to a reading lamp and sat down. He had taken a poncho liner along with him to cover himself for additional warmth, as it was getting cold out, and it was seeping into the offices. He laid down on the couch and tried to make himself as comfortable as possible, but he didn't have any sort of pillow or headrest, so he went back to the duty office and picked up his TA-50 and steel pot. He rested his head on the helmet and covered himself with the other poncho liner to relax and wait for the incoming message. As luck would have it, just as he started to fall asleep, the Duty NCO came halfway down the hall screaming, "Sir, sir, we have flash traffic." The Lieutenant woke up as best he could and scrambled down the hallway.

He sat next to the NCO as he received the numbers and letters that made up the coded message. As soon as they had received all of the characters, the NCO read them back to make sure they were correct and handed the message to the Lieutenant. He took his key ring and went downstairs to the S-2's office, which he unlocked. He then unlocked the safe to get the code book with the codes for that day. He turned on a reading lamp by the S-2's desk, sat down with the message and the book, and began to break the message. As it turned out, it was only a test message. After he had decoded it, he placed it in the safe with the book for the S-2 to review the next morning.

He walked slowly upstairs to the NCO, who had logged the receipt of the message. The Lieutenant wrote in the entry for decoding the message, put in the time, and initialed it. Once he was done, he asked the NCO if he" made a telephone check of all the CQs in the companies. The NCO told him, "No, sir," so the Lieutenant directed him to do it in the next few minutes and make entries in the log with regard to each call. Then he walked down the long hallway to the colonel's office to lie down on the couch, where he made himself as comfortable as possible, using part of the poncho liner as an additional pillow on the helmet, with the rest draped over his chest and head for warmth. Reminding himself to bring a second poncho liner next time, he promptly went to sleep.

The evening was uneventful, so the Lieutenant was able to sleep lightly but fairly soundly, like being in the field, until a little after 0200 when the NCO ran down the hallway and knocked on the door to the colonel's office to awaken him. The Lieutenant took a deep breath and looked at his watch before he ran down the hallway to the NCO to pick up the next message. The Duty NCO had already transcribed the message and had it read back for accuracy when the Lieutenant walked in the doorway, so he handed him the message, and the Lieutenant took the keys and went back downstairs where he again broke the message. It was another test message. He breathed a sigh of relief and placed that message in the codebook before locking everything back up and walking up the stairs to the duty office.

EVERY KINDA PEOPLE

Standing at the door, he asked the NCO what he could expect for the rest of the evening. The NCO told him that it would probably be quiet until 0600 because the colonel and the S-3 would come in early on Saturdays to plan upcoming exercises and make sure that orders were properly issued for the upcoming week. Since he was going to have a busy day from about 1100 on, the Lieutenant decided he was going to skip breakfast in the battalion to try to get some sleep and since he was up, he would go ahead and make his final checks, and close out his log in order to get a couple of hours of rest before the colonel came in.

The Lieutenant looked at the Duty NCO and told him, "I'm going to go ahead and make my checks. Call the CQ in each company and let them know I'm coming." He pulled off the shoulder holster and put on his field jacket. Then he slipped back into the holster, took his ring of keys, and drew the .45, making sure it was off safe. Then he stepped out of the headquarters into the cold dark of night. It was near 0300, the deadest time of night. He decided to check the arms rooms in reverse order from what he had done before, so he went to the first horseshoe set of billets and entered Charlie Company first.

He walked past the CQ and told him he would be checking the arms room as he went down into the pitch darkness of the basement. He flipped on the hallway light switch at the bottom of the stairs and proceeded to the arms room where he initialed the time he entered on the check sheet attached to the door to the arms room, then he turned slowly, walked down the hall, turned out the lights, and went up to the CQ desk to let the Specialist there know that he was leaving the company. The whole time he was jangling his keys. He did the same in Alpha Company and Bravo Company. All of them were silent and he moved quickly, especially in Alpha, that for some reason seemed more ominous.

Then he walked to the horseshoe of barracks to the west and did the same in Combat Support Company before going into Headquarters Company. There was still a lot of activity there, voices, and occasionally some shouting, echoed through the building. A couple of stereos were still cranked up and rock music filled the building. The smell of smoke, cheap cigarettes and hash was thick in the air. He also picked up the faint scent of tepid beer and vomit, not unusual on a weekend in the barracks. He went past the CQ desk after greeting the buck sergeant on duty, a gunner from the headquarters tank section. Once in the basement, he flipped on the lights and checked the arms room there as he had done in the other companies. While he was filling out the form, he thought he heard footsteps behind him but turned around and saw nothing.

He walked up out of the basement and through the doors into the courtyard. He could see there were lights on in the mess hall, so he went over to see who was up there. He saw one tall, lanky African American soldier who was busy baking bread and biscuits for the day. Occasionally, a soldier would slip in, visit with the baker, and then leave with something. The Lieutenant observed this for a while and then proceeded back to the Headquarters where he logged his checks in on the duty log before plodding back to the colonel's couch to try to get a little more rest.

A gray and foggy dawn finally came. The Duty NCO came down and knocked on the door to advise him that the colonel had pulled into the driveway in front of the headquarters. The Lieutenant thanked him for the heads-up and looked through bleary eyes at his steel diver's watch; it was 0633. He went down to the battalion washroom, across the hall from the duty office to wash his face, wake up, and make himself as presentable as possible. He also took his steel pot, TA-50, field jacket, and poncho liner back down to the duty office. When the colonel came into the headquarters, and stopped by the duty office, the Lieutenant popped up from behind his duty desk and reported to him. The commander asked him if anything unusual had happened and he replied, "No." The colonel also asked if there was any flash traffic and the Lieutenant told him that they had received two messages, but both of them were tests only. The colonel then asked the NCO if he could put on a fresh pot of coffee down in the S-3's office, as the S-3 and XO would be joining him at around 0700 and he liked to have a little quiet time with a hot cup of Joe before they started their meetings.

Chapter 11
Weekend in New England

The Duty NCO returned from making coffee, so the Lieutenant told him that he could go ahead and get breakfast in the mess hall since they were serving at 0700 and when he got back, they would wait for their relief at 0800. The Lieutenant had a breakfast bar in the pocket of his field jacket, so he ate that with a small cup of fresh coffee for breakfast while he waited for the Duty NCO to return, which he did at 0750. At 0800, the new Duty NCO took over the duty desk. The outgoing NCO wished the Lieutenant well and shot out the door to go home to his wife and kids.

The new duty officer, who was the Scout Platoon Leader, arrived five minutes later and asked if anything unusual had happened. Since nothing had, he signed in, and the Lieutenant signed out, before breathing a sigh and happily exited headquarters to go back to Bravo. He ejected the round in the chamber, returned it to the light clip, then cleared and turned in his sidearm to the armorer. Having done that, he carried his poncho liner and equipment back to his BOQ room, laid everything out on the sofa in the living room, took a quick shower, and went to bed. It was to be a short nap since his alarm was set for 1100 and it was already after 0830.

He had no sooner laid his head down when the alarm jingled. He quickly showered, dressed, and walked back to Bravo. When he got there the First Sergeant told him that the soldiers had been released to go to Thanksgiving dinner and that they were moving through the chow line. He entered the mess hall and picked up a tray to be served the standard Army Thanksgiving dinner, which was ham, turkey, dressing, mashed potatoes, gravy, green beans, cranberry sauce, rolls, and pumpkin or apple pie. Once he had his tray together, he looked across the dining facility and realized that the married soldiers had brought their families along, so the place was packed, and the room that was normally reserved for officers was full of sergeants and their wives and children.

Out of the corner of his eye, he saw Sergeant Robb and Dawn at a small table in the corner so he went over and asked her and Robb if he could sit with them. She said, “Yes,” right away, so he pulled over a chair. Sergeant Robb asked him how duty had gone and he told him, “I didn’t have any incidents, but Monday I have a couple of questions about the battalion.”

Sergeant Robb told him, “Sir, I thought you would.” The Lieutenant also asked if there was any news with regard to the filter situation. Robb told him there was no news on that front so it looked like he was going to have to borrow the First Sergeant’s Jeep first thing Monday morning to retrieve the filter and bring it back to have it installed so they could road march the platoon that morning.

Halfway through the meal, the battalion commander and his lady, Jeannie, came through the line and everyone came to their feet to greet them. Lieutenant Colonel Shea welcomed everyone there and told them that he hoped they enjoyed their Thanksgiving dinner. They had to eat it early in the month since they were going on alert Thursday morning and wouldn’t be getting back until the Tuesday before Thanksgiving, from gunnery and field exercises in Baumholder. Then he told the troops and their families to have a good weekend, and he and Mrs. Shea sat down to have dinner with their two sons.

As soon as the Lieutenant was through eating, the Lieutenant excused himself from Sergeant Robb and Dawn. He caught sight of Captain Batson and Captain Borman and their wives in a booth, so he stopped by and said, "Hello," to them and let them know that he had showed up for the formation. Then he excused himself and told them he was going to his quarters to get some sleep since he was just coming off duty. Captain Batson told him that he would see him later and to be sure and let him know if there was any progress with regard to the filter for 1-4. The Lieutenant said he would and then promptly left the mess hall to return to his quarters.

It was a pretty day, cool and sunny, like Thanksgiving back home in Kansas. He leafed through memories of those holidays and others, not as traditional as those he had enjoyed over the years. Those included recollections of leg of lamb and mint jelly at a German restaurant in El Paso, and stuffed shrimp in Bridgeport, Connecticut after traveling in cadet grays from Virginia to New York with thirty hookers and their pimp, a six-foot-four black man attired in a purple Nehru jacket and elephant bells with a matching Panama hat, six-inch high heeled platform shoes, and a four foot ebony walking stick topped with a silver skull the size of an eight ball. It was a religious experience, hurtling through the darkness, on an express Greyhound bus that left D.C. at 6 p.m. and arrived at the Port Authority exactly at midnight. Known as the "Jungle Run," he prayed the entire time not to be in an accident, since there was no way he wanted to try to explain to his mother what her eighteen-year-old son was doing with these good people. Join the Army and see the world.

TIME PASSAGES

He was dead to the world. When his alarm jangled at 1730, the Lieutenant, who had simply slept in his fatigues, got up and changed into slacks, a sport shirt, sweater, and civilian shoes for the party at Captain Coyle's quarters. After a few seconds he was completely awake and remembered the address Batson gave him. He slipped on his VMI ring and walked there as quickly as he could. It was already almost dark and cold by then. He found the address and walked up one flight of stairs to Captain Coyle's apartment on the second floor of married housing in Benjamin Franklin Village.

When he walked in the door, he heard someone shout his name. He looked in the corner of the living room and recognized Bill McCallister, who had been one year behind him in the same company at VMI. Since they hadn't seen each other in several years, the Lieutenant walked over to McCallister to get up to speed on where he'd been since leaving school. McCallister told him that after Armor Basic at Knox he was shipped to the battalion where he had been a tank platoon leader in Alpha for about a year and a half. He had done well, so he was set to take over the Scout Platoon, which was a plum assignment, even though he was from VMI. He was dating the daughter of the general who commanded the MP Brigade in Mannheim, so his personal life was going well, too.

The Lieutenant brought him up to speed and all the places that he had been over the intervening couple of years. Then McCallister led him over to meet Captain Coyle whom the Lieutenant knew was from VMI, although they were not well acquainted, because they were from different companies in the Corps. After the introduction, he asked Coyle where he had been since graduation. Like Batson, he had attended Armor Basic and was then assigned to the battalion. He had opted to attend Motor Officer Course en route while Batson had gone to Airborne and Ranger Schools, so Coyle arrived in Germany a few weeks before Batson. Both moved rapidly through positions in the battalion and had extended their stays to be tank Company Commanders.

They were both slotted for Knox and Advanced Course before being assigned in their as yet to be designated secondary specialties. Coyle finished speaking and took the Lieutenant around to meet the other officers. Since he had already visited with McCallister, they went across the room to see Ray Kowalski, formerly a platoon leader in Alpha, who had been at VMI a year ahead of the Lieutenant. He did not know Kowalski very well and, in fact, he hadn't seen before in the battalion. Ray stepped up and reintroduced himself and his wife and told the Lieutenant that the reason he hadn't seen him was because he was the Support Platoon Leader. That job had him on the move most of the time getting supply and transportation set up for Reforger, then serving as an umpire in the exercise. Now that he was back, he spent most of his time circling back and forth to Baumholder to make sure that everything would be ready for the battalion's road march on Thursday. He was responsible for all of the food, fuel, and ammunition in the unit, and also for arranging for distribution of ammunition from the preposition stock point in the event of a full alert.

From Kowalski's description, it was a thankless job, requiring grueling work and many hours in the field. The Lieutenant thought to himself that if he could help it, he would not be assigned that job. He looked over into the far corner and saw that Lieutenant Marley was there as well. He didn't acknowledge him but moved on with Captain Coyle to meet Alpha's newest platoon leader, First Lieutenant McCrae. He actually knew McCrae a little. He had also been in the Fourth Training Brigade at Fort Knox, although not the same battalion. He was a friend of his buddy, Scotty, and, like him, was from western Pennsylvania. He had not gone to Fort Bliss with the Lieutenant, so the Lieutenant did not know him very well, only that he was a West Pointer of the same year group as Scotty and him. They visited briefly, until Captain Borman and his wife entered the room. Harry Coyle welcomed them loudly and then went around the horn to introduce them to everyone. As he did that, he hailed the new arrivals, and all said farewell on behalf of himself and Batson and their wives.

When Harry finished, his wife came out and announced that supper was ready. Like any good Army wife, she could feed four or forty on an hour's notice. So, in this case since she was moving and was trying to pack up most of her utensils she made tacos. Everyone broke into a grin because the one thing you couldn't get in Germany was Mexican food. She had set up a buffet line in the dining room with bamboo trays and paper plates, beer, tea, and ice water. To top it off, she had managed to get her hands on a large bag of taco flavor Doritos, the most craved snack food in Germany, because the PX didn't import them. The Lieutenant took a German beer made himself a plate with a couple of tacos and a few chips. Then he stood in the corner of the living room to visit with the Alpha platoon leaders. McCallister told them a few more stories from the battalion based upon his experience there. He also told them that if you wanted to know whether or not the colonel was mad you should watch his right foot. If he started stomping it, it meant he was angry and the more he stomped it, the angrier he was. "For those in the know, we keep an eye out for him and are wary of him if he appears to be upset. He's Blackhorse and black Irish, with a terrible temper. Other than that, he's a good commander; good to work with and he takes a keen interest in developing his young officers."

TEAM YANKEE

The Lieutenant finished eating and made the rounds, chatting with everyone there, even Marley and his wife, at least briefly. Having filled the social dance card, he went to Harry and Mrs. Coyle and told him that he would be leaving shortly as he had been on duty all night and really needed to catch up on his sleep. They thanked him for coming and he thanked them for having him. At the door, he took Coyle aside and asked him how it had been serving in this battalion for almost five years. Harry told him, "It was a great experience. My company is cross attached to the infantry to be part of Team Yankee. I have a little more leeway with regard to field training since, if push comes to shove, I'll actually be working for a different battalion. I really enjoyed being a Company Commander. Someday I hope to write a book based on it, titled *Team Yankee*, about what it was like to serve in the Cold War in a tank battalion. It will have to be set in the future though. Nobody would believe how it is now."

"I would. I've only been here three months, and I'd believe anything."

"Exactly," Harry laughed, "we're nobody."

"Anyway, it sounds fantastic and I hope I can read it someday. Godspeed." With that, the Lieutenant left the apartment and walked briskly back to his room under the dim light of the moon flickering through the fog.

NOBODY

When he got back to his room, it was early in the evening. He took another hot shower to relax and try to get to sleep before any excessive noise from next door came through the wall or the music streaming out from a stereo playing in a party upstairs got any louder. Lying in bed, the black dog of loneliness came in the room. He thought about what Harry said. Like everyone else in uniform right after Nam, they were shades, nobodies. Ignored, if not snubbed, in bus stations, airports, and motel lobbies, even travelling in civies.

Getting pelted by rocks, empty beer cans, and liquor bottles, called "dumb fucks" and "baby killers" by a bunch of drunk college kids from Georgetown as the Corps came off Pennsylvania Avenue at the end of President Carter's inaugural parade in January of '77 was bad enough, but looking back, he figured he bottomed out at a diner in Missouri where he stopped for lunch after driving most of the day across the plains on his way to New Jersey to get shipped overseas. The manager refused to cash a twenty-dollar American Express traveler's check for him because he was in military service. So, he left his Army ID as collateral and drove a quarter mile back across I-70, got five bucks worth of gas, and came back to pay and tip the waitress out of the change from that same check. As he put the ID card back in his wallet, he couldn't help himself, saying, "Well, if the Russians ever come through here, I bet you'll cash their checks."

With that, he climbed back in the car and drove all day and into the night to get to Scotty's folks house in western Pennsylvania where he could stay without the hassle of getting a room. It made him wonder why he ever signed up; then he thought of his girl and the future they planned, if he could ever get to law school. He curled up and came back awake to the sound of running water from the quarters above him. The shower upstairs was running constantly. After a bit, he got used to it, took a deep breath, and fell fast asleep, promising himself he wouldn't get up until at least the middle of the morning.

GOOD TIMES

He woke up to complete silence. Peering across the bed, his clock showed it was well after ten. Although he was Methodist and religious, he wasn't churchy. He determined that he would not be going to church but instead focus on getting ready for the day and the week ahead. He rolled out and turned on his hot pot full of water to make some coffee. After an extra-long shower, he sat in the club chair in his living room reading a few pages from the Army Times, that was always at least two weeks behind the times, a chapter of the Bible that he picked at random and finished two cups of what passed for coffee with a breakfast bar. Feeling a little more alive, he read a couple more chapters of *The Dogs of War*. The business of war never changed.

By 1130, he was awake and bored enough to begin making preparations for the field. He went to the PX and bought a small beef roast, fresh potatoes, a packet of dry onion soup mix, and canned vegetables for the communal dinner as he wanted to eat with someone that evening. He also stocked up on red licorice, breakfast bars, and a large bag of Snickers, so he wouldn't starve during gunnery. He returned to put the roast in the refrigerator and the rest on the counter, and then he put all of his other goodies back in his room and walked the hallway to see if anyone was up and around. From the looks of it, everyone in the BOQ was sleeping in as late as possible so he went back to his room, turned on the radio to listen to some soft rock music on AFN, and took a mid-afternoon nap.

Later that afternoon he came to when he sensed movement in the hallway, so he looked out his door and saw Buzz and Elizabeth heading for the kitchen. He walked down to see them and told Buzz that he had picked up a roast for dinner. Buzz grinned and began preparing it. The Lieutenant sat down next to Elizabeth and asked how her year was going. She told him that things it was going very well; she really liked working for the American government and visiting in Germany as it wasn't as dreary as England. Only having been at Heathrow himself, the Lieutenant had no knowledge of the weather there, but could sympathize with her, and from his limited experience, the food there was also terrible.

Buzz mixed the soup with a little water, and poured it over the roast, potatoes, and vegetables. When he had dinner in the oven the Lieutenant asked him if he could retrieve his greens and also asked how the promotion ceremony had gone. Buzz thanked him again profusely for loaning him his greens as they were tailored and fit him perfectly. He had the Lieutenant follow him across the hall to his room where he removed his rank and branch insignia and handed the greens back. The Lieutenant said he was happy to be of help, and although he hoped he would never need a dentist while he was in Germany, he was glad that he knew one.

He asked Buzz how things were back in the States. Buzz told him that everything was fine and that he planned to take two weeks of leave after the Christmas holiday, which would put him back at Fort Hood until New Year's. After that, he would come back to spend a few days sightseeing in Europe. He was covering for other dentists in his office who wanted to have time off up to Christmas day with their families and he was happy to assist them in that regard. Buzz asked him if he would like a beer and the Lieutenant said he certainly would, so Buzz handed him one before they walked back to the kitchen together.

As soon as they were back in the kitchen, Pavel came in and sat down with them at the table. Buzz asked him how his week had been. Pavel said, "I've been in the field the whole week, training the Scout,

Radar, and Mortar Platoons for their field qualification tests to be given at Baumholder." He looked bone tired as he continued, "I was on the move constantly, day and night, so I'm looking forward to a couple of quieter days in garrison before we road march to gunnery."

Buzz turned to the Lieutenant and asked him the same question. He replied, "It was pretty uneventful. I pulled my first night of staff duty and nothing really happened. I ate Thanksgiving dinner with the battalion yesterday and after that went to a little hail and farewell for Captain Coyle and Captain Batson, who are rotating out, and Captain Borman who is my new CO. Due to a parts problem, I wasn't able to get all my tanks off deadline before the weekend so I'm on Batson's shit list. With any luck, I'll get my last tank up in the morning and spend the day shaking the entire platoon down with a short road march and cross-country driving. Then we'll bring them back and get then cleaned up so we can go over to brigade Tuesday to fire at a mini tank range."

Never having been on one, the Lieutenant asked Pavel what the mini range entailed. Pavel said, "Well, you bolt an M16 to the gun tube of one tank and pull it into a large garage-like facility with a miniature range that has small metal pop-up targets. Once you have that tank zeroed, your crews will rotate through it in a simulated Table 8 for individual gunnery against various target types at various ranges."

The Lieutenant asked, "What is the rest of the platoon supposed to be doing while that one crew is on the vehicle?"

"They will do individual training, like friend and foe aircraft and helicopter identification, and friend and foe armored and wheeled vehicle identification using flash cards. They can also practice fire commands. The platoon sergeants will have ammo cans full of cards and other training devices to use, so don't worry about it."

The Lieutenant nodded and told him, "That's similar to what I did at the training center at Knox where I spent a year running rifle ranges with drill instructors and I did a lot of time on tank gunnery ranges at Bliss."

Pavel smiled a little and continued, "It's actually a pretty good time and gunnery is fun, too: you get to shoot things and have a break from the chicken shit in garrison, while the troops are too busy to blow dope or hit the needle, at least as often."

THAT'S ROCK 'N' ROLL

The Lieutenant then asked Pavel what Baumholder would be like. He answered, "Since we are road marching, after we get there, we'll spend the better part of two days fixing broken stuff, locating parts in the division supply area and if necessary absconding with them to get all of their vehicles in better condition by the next Monday. The line companies will rotate out into field exercises and basic maneuvers: movement to contact and fighting from successive defense positions. You will have an exercise, company on company, with an infantry company from 2/13 and a tank company from the battalion cross-attaching platoons and attacking and defending against each other. You'll also get a full day exercise for platoon leaders; just you and your tank with an infantry platoon leader in his track. You'll go to a mini-range that simulates the Fulda Gap and take up a firing position using the M16 attached to your main gun, and you can call for artillery from small tubes that put out little smoke rounds. Due to the limited amount of time at Baumholder, as opposed to Grafenwöehr, and ammo constraints, individual tanks won't fire Table 8 with the main gun. Tankers will fire their forty-fives, grease guns, and fire the machine guns on the tanks when they are attached to them on the firing line. At the end, your platoon will fire Table 9. With all of your tanks and the COs on line, you'll defend against a rolling tide of various armored vehicle targets. If you manage to kill enough of the approaching targets, you'll qualify your platoon."

The Lieutenant thanked him and Buzz asked how Baumholder itself was. Pavic told him, "We're getting there in November, with winter coming on, so we expect to have foggy and dreary weather at the base of the mountain, known as the 'Rock,' and as we drive higher and higher, we'll go through cold rain, freezing rain, snow, and then finally, windblown ice up on the ranges. The upside is you don't sleep on or inside your tank. There are good billets with officers separated from enlisted men and there is a fantastic club there with half of the building serving the enlisted and the other half serving officers. You can get really good Jaeger schnitzel, chicken schnitzel sandwiches, French fries, and beer, so other than night firing exercises and night road marches, if you were lucky enough to get back to the barracks and haven't frozen to death, you can look forward to a great meal and some camaraderie. Kind of like here."

Buzz nodded silently and asked the Lieutenant what he had coming up in the next week. This was always of interest to him because all of his weeks were pretty much the same. The Lieutenant told him that after "mini tank range Tuesday," on Wednesday there would be a change of command in his company and Captain Borman would be taking over for Captain Batson, who was leaving in the next week or so. Thursday morning, there would be a mock alert, in preparation to road march the ninety miles to the Rock, in rain or freezing mist.

Buzz looked at him, shook his head, and said, "I'm glad it's not me."

The Lieutenant just laughed, "That's life in the Cav. They like to ride you hard, put you up wet, in the desert, with an open Jeep and a driver. They'll toss you a case of Cs and ten gallons of water, pat you on the ass, and tell you they'll come back and get you in a week. It's just the way it is. But I have a great platoon sergeant, so I'm actually looking forward to it."

THE YEAR OF THE CAT

From the smell in the room, he could tell that the roast was ready. With Buzz's approval, they made their plates and sat down to eat. The Lieutenant glanced to see if the female signal officer was still engaged in her attempt to try to get Bernie back. Although it was early in the evening, dense clouds were overhead and there was a cold damp mist in the air. She did not appear, so as Buzz predicted, it was too cold and she'd given up or perhaps had moved on. Buzz asked the Lieutenant if he had any plans to go home for the holidays. The Lieutenant said, "No, I'm saving my leave to go back in March to get married in Abilene, Kansas."

Elizabeth said, "I'm sorry you will be away from home, but I'm excited for you getting married. I have to go back home to England to see my family for at least a couple of weeks, or it just wouldn't be Christmas."

She was finishing discussing her holiday plans when Lieutenant Storch came in behind them. When he saw that they were having supper, he fixed himself a plate and sat down next to the Lieutenant, who asked him what he had been up to. Storch said, "I've been sleeping in after working all Saturday night cleaning up the records for my ordnance company and the battalion. We have the annual general inspection coming up at the end of the week, so everyone is in panic mode. Big Foot was still working when I left this morning, so I don't expect him to be up and around until tomorrow."

The Lieutenant finished eating and excused himself, leaving Storch looking over at Elizabeth with the sad puppy dog eyes of the lovelorn. He went downstairs and grabbed a Coke for a treat since it was Sunday. He couldn't call his girl, so he wrote her a long letter to tell her about his upcoming week and that once he was in gunnery, he probably wouldn't be able to write or call her for at least a couple of weeks. He would try to send a couple of short notes out during the week because he knew she loved getting mail. When he was done, he sat the envelope on the table in the living room and tried to get to bed early. He couldn't sleep, so he read a couple of chapters of *The Dogs of War*.

The protagonist in that novel, Carlos "Cat" Shannon, was his personal literary hero. The Lieutenant aspired to be an officer like "Cat," both competent and highly regarded to the point that he could assemble a unit, with non-coms and soldiers, who would willingly follow him. He also had an affinity for "Cat," because as a mercenary, he was neither respected nor appreciated by the society that nevertheless needed his skills. He reached a good place to stop, went back down to the basement, and bought another coke from the vending machine to take up and put in his refrigerator. He would drink it first thing in the morning to wake up for PT.

THE FRENCH CONNECTION

Early the next morning, the Lieutenant heard Storch's door slam closed, so he jumped out of bed, went to the refrigerator, and drank half of the Coke before the alarm could go off. He wanted to get to his company at least an hour early to see if he could meet Sergeant Robb and solidify the game plan to obtain the filter from Brigade Maintenance, get it installed, and shake down the platoon so that they would have everything prepared to road march over to Brigade for the mini-gunnery exercise. He also had several questions for Robb about some of the things he observed while on duty. He dressed right away and once he had his PT gear on, slipped out into the drizzle of the morning and he jogged over to the company area.

Not seeing Robb, he went downstairs to the orderly room to see if he was there. As usual, Robb was there early, talking to the First Sergeant. When they were done, Robb followed the Lieutenant back outside the barracks to wait for the PT formation. The Lieutenant asked Robb the status of the filter and Robb told him that there was no word at battalion maintenance yet, but he was going to borrow the First Sergeant's Jeep and drive over personally to retrieve it and bring it back by 0900. With luck, it should take about half an hour to get it installed and another half hour to get the platoon assembled and ready to march.

"Sir, we should be able to get out the gate by 1000, shake the tanks down for a couple of hours, bring them back for lunch, and spend the afternoon on post march maintenance, fixing anything that broke down in the training area. After that, we'll be cleaning them so they will be looking good at the mini gunnery range."

The Lieutenant thought a minute and said, "I have a couple of questions about duty."

"Yes, sir."

"The first thing I noticed was a large crowd around the aid station early in the evening. What was going on?"

"Alright, the medics in Headquarters Company are the source of clean needles, syringes, and codeine cough syrup, the type drank by the crew of bridge tank when we were on Reforger. They had a lock on that part of the drug market and it's almost impossible to catch up with them, as if anyone wanted to, because as soon as the heat is on one of them, he'll arrange to be assigned to the main USAREUR hospital in Heidelberg for training. Then he'll return with a letter of commendation, a certificate of appreciation, and probably a recommendation for promotion, and even more syringes, needles, cough syrup, and anything else they can lift. The doctors there all think they are in a MASH episode and really don't like the Army much. The medics can also write sick slips for troops who are feeling ill."

"Thanks. How about the night baker? He seems to have a lot of visitors through the night."

Robb lowered his voice and told him, "He's the distribution point for the heroin operation, and he sells Turkish hash that the main dealers bring up from the city. The soldiers in the PAC, the Personnel Admin Center, who are brighter than most, handle procurement. Every other weekend, they get in their new BMWs and drive across Germany and Belgium to Amsterdam. Members of the terrorist cells from Germany and France guide them, for safe passage along the old champagne route. Amsterdam is a free port and everything is legal. Once they are loaded up with drugs, they wait to be smuggled back across the borders again to Mannheim. When the drugs get here in Mannheim, they are distributed in Headquarters Company by the cooks, and transported on ration trucks with supply sergeant drivers, to the big supply and transport long-haul battalions that operate out of the city. These long-haul trucking companies distribute parts, food, fuel, and ammunition all across USAREUR so the drugs go out and after paydays the drug money comes back on the trucks to be divided up by the network in Mannheim."

Robb continued, "A few Company Commanders have tried to fight the trade. It's really hard because the MPs are on the take and will lose the evidence. The other piece of the puzzle, sir, is this: on the weekends, when the PAC troopers are not picking up drugs in Amsterdam, they run to Switzerland. After paying the terrorists their share from the sales, they deposit their balances into numbered Swiss bank accounts that it can't be reached by law enforcement. So even if you court martial one of them, as soon he's served his time and gets out, he will still be able to get his money."

By now, the Lieutenant's eyes were wide, he thought for a little and then asked, "How long has this been going on?"

"Sir, I'm on my third tour in Mannheim and the same system or one similar to it was in place when I rotated out of Vietnam in '69. There was one like it operating out of Saigon during the war. In '72, when the majority of U.S. troops left Vietnam, the trade was transferred in part to Thailand. Drugs and sometimes people were flown into Germany from there by the Air Force. After '75, the level of drug activity in Europe increased significantly because of the lucrative return it paid everyone in the system. The Red Brigades in Italy, Action Direct in France, the Red Army Faction in Germany, formerly the Baader Meinhof Gang, ramped up their activity after the 1972 Olympics. They are also allied with the IRA and Black September, due to their demand for more and better weapons and higher-grade bomb making materials."

The Lieutenant took a deep breath, exhaled, and asked, "Is anything being done about it?"

"Sir, CID is full of investigators who were trying to make their careers on low level drug busts and informants to get convictions of at least some of the dealers. But there's no real effort to crack down on the main organization due to the difficulty of the task and the fact that commanders of Army units generally don't want to acknowledge any kind of bad news in their units. It might reflect badly on their abilities and NATO's readiness. With recruiting so bad, we can't court martial too many troops out of the Army or we'll be combat ineffective for lack of personnel. Oh, and sir: the East German Stasi and the KGB are active here, too. I In some cases, supporting the terrorists. So, always remember: one out of every three Germans you meet is probably a spy, for our side, the other side, or both."

NEWS

Robb thought a bit and then added, "By the way, sir, do you keep up with the news?"

"As little as possible. I'm just concentrating on surviving, getting married, and finishing this tour to get back stateside to law school."

"Well, sir, I live with Dawn on the economy, so we get local TV. The German stations are reporting that the Iranians have attacked our embassy in Tehran, taking an unknown number of Americans hostage. So, it's probably going to get worse. Just a heads-up."

"Thanks, Sergeant Robb. We'll just have to see how it goes. Stay alert and stay alive, as they say at Bragg. One other thing. Why do you keep coming back here?"

"Sir, it's not a bad place and I met Dawn here during my second tour. She likes it here. That's enough for me."

SHAKE YOUR GROOVE THING

The First Sergeant came out and called the company to attention. The captains and the executive officer were still busy finishing the inventory for the change of command, so the First Sergeant turned the company over to the platoon sergeants for PT with their platoons individually. The Lieutenant told Sergeant Robb to go ahead and take charge, and he would run at the rear of the formation and do PT with them.

Once that was done and the platoon was back in the company area, he fell out and jogged back to his quarters. He finished his Coke, heated water in his hot pot, and took a shower. When he was done, he took the P38 can opener on his dog tags chain and cut the top out of the Coke can. Then he filled it with hot water and made instant coffee, ate a breakfast bar, and put some licorice in his field jacket before dressing and returning to the company area.

He made it back for morning formation and as soon as the platoons were released to their platoon leaders, Sergeant Robb went to company maintenance and took the First Sergeant's Jeep. With him gone, the Lieutenant marched the platoon to the ready line and released the tank commanders and his gunner to check their vehicles over and do whatever maintenance needed to be done in preparation for the road march at 1000. True to his word, a little after 0900, Sergeant Robb drove into the company maintenance bay, snatched the filter out of the back of a Jeep, and walked briskly over to his tank. He gave the filter to his gunner and driver, who began installing it. The Lieutenant then went from tank to tank and gave the order to all tank commanders to set their radios on the first frequency on their sets and be prepared to move at 1000.

After half an hour, Sergeant Robb gave the Lieutenant the thumbs up, so he mounted his tank slipped into the commander's seat, plugged his combat vehicle crewman helmet cord into the commo fitting, flipped the selector toggle on the helmet to radio and asked for a commo check. All of the tank commanders answered in order, from 1-1 through 1-5 followed by 6-6. Each tank then dismounted the loader who served as a ground guide to move the tanks from the motor pool down the street to the east and then south through the main entrance to the maneuver area.

When all of the tanks were through the gate, the loaders mounted and the Lieutenant gave the order for the platoon to proceed to the first intersection of the main road that they were traveling down. At that intersection, he ordered his tank, with 1-2 and 1-3, to move off the road to the left and the light section, 1-4 and 1-5, with 6-6 to the right, and form online in an attack formation for five hundred meters to exercise the shock absorbers, torsion bars, and tracks to see if anything was loose. When that was done, he ordered them to reassemble on the road, and with 1-2 in the lead, they picked up speed and drove for twenty-five minutes to Brigade and back to the battalion. The commander of 1-2 made sure to keep his head below the machine gun mount in case of wires. Once they were back at the gate to the main street, the loaders dismounted again and guided the tanks back to the ready line.

It was time for lunch, so Sergeant Robb released the platoon for thirty minutes to eat lunch and return to him. The Lieutenant turned to Sergeant Robb and said, "I'm going to go down and see if the old man had anything for me to do, with my luck, I'm sure something's come up."

THE NAME OF THE GAME

He walked into the orderly room and asked the First Sergeant if the CO had left any instructions for him since he was back from road marching his platoon. Guerra said, "Yes, Lieutenant, I have a note from Captain Batson regarding a mission he needs you to finish right away."

The Lieutenant took a look at the note and then reread it. Apparently, Batson wanted him to write the citation for an Army Commendation Medal to be awarded to Staff Sergeant Jackson, the Assistant Platoon Sergeant in Third Platoon. Since the Lieutenant barely knew the man, this could prove to be challenging. Fortunately, however, in his first assignment at Knox, his first Company Commander had actually given him a field manual from the Vietnam War on how to write citations, so he excused himself, went back to his quarters, found the manual among the other books in his small professional library, and brought it back to the company area.

When he was back, he asked the First Sergeant where he could find Jackson. "In the day room, on the sand tables, Lieutenant." He went up to the fourth floor to the sand tables and saw Jackson and Ramsey in the corner. He went over to Ramsey and told him that he needed to speak with Jackson in order to write up a citation for an ARCOM since he was rotating back to the United States for a CONUS assignment and Captain Batson wanted to have this done before he left command. Ramsey nodded his head in assent and looked away to engage with one of his tank commanders.

The Lieutenant walked up to Jackson and explained what he needed. The two of them went over to the far corner of the room and sat down. The Lieutenant took notes –on when Ramsey had arrived in the battalion, when he expected to depart, the various jobs that he had while he was there for the last three years, any letters of commendation and certificates of appreciation he had received. Jackson was an excellent NCO and rapidly provided all of the information. The Lieutenant thanked him and went down to the orderly room where he asked the First Sergeant if he had a couple of pieces of typing paper and if he could use the Supply Sergeant's IBM electric typewriter.

The Lieutenant sat down at the clerk's desk, took his pen from his pocket, and wrote out the citation in longhand. Then he typed it up. Based upon the field manual, it always helped to add some bit of poetic license to citations, so he inserted at the end that Staff Sergeant Jackson valiantly served on the front lines of freedom. That being done, he gave the completed typewritten citation to the First Sergeant and asked him if there was anything else. Guerra said, "No, but your platoon must be cleaned, prepared, and ready to go to brigade no later than 1000, as Third and Second Platoons will be going over for gunnery ahead of you. Your people will be the last to fire and the last to come back. And, Lieutenant, I expect your platoon to be spotless and ready to roll Wednesday morning for the change of command ceremony. All of your tanks will be on line with the crews in perfect uniforms in front for the photographs of the ceremony, with the battalion and brigade commanders and their staffs."

The Lieutenant thanked him for the heads-up and went back to his platoon in the motor pool to close the loop with Sergeant Robb. He told Robb what he had been up to and Robb told him that the citation sounded like it should be more than sufficient, since he written several of them himself over the years. The Lieutenant then asked, "Do you need me for anything?"

"No, sir. I'll see you tomorrow at PT. From what I hear, the inventory of the company is almost done so we can expect to see Captain Batson and Captain Borman at formation. Both officers will be busy getting ready for the change command Wednesday, so this will probably be the last time we have PT with Captain Batson."

"Thanks, sergeant," the Lieutenant noted, then saluted, and walked thoughtfully in the late afternoon sun, to his room, where he slipped a TV dinner into the toaster oven and took a quick shower.

WITH A LITTLE LUCK

When he had finished eating, he listened to the radio. In the endless stream of gray days and dreary nights, music was his one comfort. He jotted a quick letter to his girl. He told her that things were happening fast now, as he was getting a new Company Commander and going to a Baumholder with him. He would try to again Wednesday night and put a letter in the mail to her Thursday morning before the alert, but if not, he would have to catch up with her when he returned, the week of Thanksgiving.

As he set his alarm for 0500, his thoughts drifted back to his last assignment at Fort Knox. He was the executive officer in a basic training company. He was also the training officer and mess officer, so he would usually eat breakfast in the mess hall. He also had a newspaper account and an account for the Army Times. The children of noncommissioned officers delivered these papers to make extra money. He remembered the first time he saw the newspaper delivery boy in his new company after he returned from Texas to the brigade.

The boy was about ten years old. When he came in, he introduced himself and so did the Lieutenant. The boy asked him where he was in from and he told him that he had been at Fort Bliss but had previously been in the 17th Battalion which was stood down while he was away. He was now newly and temporarily assigned to Echo Company in the 15th Battalion but had orders to go directly to Germany.

The boy looked to him and said, "Germany isn't bad. I was born there and have been there twice. Just hope you're not going to the 8th ID."

The Lieutenant looked up somewhat startled and said, "As luck would have it, I'm going to the 8th Infantry Division."

"Oh," said the boy, "where are you going?"

"I'm going to Third Brigade in Mannheim."

The boy's countenance brightened at least a little and he said, "Well, at least you're not going to Baumholder. My dad was assigned there his last tour, and Baumholder is the armpit of the universe."

The Lieutenant looked at the boy and said, "Really?"

"Yes, it's in the middle of nowhere and the weather's terrible. You can get four seasons in the same day most days, and it's freezing cold in the winter. The only good thing about it is there's a Kentucky Fried Chicken restaurant at the bottom of the mountain when you come off the Autobahn, that's it."

Chapter 12
Right Down the Line

With that happy thought about his upcoming trip, the Lieutenant passed out and woke up a couple of minutes before the alarm rang. He turned it off, because he hated the noise, and out of respect for the neighbors next door. He slipped on his dark blue PT gear and jogged over to the company area.

Once he had arrived, he hurried down to the First Sergeant's office for a strong cup of coffee, the elixir that made early mornings survivable. He could hear the two captains talking in the Company Commander's office and when the First Sergeant went up the stairs to call the company to attention for morning formation, the Lieutenant walked up behind him casually in the back of the formation. He said good morning to Sergeant Robb, who moved to the front of the formation so that he was in position with the other platoon sergeants.

First Sergeant Guerra called the company to attention and turned it over to Captain Batson, who put the company at parade rest and then at ease so that he could speak to them. He told the soldiers that he would be changing command and introduced the new incoming Company Commander, Captain Borman, who was one of his classmates from West Point. He also told them that the change of command ceremony would be the next morning so the company area, billets, and vehicles must be in top shape as the brigade commander and other high-ranking officers would be attending. When he was done, he turned the company back over to Guerra, who released the platoons to their sergeants for individual platoon physical training.

Sergeant Robb had the platoon execute a left face, march forward ten paces, execute a right face, and open ranks for calisthenics. When they were finished with the exercises, he quickly closed the platoon up and started off on a two-mile run. At the halfway point, the Lieutenant jogged up beside Robb and asked if he could call cadence. Sergeant Robb looked at him a little quizzically and said, "You, sir?"

"Yes, I just spent a year in the training center pushing BT, before that I was a cadre sergeant at VMI, and I did my time in Airborne School in '75, right after Saigon fell. I'll be alright."

Once he had the rhythm of the platoon, he started and the troopers called back as they ran:

"*A – I – R – B,* **A – I – R – B**
O – R – N – E, **O – R – N – E**
Airborne, **Airborne**
Infantry, **Infantry**
Death from, **Death from**
Above, **Above**
Death from, **Death from**
The Sky, **The Sky**
Angels, **Angels**
of Death, **of Death,**
From tyranny, **From tyranny**
Deliver me, **Deliver me**
Yaa, **Yaa**
YaHaa, **Yahaa"**

The soldiers were responding well, so he continued with a few of the other cadences he had picked up over the years. and they completed the return mile quickly. As they came into the company area, he ordered them to quick time, to sound off, and take it on down, followed by skip cadence, delayed cadence, count cadence count. Having all been through basic training, this cadence sequence was well known to all of his soldiers. The Lieutenant then halted the platoon and had them execute a right face. He turned the platoon over to Robb and told him that he would be back in the company formation at 0800 and that he would inspect the vehicles and the troops to make sure that everything was ready for mini gunnery at Brigade.

After a quick shower, he slammed a breakfast bar and a cup of instant coffee before he dressed in fatigues, put some licorice in his lower field jacket pocket, and walked back to the company. After the First Sergeant had formed and released the platoons to their platoon leaders, Sergeant Robb marched the platoon to the ready line and the Lieutenant followed. When they were there, Robb had the tank crews stand to attention in front of their vehicles while he and the Lieutenant inspected the tanks climbing up to the turrets to look inside them, then inspecting the soldiers' uniforms and personal appearance. They were a little sleepy, but not buzzed, so it looked to be a good day.

Sergeant Robb noted some corrections to be made on the vehicles and directed the crews to take care of them right away. Then the Lieutenant went up to the S-2's office and drew his CEOI for the day. When he was back, he gave the platoon the company frequency and platoon frequency so that they could set the radios on their tanks. While they were doing this, Second Platoon and Third Platoon had both departed to drive through the training area to Brigade. At 1000, an hour after Second Platoon departed, the Lieutenant ordered his platoon to move out. The commanders dismounted their loaders as ground guides to get the tanks into the training area. Once the loaders had climbed back into their hatches, the drivers hit the gas hard and they all took off for Brigade. They had not installed the M-85s, since they would not need them, so the tank commanders crouched down behind the machine gun mounts to avoid any possibility of hitting wire in the trees.

They pulled up at Brigade and dismounted their loaders again. Sergeant Webb climbed down off 1-2 and led them to a large vacant lot that was the holding area for the range. Third Platoon had already finished and was heading back to the company. Second Platoon was rotating through their platoon leader's tank 2-1, that was parked in the bay of what looked like a large garage. Since he had never been to the mini gunnery range before, the Lieutenant slid down off the front of 1-1 and walked into the building to observe. He knew he could count on Robb to have the crews busy adjusting the tracks, tightening connectors, and reviewing the training materials that they had brought with them while they waited.

Once he was inside the building and adjusted to the light, he could see that 2-1 had been pulled up into a position with the gun tube oriented toward the far end of the garage. At that end there was a metal structure with a bullet trap and large sandbox with small metal targets that would fall when struck by a bullet. Attached to the main gun was an M16A1 rifle with a device hooked to the gunner's trigger so that he could fire or the tank commander could fire from his position, using the override at his position once the target was acquired. It was a good exercise and a lot of fun to watch. The Lieutenant had been on numerous rifle ranges but never one quite like this.

THE RUBBERBAND MAN

When the crew on 2-1 had finished firing, the tank commander took off his helmet, and the Lieutenant noticed that he had been observing an African American second lieutenant, who apparently was the new Second Platoon leader, and must have just arrived. As he dismounted the tank, the Lieutenant walked up to him and introduced himself.

The new officer introduced himself as Charles Matthews. The Lieutenant asked him where he was in from and Matthews told him he had just finished basic course at Fort Knox. Before that, he and his wife were born and raised near Fort Sill in Oklahoma. The Lieutenant welcomed him to the company.

Remembering his rocky start, he told him where he was living in the BOQ and then added, "If you need a place to stay, I can put you up."

The new lieutenant told him, "That's not necessary, I have married quarters assigned and I'll get the keys after Thanksgiving. In the meantime, I have a friend from ROTC in college who has an apartment on the economy where I am keeping some of my stuff. I am sleeping at the VOQ where I have a room, to be near the company, so I'll be okay.

The new lieutenant continued, "All of my free time is being taken up by Sergeant Wilson, who is trying hard to get me and all my gear ready, as well as familiarize me with the platoon in the few days we have left before gunnery."

The Lieutenant nodded and told him, "Maybe we can grab lunch somewhere after the change the command tomorrow, and if not, we'll be seeing each other pretty often at Baumholder."

Lieutenant Matthews looked to his right and saw his platoon sergeant frantically motioning to him, so he shook his head and took off to meet him. He had a bee-bop sort of gait and a cool attitude, the Lieutenant thought to himself: *We're going to have a good time.*

GET USED TO IT

As the new lieutenant and his platoon sergeant left the building, Sergeant Robb came in with the crew of 1-2 to fire the gunnery table. The crew of 1-2 climbed aboard and the Lieutenant and Sergeant Robb observed them as they fired five practice rounds, at a large zero target in the center of the range, to make sure that they could properly use the devices attached to the vehicle. The gunner fired three rounds to the tank commander's two. When that was done, the NCOs running the firing range gave the "all clear" to begin the exercise, and the farthest row of targets appeared. When possible, the crew tried to shoot the targets on the flanks first, then focused on those closest to them; bank after bank after bank of targets rose up and they fired as quickly as possible. When that crew was done, Sergeant Robb had the crew of 6-6 come in and fire. This was followed by the crews for 1-3, 1-5, and 1-1.

When his crew came in, the Lieutenant climbed aboard the tank and slipped down into the cupola. He quickly manipulated the tank commander's fire control and asked Blaylock to come on the intercom to announce, "Up," when he was ready. As the tank crews before them, they waited for the larger zero target to pop up in the middle of the field and Blaylock fired three times, hitting the target every time. The Lieutenant grabbed the pistol grip at his firing station and trained the site on the target. He fired and missed high to the left a little bit, so he compensated and hit it the next time – dead in the center.

When he was done, the crew got the order to begin firing on the targets as they came up on the range and Blaylock went to work. The Lieutenant would slew the gun toward the target he wanted hit, and the gunner would fire. Then the Lieutenant would move the turret back in the other direction and the gunner would fire again. He kept a constant conversation going with Blaylock. The driver and loader would come in too if they identified targets that the officer was missing.

Time flew and the cease fire command came over the radio before they knew it. They all dismounted and walked over to Sergeant Robb, who critiqued them. They had done pretty well but had not hit all the targets. They did not have their rate of fire high enough to shoot them all in the time given. The Lieutenant would have to pick up the slack by firing from his position as the targets came nearer to reduce the time between acquisition and getting the shot off. The Lieutenant nodded, thanked Sergeant Robb, and told him he would stay at the range to watch his crew and see how it was done.

He did. Robb and his gunner hit the zero target five times in a row, then as the banks of targets appeared before them, Robb and the whole crew on the tank operated smoothly, as a single being. They hit every target except one in the allotted time. Robb and his crew dismounted the tank and turned it over to Sergeant Webb, who backed it out of the range. Sergeant Robb talked to his crew briefly about their performance and they all left to go out to prepare 1-4 and the rest of the tanks to head back to the company area. When the Lieutenant came back to his platoon, he was happy to find that the crews were still training on their flash cards for vehicle and aircraft identification and playing hearts with a special set of cards that had silhouettes of vehicles on them. When the crews were mounted again, the Lieutenant called for a quick commo check. When that was done, they took off at over twenty miles an hour back to the battalion area where the loaders dismounted to lead them back into the motor pool.

HOW YOU GONNA SEE ME NOW

When the tanks were back on the ready line, the Lieutenant saw Batson's driver headed toward him. The driver told him that Batson needed to see him right away. The Lieutenant went over to Sergeant Robb to let him know that he was ordered to see the "Old Man." He asked if he needed to do anything to assist when that meeting was over and Sergeant Robb said, "No, sir, we'll be all right."

The Lieutenant told him, "After meeting with the captain, I'm going to the bank to get enough marks to see me through Baumholder since the officer's club there only takes marks. After that, I'll check in with you at evening formation."

The Lieutenant hurried to the company and downstairs to the Batson's office. He knocked at the door and when asked to enter, saluted, and took a seat. He sat in the chair beside the desk as Batson told him that he had received the citation and it was excellent. Batson especially liked the line about "the frontiers of freedom" and he planned to use it on some of the citations that he was writing up as well. He then told the Lieutenant, "From what I hear, you did an excellent job at the mini gunnery range in training your soldiers while they weren't shooting and during the firing exercise as well."

The Lieutenant replied, "I've been running rifle ranges at Fort Knox for a year and I spent the better part of seven months on tank firing ranges at Fort Bliss, so it's second nature by now."

Batson appeared surprised at this, although the Lieutenant swore he had told him this when he initially came into the company and was asked about his background. Nevertheless, the captain told him, "The change of command ceremony will be at 1030 followed by a short reception and early lunch. As is the custom, I expect that Captain Borman will release the company for the rest of the day, so everything has to be ready for the road march to Baumholder and the alert Thursday. And, Herr Leutnant, as soon as lunch is over, come and see me. I'll be using this office for a few hours in the afternoon finishing up company business before I depart. Dismissed." The Lieutenant rose, saluted, and took off for the American Express bank as fast as he could. He just hoped that he wouldn't have to stand in line for over an hour there trying to get a check cashed and exchanged for the two hundred marks he needed to get through the two and a half weeks to come.

Once he had his German marks, he stuffed then in his right breast pocket, next to his VMI ring, and walked quickly back to his tanks on the ready line, since it was almost time for evening formation. The crews had washed them and tightened up any track components that had come loose on the trip back from Brigade. He congratulated Sergeant Robb on having them looking so good and asked, "What's on the schedule was for tomorrow before the change of command?"

Robb answered, "I'll have the crews come and wipe the tanks down after dinner to make sure they're completely clean and dry. In the morning, I'll have an inspection in ranks, to make sure that they are all up, and their uniforms are squared away, so they can stand an inspection by you at 0900."

By that time, the First Sergeant had called the company to assemble, so he and Robb followed the platoon to the courtyard and waited for announcements from the First Sergeant.

Guerra laid out the schedule for the rest of the evening, which was to be spent preparing for the change command ceremony. Once the soldiers had breakfast, they would make final uniform adjustments and post in front of their vehicles. He expected that the brigade commander and the battalion commander and his staff would inspect the company on the ready line. After that, they would form up in the company area, where the battalion commander would pass the company guidon from Captain Batson to Captain Borman, with the brigade commander in attendance. He expected that the brigade and battalion commanders would say a few words as would Captain Batson and Captain Borman. He then released the troops to their platoon sergeants.

The Lieutenant told Sergeant Robb, "I'll see you in the morning," and saluted. The sun was setting and damp fog set in as he walked, lost in thought, back to his room. He slipped on a V-neck sweater over his t-shirt and pulled on his well-worn jeans. Then he listened to a couple of pop songs on the radio, remembering happier times and relaxing before walking down to the dining room for dinner with his new band of misfit compatriots.

NEW YORK GROOVE

As he came around the corner into the dining room, he noticed that Buzz was checking on the progress of two chickens he was roasting with herbs, red potatoes, onions, and various spices. He had made a large green salad that was setting out on the counter next to a six-pack of German beer. The Lieutenant said, "Hello," and asked, "when do you think dinner will be ready?"

"About twenty minutes. I made this meal special because it is one of Elizabeth's favorites. I expect that Susan and Chris, the ladies who room next door to you, will be joining us. They are supposed to bring some white wine with them. Bigfoot can't make it. He is entertaining, but I think Tom and Bernie will show up. I haven't seen Carey Pavel, so I don't know about him."

"My guess is he is busy up in headquarters working on the movement orders and final coordination for the road march to Baumholder; he's been helping in the ops shop while he waits to take over his company."

The Lieutenant helped himself to a beer and opened it with the Swiss knife he had purchased at the PX after his return from Reforger and now carried everywhere. It was perfect for the field; not only could it open beer and wine bottles but cut coaxial cable as well. After he sat down in the corner and looked out the window into the gloom and mist of the early German evening in Mannheim, he asked Buzz, "What's it like, being a newly minted colonel?"

Buzz replied, "It's great! For once I can avoid the crappy duty schedule while younger dentists fill in." As he was saying that a dainty AG captain, named Wells, still in uniform, came through the door and introduced herself as Susan. She was followed in by an admin Specialist Four, named Sims, who was in fatigues, too. She had a slightly larger build, and said, "I'm Chris," as she sat down at the table.

Susan had a bottle of white German wine that she opened with the communal corkscrew on the counter. Then she retrieved four glasses from the cabinet and sat them down. Buzz told her, "I'll take one," so she poured one for him and for herself and her roommate. She looked over at the Lieutenant, who declined, explaining, "I'm sticking with beer for the evening because I have to get up early tomorrow." He then asked them, "How long have you been living in the BOQ?"

Susan replied, "I rotated in about a year ago in August."

Chris chimed in, "I decided to move in about six months ago.

We're lesbians in a committed relationship, and we're trying to get assigned together to West Point. We think that has the best possibility for entering into some sort of domestic partnership, like marriage, since we will be in New York."

The Lieutenant nodded his head and told them, "I understand your frustration. I'm busy trying to get married myself and Army certainly hasn't made it any easier. Nevertheless, as it stands now, I'm hoping to go back to the States to marry my fiancée in March and bring her back to Germany."

Both ladies were interested in the details of the wedding arrangements, but the Lieutenant simply didn't have any information as he was leaving that all up to his girl. He was more than busy in the battalion. Curious, he asked, "Do you two have a dog? I could swear I hear one sometimes, but I could be losing my mind from sleep deprivation."

Susan answered right away, "We sure do, a little Yorkie named Ricky."

"Good – now I can sleep knowing I'm still sane, except for one other thing."

"What's that?" asked Chris.

"It seems some nights I hear water running constantly, like there's a leak or something."

"Oh, that's Meghan."

"Meghan?"

"Yes, she's a transportation butter bar with a long-haul truck platoon. Been here about a year. She used to come down and eat with us regularly, but she got into a relationship with three of her drivers and some nights they all get together to party. You got your place because the finance LT who was there before moved out on the economy. He couldn't take the noise on the weekends – like living under Niagara Falls."

"Oh, well. Good to know. But after four years at VMI with a thousand cadets, I can sleep through anything for the next five months."

Susan added, "Cheers" as she lifted her glass and gave Chris a hug.

He raised his beer, touched her glass, and said, "Prost," in reply.

The smell of dinner was fabulous and Buzz rose to announce it was ready as Elizabeth came into the room. When she saw the dinner that he had made for her, she smiled and thanked him. Buzz made her a plate and sat it down next to where he had been sitting. He also poured her a glass of the white wine. While he was doing that, Bernie and Tom came in together. Each of them took a bottle of beer and stood to wait their turns as the others made themselves plates and sat down.

Once they had eaten and praised Buzz for his culinary skills, the Lieutenant turned to face Bernie. Since he had not had an opportunity to visit with him before, he asked him, "What will you be doing when we go to gunnery?"

"The division artillery is in Baumholder so the other fire support officers and I will train with the artillery batteries to work on fire support coordination. I'll also do the fire support coordination for the Scout Platoon and work with the mortars in getting their tubes laid and fired correctly. There is also a miniature platoon fire control exercise that I'll be plugged into as FSO to take calls for fire from the lieutenants on that firing range."

Of all the people the Lieutenant had met so far, Bernie seemed to be the only one who was excited about going to Baumholder. The Lieutenant thought he was probably happy to be going where he would be surrounded by other artillery officers and, in addition, he would have more markets for his secondary business up on that mountain in the middle of nowhere.

Bernie asked Elizabeth how she was doing. She told him, "Things are going well and I'm looking forward to the end of the semester when I can go back to England for the holidays. Even though Germany is exciting and it's a lot of fun to be with Americans, I still miss London and my old neighborhood. And I look forward to seeing my parents and sisters when I get back."

Tom Storch listened to her quietly, hanging on every word. He was still infatuated with her and frustrated beyond belief. The Lieutenant sympathized with him. He understood only too well what it was like to be a lonely, single Army officer in a garrison far from home. He asked Tom if he had any plans for vacation over the holidays. Tom said, "I plan to go back to Indiana for at least a week, since the Army generally shuts down on a half-day schedule and there's not much to do. I've been in country almost two years, so there are officers junior to me who will be pulling duty in my absence."

"Well, I sympathize with those officers. I've already been promised that I will be pulling duty on Thanksgiving Eve, Christmas Eve, and New Year's Eve, with the last being at Brigade. I don't mind though. It breaks up the monotony."

Buzz asked, "When are you leaving for your vacation on the Rock?"

"I have to be up early in the morning for a change of command and final preparations. At 0400 the day after tomorrow, we have a mock alert, followed by the road march to Baumholder that will take the rest of the day. Friday we should start shooting." Looking down he saw the time and saw it was getting late, so he rose, washed the silverware and dishes, and sat them beside the sink to dry. Then he turned, excused himself, and said that he was going to bed down early since he was going to be kept pretty busy for the next few weeks.

Halfway to his room, he changed his mind and ran downstairs to buy two Cokes from the vending machine so he would have one to drink first thing in the morning to be ready for PT and the change of command, and the second to drink at 0300 on Thursday before the alert. When he got back to his room, he put both in the refrigerator, got ready for bed, and relaxed. Fortunately, the ladies next door had had enough wine to be pretty tired, so it was quiet. There wasn't any noise from upstairs either, so he curled up in the poncho liner on top of his bed and faded out of consciousness.

JUST A SONG BEFORE I GO

At 0430, he came wide awake and drank his first Coke before dressing and going into the company area for physical training. When he got there, Sergeant Robb stopped him and told him that the First Sergeant was turning the platoons over to the platoon sergeants for PT. They would only run one mile so they could get through breakfast as quickly as possible and get dressed to make sure the vehicles were cleaned up and in perfect condition for the change of command. The Lieutenant told him he would run with the troops, then get squared away and meet him and the platoon on the ready line a little after 0800.

After PT, he went back to his quarters, showered, and had a breakfast bar and a cup of instant coffee. Then he laid out his fatigue uniform and shined his Airborne jump boots to a high gloss. He also made sure that all of the insignia and his parachute wings were properly placed on his baseball cap and the field jacket that he would carry with him in case he needed it. After dressing, he walked to the courtyard where the First Sergeant had formed the company and was giving the troops their detailed instructions for the day. He was running a little early, so the Lieutenant walked to the back of the formation and waited until Guerra released the platoons to the platoon sergeants to march them to the ready line and prepare their vehicles for review by the commanders.

He trailed the platoon. When it reached the ready line, he watched as Robb had the crews wipe down the tanks with rags to make sure that they were perfectly clean and dry. Robb also inspected the soldiers' uniforms to make sure that they were squared away. When that was done, he and the Lieutenant inspected the tanks, outside and inside, as well. Then they both inspected the troops, who were at least ninety present alert. By the time they were done, it was a little after 0930, so the Lieutenant had Sergeant Robb put them at ease as they waited for the company, battalion, and brigade commanders. A quarter of an hour later, they saw the battalion commander arrive in the courtyard. The brigade commander pulled in behind him and they were met at the entrance to the barracks by Captain Batson and Captain Borman. After salutes and handshakes, the entourage formed and crossed the street toward the motor pool. Sergeant Robb called the platoon to attention, and the Lieutenant took his place in front of his men, while the other platoon leaders did the same. The Lieutenant could tell that Lieutenant Matthews was a little uneasy since he had limited experience with the command leadership.

The Lieutenant watched out of the corner of his eye as the officers approached Second Platoon. Lieutenant Matthews came to attention, saluted, reported, and then followed Captain Borman as they made their way down the line introducing themselves to the troopers and asking a few questions. When they were done, Matthews saluted the senior officers and they returned to salute before turning to First Platoon. As they came near his platoon, the Lieutenant reported and the brigade commander, Colonel Means, came up to him. Colonel Means asked how the Lieutenant had liked his first Reforger and that he had heard a report from higher that the Lieutenant's platoon had performed well.

The Lieutenant told him it was just fine and he had a good time in the field. The colonel asked, "Are you related to the German general?"

The Lieutenant answered, "Yes, sir, distantly."

The senior officers then began their inspection. The Lieutenant followed the senior officers as they made their way down the line of his troops visiting for a short period of time with Sergeant Robb. When everyone had reached the end of the line, the Lieutenant saluted and the reviewing party moved on to Third Platoon.

After Third Platoon and Headquarters Platoon had been inspected, the captains and colonels turned and made their way back toward the company area. Guerra called the company to attention, and the sergeants marched the platoons back to the courtyard, where the company was formed in preparation for the change of command.

The platoon leaders took their places in front of their platoons and the Colonel Means stepped forward to say a few words about how happy he was to see a new captain coming in to take the company over from Captain Batson who had done an outstanding job with it.

Lieutenant Colonel Shea echoed those remarks and took his place in the middle between the two captains. The First Sergeant took the company guidon from the guidon bearer and presented it to Captain Batson, who in turn passed it to Shea, who then passed it to Borman. Guerra took it from Borman and handed it back to the guidon bearer. When that was done, Captain Batson had the company stand at ease as he spoke to them briefly about how proud he was of them and that he wished he could stay longer; he then introduced Captain Borman.

Borman introduced himself and told the assembled troops that he was proud to be there and pleased to be taking over such an excellent company from an officer that he had known for many years from their time as cadets in New York. He then turned the company over to the First Sergeant, who informed the troops that they could assemble in the mess hall for tea, lemonade, and cookies. Also, at the direction of Captain Borman, they would be released after lunch to attend to their personal needs and prepare for the movement to Baumholder.

The Lieutenant looked over at Lieutenant Matthews and moved his head slightly to indicate that he should join him. The two of them walked into the mess hall where they were met by the Adjutant, who directed them back to the officer's dining room. They formed up behind the other officers in the battalion, after the brigade and battalion commanders, in the receiving line for the new Company Commander and his wife, who had just arrived on post shortly before the change of command ceremony.

Once the reception line was completed, the officers mingled for a few minutes, and the Lieutenant made sure that he spoke to Captain Borman and his wife as well before grabbing Lieutenant Matthews and preparing to leave for lunch. As he was headed toward the door, Captain Batson stopped him and said, "I need to see you in my old office after lunch. I'm working there until 1600, when I leave the company for the last time."

The Lieutenant said, "Yes, sir," and then turned and walked out of the dining facility with Matthews. Since he had a limited amount of time for lunch now, he suggested that they just walk over to the snack bar and grab a sandwich or a hamburger so he could come back to see Batson.

As they walked to the PX, he asked Matthews if he was ready for the road march in the morning. Matthews said, "I have no idea, man. Things are moving so fast."

"I understand; I was in the woods the morning after I got here, and I just got back from three weeks in the field on Reforger." When they got to the snack bar, they ordered cheeseburgers with fries and American Cokes and sat down. He told Matthews that the food wasn't bad, but he couldn't handle snack bar Mexican food after a really bad experience he had at Bliss.

They took a few bites, lost in their own thoughts, before he asked Matthews how things were with his wife and getting moved to Germany. Matthews told him that they were newly married and their household goods were back at his wife's parent's place in Oklahoma. They had lived in married quarters at Fort Knox and had their hold baggage packed out of there. She was waiting for a phone call from him that would tell her when she could make arrangements with transportation to ship their things and fly her to Germany once he had quarters arranged. He and his wife had known each other since high school and went to the same college. Once they graduated, they were married, and now they were off on the adventure of a lifetime in Europe.

The Lieutenant told him, "I'm glad to hear that. It will be great to have a newly arrived wife as a friend for my wife when I bring her back in March. By the way, do you understand what you'll be doing at 0400 when they call the mock alert?"

Matthews looked at him quizzically, so the Lieutenant told him, "You and your senior NCOs will draw sidearms and grease guns and then hump the M-85s and coaxial machine guns up to your tanks. It will probably take two trips."

Matthews looked at him even more confused, so the Lieutenant asked, "Have you read the Battalion SOP?"

"Yeah, it was in a binder in the Three's office. Some sergeant gave it to me."

"OK. Have you seen the real one from the Three's safe?"

"No."

"Then let me give you the lay of the land." The Lieutenant quickly briefed Matthews on what was expected and why.

Lieutenant Matthews rolled his eyes, after which the Lieutenant told him, "Yes, brother, it's a whole 'nother world. Stick with your platoon sergeant and listen carefully to what he tells you to do because this place is nothing like a CONUS unit. I've been assigned to a tank battalion, two infantry battalions, and an ACR stateside plus you have a lot going on here. For one thing, hash and heroin are everywhere so at least half of your troopers will be stoned or recovering when they fall in for PT. Just look into their eyes and you can gauge how things are going. And if you need anything from me, just ask. We have to look out for each other." They finished lunch, Matthews thanked him, and they walked back to the company area where the Lieutenant broke off and quickly shot down the stairs to the commander's office.

STUCK IN THE MIDDLE WITH YOU

He stopped at the door and knocked; when asked, he came in, saluted, and took a seat beside Batson. The captain gave him a draft of his Officer Efficiency Report (OER) and told him to look it over. The narrative portion of the OER read very well, but Batson had scored him ninety-seven out of a hundred points for proficiency. He asked Batson why that was, since he had never received less than one hundred in any of the previous four reports he had received in the past two years. Batson explained that he was new to an actual tank battalion, he had a lot to learn, and he needed room to improve in his next OER. The Lieutenant shook his head slowly and said, "Very well."

The Lieutenant thought to himself, *It's okay*. He was not in the habit of arguing about efficiency reports, and the "room for improvement" line was pretty standard for officers who wanted to make it look like they were supporting the myth of the HR system at the expense of their subordinates. And in any event, he was planning to go to law school as soon as he had done his time in Europe. Batson motioned for him to sign and return the draft report, which he did. The Lieutenant then rose, shook hands with Batson, wished him good luck on his next assignment, and safe travels. Batson wished him good luck in his tour in the battalion as well.

Sensing that they were done, the Lieutenant saluted, left the office, and never saw Batson again. The Lieutenant thought Batson did teach him several things in their short time together and exemplified the "three As" required for success in the Army: "Kiss Ass, Cover Your Ass, Be an Ass," a mantra he personally was determined not to live by.

As he was walking out of the office and passed the orderly room, he saw Sergeant Robb out of the corner of his eye, so he went in and asked Robb if there was anything he needed him to do. Robb said, "Sir, just be sure to have your A and B bags packed with all of your cold weather gear and be in the company area by 0330, so we'll be ready when the alert is sounded at 0400."

"Alright, I'll be there. If you need me, I'll be in my quarters." Then he headed back to Benjamin Franklin Village. The sky was lead gray, overcast and foggy; the perfect start to an adventure. In the back of his head, he played some of his favorite Kansas songs. He was a child of the plains, and they reminded him of home.

Chapter 13
Carry on Wayward Son

As soon as he stepped into his room, he put a frozen dinner in the toaster oven and took a shower. Once he was dressed, he put his A and B bags in the living room and laid out his TA-50 items to make sure he had packed everything and knew where all his gear was. He put the Mickey Mouse boots and overshoes in the B bag with extra uniforms and put his rubber duck, poncho liners, an extra uniform, plus a field shirt and pants, in the A bag with his sleeping bag. Then he packed a couple of boxes of breakfast bars and five packages of red licorice, as well as his cold weather headgear, and a scarf to use as a face covering; it was bound to be freezing on the road march.

Once he was done with dinner, he sat in the easy chair in the living room for about an hour relaxing, reading a copy of the Army Times and listening to some mid-seventies rock on the radio. Then he turned into bed early after setting his alarm for 0300. He hoped that he would wake up before the alarm went off so it wouldn't disturb his neighbors next door. After thinking about his girl and praying that she would have a good day at work he fell asleep on top of his red comforter, until he checked the luminous dial on his watch that showed it was 0220.

He got up, turned off the alarm clock, shaved, and washed his face. Then he packed the clock, razor, shaving cream and soap for the field. He put on his uniform and made sure that his sparkling sapphire VMI ring was in the right breast pocket. It was way too large to wear safely in the turret of a tank. Looking out the window, he could tell that it was cold, foggy, misting and pitch dark so he put on his field jacket, picked up the A bag placed it on his shoulders like a backpack, then slung the strap for the B bag over his left shoulder made his way to Bravo.

When he came into the courtyard, he saw Sergeant Robb standing next to a tree by the sidewalk, drinking a cup of coffee. He told the Lieutenant that he had already visited with the crews in barracks and that Blaylock was already on the ready line with 1-1, so he could take his bags over and stow them before coming back for the alert and issuing of the heavy weapons, sidearms, and grease guns.

The Lieutenant asked, "How are the drivers this morning?"

"Fine except for 1-5. Ames' gunner will drive until we get at least halfway – he had a rough night. I'll check him again before I let him get on the Autobahn. The companies will be rotating through the mess hall for breakfast beginning at 0430, so the cooks can pack up to go along with the battalion."

The Lieutenant thanked him for the information and carried his bags over to the motor pool where he greeted Specialist Blaylock. He liked Blaylock a lot. He was usually happy, competent, and he didn't do drugs. When he got to 1-1, he tossed his bags up to Blaylock, who stowed them in the bustle rack. Blaylock had already checked the radios to make sure they were operable and as soon as they had received the CEOIs from battalion, they could set the frequencies to the platoon and company frequencies and do a commo check before following Alpha Company to the Rock.

Through the fog, the Lieutenant could make out the entrance to Bravo by the outside light over the front steps. As he watched, Guerra came out. The NCOs and troops that were still in the barracks fell out into formation. Robb stepped into his place at the far end of the courtyard and the Lieutenant hurried to join him. Guerra told the troops that they were on alert as of now and that in accordance with the SOP, First Platoon would draw heavy weapons and sidearms right away. Robb called First Platoon to attention and led them down to the arms room where he had the senior sergeants and the Lieutenant come to the front of the line to pick up the heavy machine guns and their pistols with clips on the first trip. As the Lieutenant placed the heavy machine gun over his shoulder, he caught sight of Lieutenant Matthews who was watching him, still in amazement. The Lieutenant caught his eye and told him, "Just as I said, you'll be humping these things up the stairs into the motor pool before you know it." Lieutenant Matthews shook his head again, but he was smiling, a wiseass grin, because he was an adventurous sort who just wanted to get going.

When the Lieutenant got to 1-1, he handed the heavy machine gun up to Blaylock, who wrestled it into the turret. After he climbed up, the Lieutenant helped him set the machine gun in its mount in the tank commander's cupola. After pushing in the locking pin, he told Blaylock that he would be coming back with the coaxial machine gun and when that was mounted, the gunner could go to the arms room to draw his forty-five and a grease gun. The Lieutenant dismounted, jogged back to the arms room, made his way to the head of the line again, picked up the coaxial machine gun for his tank, along with a grease gun with its sling. Back at the ready line, he had the loader take the coaxial machine gun from him and place it up on top of the turret so the Lieutenant could climb and hand it down to Blaylock to put it in position.

As soon as Blaylock had functioned checked the machine guns, the Lieutenant asked his driver and loader if they had been issued their sidearms and grease guns. They said that they had so he sent Blaylock to the arms room to pick up his. Having done that, he climbed up into the tank commander's seat and looked out over the motor pool. It appeared that all of the tank commanders were in position, but he still had not seen the Company Commander yet. Since everything was loaded, the Lieutenant settled in, reached in his A bag, pulled out a bag of licorice, and put it in the ammunition tray of the M-85. He told the crew that as usual, he had licorice with him and if they got hungry to help themselves. As you looked into the company area, he saw that the S-2 had pulled into the courtyard to issue the CEOIs. He signaled to Robb, who nodded and both of them went to the S-2, where they signed for the code books and slipped the lanyards over their necks.

When they were walking back to the motor pool, Robb asked him if he had packed his NBC gear and gas mask. The Lieutenant said, "Yes, they were on top of my A Bag, just below my rubber duck and a poncho liner."

Satisfied, Robb turned back toward 1-4 and said, "Sir, go ahead and eat breakfast first, once the commo check is done."

The Lieutenant looked down at the book hanging from his neck. He wrote down the call signs and frequencies and then went up and down the "P line," talking to each his commanders and giving them that information. He also directed them to get those frequencies set on the radios right away. He then climbed back up in the commander's cupola and when he got the thumbs up from all of the tank commanders, he called for a commo check across the platoon. All the radios were up and working so he tried to do a commo check with Borman, but no one answered. He assumed that the new commander was simply beyond busy getting ready for his first road march across Germany, with a new company, during an alert, in the middle of the night.

The sky was beginning to lighten just a little and the Lieutenant caught movement from Sergeant Robb's tank in the corner of his eye. He looked over and Robb was pointing toward the mess hall. The Lieutenant picked up the combat vehicle crewman helmet, put it on his head, and pushed the toggle switch to talk to his tank commanders by radio. He directed each one to release half of his crew for twenty minutes to get breakfast, after which the other half would grab their meals. Sergeant Robb shook his head "no," so the Lieutenant and his driver went to the mess hall together.

After walking through the door, he went through the serving line and made his way toward the officers' dining room. Looking back over his shoulder, he saw Matthews coming in behind him, so he motioned for him to join him as soon as he had his tray. After sitting down, the first thing the Lieutenant did was drink one cup of black coffee about as fast as he could toss it down. Then he excused himself and went to get another. After sitting down again, he began to eat and asked Matthews how things were going.

Matthews said, "Okay, I guess. I'm jet lagged; it's amazing to see so many people busy at the same time, frantically trying to get things done."

The Lieutenant nodded and asked, "Have you seen the CO yet?"

"No, but the executive officer is in company maintenance getting headquarters platoon ready to go. When I spoke with Borman yesterday at the reception, he told me that Marley was one year behind him at West Point, so he knew him, but not particularly well."

Looking at his watch, the Lieutenant said, "Hmm, we better get a move on." They both ate as fast as they could, so they could get back to their tanks and release the balance of their crews to eat. As they walked back to the motor pool together, they saw the Company Commander's Jeep pull up into the motor pool, with Captain Borman inside. It was going on 0630 - pretty late for his first alert, thought the Lieutenant. He had already seen Colonel Shea driving through the battalion area with the operations officer, so the Lieutenant assumed that the battalion commander would certainly be unhappy with Borman's late arrival. Borman had not seen Shea "lit" by his Black Irish temper, so it should be an experience.

The lieutenants parted company and jogged over to their platoons. When the Lieutenant reached 1-1, he released Blaylock and his loader to get breakfast. Then he walked up and down the line. He saw that Robb had gone to breakfast to eat and make sure that the first half of their crews were done eating and sent back so the second half could go to the mess hall. Once he settled back into his tank commander's cupola, he looked over to the company area and saw Captain Borman reaching for his radio, so the Lieutenant made sure his headset was on and listened for the order that he assumed was coming.

Sure enough, the first order was for a commo check of the company. Marley came on for Headquarters Platoon. The First Sergeant checked after that, followed by the lieutenant for First Platoon, Matthews for Second Platoon, and Ramsey for Third Platoon. Borman completed the check and relayed the order he had just received from Battalion that was they would be pulling at 0900, as opposed to 1000 as there was heavy weather coming in from the west into Baumholder and they wanted to try to get to there before they were caught by freezing rain.

Sergeant Robb returned from breakfast with the rest of the platoon. He ordered the crews to do one final complete maintenance check and crank the tanks up to make sure they would start. Once that was done, they let them idle and tried to keep warm. At 0805, Shea, the S-3, and Major Reardon, the executive officer, all made the rounds of the companies to make sure that everyone was ready. With all of the tanks and other tracked vehicles running, the noise was deafening.

At 0830, the scouts from Combat Service Company pulled out of the front gate, followed by the mortar, air defense and ground surveillance radar platoons. Once they were cleared, the staff left, and Alpha Company came in behind them, with its blade tank leading and the platoons in order. Borman followed in his Jeep with Bravo 5-5, the blade tank, close behind, trailed by First Platoon and Bravo 6-6 with Second and Third Platoons, Marley and the Headquarters vehicles falling in behind the convoy. His tank rocked forward, bucking a little as the treads gripped the roadway.

The Lieutenant strived mightily to keep up with where he was on the map sheet that he had marked in his large cover, for the road march while maintaining his interval with the other vehicles. The temperature was dropping fast, and the mist was beginning to turn to light rain as well, as they moved up onto the Autobahn. He fished his rubber duck out of his A bag and pulled it on over his field jacket to try and keep most of the rain off. It was going to be a long march to Baumholder in the soup.

DEACON BLUES

Once on the Autobahn, the battalion increased speed to forty kilometers per hour which was a good cruising speed for the M60 tanks. After they had been on the march for about an hour, Borman came on the company net and told them that they would be taking a rest break about halfway to Baumholder where there was a large German rest area with a restaurant where they could get hot beverages. Borman relayed the exit number for that stop and the Lieutenant wrote it down in his notebook.

After two hours of driving through ice cold heavy mist and rain, the Lieutenant saw the sign for the exit to the rest area. He keyed his headset and reminded Webb in 1-2 that he and the rest of the platoon would be turning into the rest area. He received a "roger that" from Webb and a few minutes later, his tank took the exit toward the rest area. At that time, Sergeant Robb came on the net and reminded the tank commanders to pull their vehicles in as straight as possible on the far side of the parking lot away from the rest area itself. That way they could back straight up and pull out to get back on the Autobahn later.

When all of his tanks were parked in line, the Lieutenant dismounted, went over to Sergeant Robb, and thanked him. Robb looked over his shoulder and pointed to a large pile of square granite cobblestones that looked like large dice. He asked him what those were and Robb he said, "That's where a tank from Alpha must have tried to neutral steer in this parking lot. The lot itself is designed to last forever because it's paved with loose cubes of granite. If a tank or other armored vehicle tries to execute a tight turn, it tears them up; breaks them free from their underlying foundation and the battalion will have to pay to have them replaced and the parking lot resurfaced wherever it has been damaged. Alpha Company has a new commander who came over from Headquarters Company. He just took command and didn't publish proper movement instructions. He also had a new platoon leader since the First Platoon leader is now the Scout Platoon Leader." Again, Sergeant Robb had proved himself worthy of all the praise that his Lieutenant had given him in the preceding months. Robb acknowledged the thanks the Lieutenant gave him and said, "Sir, we need to talk about one other thing."

The Lieutenant turned and asked, "What?"

"When we pull off the Autobahn to take the exit for Baumholder, at the bottom of the mountain, you will see a Kentucky Fried Chicken restaurant. It's the only one that I know of anywhere near our division in Germany. Customarily, the platoons will pull over and buy some buckets of chicken to take with them to garrison on top of the mountain so they can eat it while they do maintenance on the tracks."

"Really?"

Robb looked at him as seriously as possible and told him, "If we don't do this sir, we'll have a mutiny on our hands."

The Lieutenant looked at him and said, "Having survived an attempted mutiny in the desert at Bliss, I'll tell you - I don't want to go through that again, so I'll be glad to pull over and let the troops get their chicken. I imagine it's the only real upside to go to gunnery on the Rock."

As usual, the Lieutenant and half of the crews used the restroom first, and when they returned, Sergeant Robb and the remaining crewmen used the facilities. Once everyone was back on line in the parking lot, the Lieutenant looked to see if Borman had pulled in. He found him and his driver combat parked in his Jeep up close the building, so he ran over and told Borman that his platoon would need to pull over at the base of the mountain at Baumholder to allow the troops to buy lunch at KFC, as it was customary and they looked forward to it. Borman thanked him and relayed that message to the entire company so that all of the platoons of Bravo would have an opportunity to eat food that at least sort of reminded them of home.

Scrambling back, the Lieutenant mounted 1-1 and called for a quick commo check from the tank commanders. Having received confirmation of good commos, he waited for the order from his Company Commander to move out. Third Platoon and Second Platoon had already departed the parking lot, so First Platoon took its normal position in the rear of the company, followed by the headquarters, supply, and maintenance vehicles. As his tank pulled out of the parking lot, the Lieutenant looked at all of his tank commanders and saw that they were smiling, which was a further indication that picking up chicken before gunnery was the right thing to do.

As the tanks accelerated to their cruising speed of twenty-five miles an hour, the rain fell heavier and the wind came up as well. Soon they were being lashed by rain coming down in sheets so they slowed to fifteen miles an hour so that the drivers could see through the water that was running off their goggles. After two hours of driving in the rain, getting colder and wetter the entire time, the members of the battalion approached the exit for Baumholder. The Lieutenant gave the order for his platoon to pull off to the right side of the road opposite the Kentucky Fried Chicken restaurant and directed that one man from each tank be delegated to go and get chicken for the crew members of that vehicle. His loader volunteered and the Lieutenant said that he'd be happy to chip in so he took two five-mark coins out of his pocket and gave them to the volunteer.

It took less than ten minutes for his crewmen to get their buckets of chicken and return to their vehicles. The restaurant was well aware of the oncoming rush since Alpha had already been through, and he could see that Third Platoon and Second Platoon had already finished and were preparing to move out. Once his tankers were aboard and the commanders signaled that they were ready, he gave the order to follow Second Platoon and Bravo 1-2 took off up the mountain.

THE LONG AND WINDING ROAD

His platoon slowly climbed the Rock and the Lieutenant became aware of why the soldiers disliked Baumholder as much as they did. Although the rain stopped after ten minutes when they reached the lee side of the mountain, they were still in fog and drizzle for another half an hour. Then, as if by magic, the sun came out and it was actually hot on top of the tanks and soon they were covered in fine dust and swirling diesel smoke. Twenty minutes later a light rain began to fall, and they moved into another bank of overcast. Shortly after that, they experienced snow flurries and then heavier snow until finally, when they reached the Baumholder motor pool and hit garrison area, there was freezing rain.

The motor pool was a pleasant surprise. Since the Germans fired there, too, it had a well laid out hardstand, heavy concrete for the tracks, and a smooth lighter concrete for the wheels. The platoons pulled into position similar to those that they held back in Mannheim. When the tanks were in position, Borman ordered his platoon leaders to come to his Jeep, where he gave them their billet assignments and keys. When that was done, the First Sergeant met with all of the platoon sergeants and gave them the billeting instructions for their crew members. The Lieutenant caught Lieutenant Matthews' eye, came over to him, and asked, "What room have you got?" When he heard the number, he said, "Good, that's mine too. Looks like we'll be sharing a room with two beds and a shower."

This made the Lieutenant happy since he would finally have someone to talk to while going through the next two weeks. Both lieutenants went back to their platoons when their men were doing maintenance on the tanks. The Lieutenant helped check the track on 1-1 and found a couple of wedge nuts that needed to be tightened. He took care of those and when he was done, the commander's driver came down and requested that all platoon leaders meet Borman at Bravos headquarters offices.

The Lieutenant gave Sergeant Robb his room number and told him that he would be going to a meeting at the Company Commander's office and gave him the building number for that as well. Robb already had that information through the NCO network from the First Sergeant but appreciated the Lieutenant coordinating with him anyway. Sergeant Robb told him, "Sir, since it's Thursday, we'll probably spend tomorrow morning doing maintenance. The battalion and company maintenance platoons will fix any vehicles that broke down along the way. Starting in the afternoon, and through the weekend, the companies will download most of their live ammo to make room for some training ammo to shoot on the ranges. We'll take our tanks up to some zero ranges to test fire the co-ax machine guns, and the M-85s and calibrate the main guns. We'll practice zeroing M-2 machine guns attached to the tubes, because we'll use them to fire most of the tables before Table 9 platoon fire for qualification. It saves a lot of money on ammo and wear and tear on the tubes. As part of the zero exercise, the tank platoons will have to rotate to the support platoon to download our main gun ammo and upload training ammunition for firing as well. Everything together will take the whole weekend since most of the time at Baumholder is spent on fog watch. Visibility conditions are really bad, especially this time of year. On Monday, we'll go into full gunnery, so we'll be moving and shooting day after day and some nights, too. Meals for officers and enlisted will be served in the mess hall by the battalion's cooks, but officers and senior NCOs normally eat at a two-sided guesthouse that serves them on one side and junior noncommissioned officers and enlisted men on the other."

By this time, the freezing rain had stopped, but a light mist continued to fall so the front slope of the tank and turret were, according to his driver, "slicker than owl shit," whatever that meant, so he had Blaylock toss his A and B bags down to him to carry to the billets. He gave Blaylock his room number and the number for the company headquarters, and told him that if he was needed, he would be in one of those two places. Blaylock had been to many gunneries there and was familiar with the layout. The good news was the tanks were at least parked on concrete so they could walk to the barracks without having to traverse a sea of mud.

Out of the corner of his eyes, the Lieutenant could see Lieutenant Matthews with two bags that were almost as big as him heading for the barracks. He yelled at him over the sounds of the tanks to let him know that he would be falling in behind him to go to their room. Once they made it into the building and got out of the drizzle, they realized that the barracks were much warmer than the outside temperatures they had been enduring since the morning. They went into the room together and stacked their bags at the foot of the bunks and closed and locked the door behind them.

They kept on their rubber ducks and walked over to the small one-story building that would serve as company headquarters for their gunnery. A slightly larger building near it served as battalion headquarters and the brigade staff and commander had their own larger building set apart from the others. The Lieutenant could tell this by the bumper numbers on the Jeeps and other vehicles parked in front of the buildings. He also saw the headquarters for 2/13 Infantry their cross-attaching battalion that had arrived a day earlier. Apparently, 5/68, the sister battalion, had stayed back. It had M-60A2s that fired missiles, and for that reason, did its own thing. This party was just for DATs, "Dumb Ass Tankers," and "Grunts," the close-in cannon fodder.

The lieutenants walked into the company offices together. Sergeant Ramsey, Sergeant Robb, Sergeant Wilson, Lieutenant Marley, and the maintenance sergeant were already there. Guerra had already briefed the noncommissioned officers on the schedule, but they all listened attentively when Captain Borman came out of his office and reviewed the training schedule and plan for the upcoming week with them. Robb had said they would spend the weekend zeroing weapons and engaging in individual tank gunnery.

Since visibility was so bad in the mornings, they would finish maintenance on the tanks early in the morning and then road march the company to the zero range to calibrate weapons, both the M-85s and main guns, and practice with M-2 fifty caliber machine guns that would be used for simulated gunnery. When they arrived at the range, the Support Platoon would be set up to help provide machine gun ammo and some limited training rounds for each tank as it proceeded to the firing line. Captain Borman then gave all of them his room number and turned the meeting over to the First Sergeant, who let them know that breakfast would be served in the mess hall at 0600 and the company would be expected to stand-to at 0700 and to be prepared to road march by 1100 to the firing range. He would bring warm C-rations to the range for lunch, and they would road march back around 1630 because the fog would set in after that.

MAKIN' IT

Lieutenant Matthews and the Lieutenant coordinated with their platoon sergeants letting them know then they would see them at breakfast and could be found either in the officers' club mess or their room. Both of them then went back to their room, took a shower, and put on some clean, dry clothes before they went to eat. Once they were cleaned up and dressed, they quickly walked over to the large club building in their winter headgear and field jackets.

It was a building about the size of a bowling alley back in the United States with a wall down the middle. Both halves of the building shared a kitchen, so the food was the same, but the junior noncommissioned officers and enlisted were on one side of a partition wall and the senior noncommissioned officers, E-7 and above, and commissioned officers ate on the other side. The Lieutenants walked up to the counter to order and were met by a German lady who took all of the orders on the officer's side. The two mainstays of the menu were the chicken schnitzel sandwich that was essentially a chicken sandwich, and a jaeger schnitzel that was a pork tenderloin in mushroom sauce with French fries.

Both lieutenants decided to try the chicken sandwich and French fries with a beer. When they ordered them, the German lady sang out in an unmistakable voice, "Zwei chicken schnitzels mitt pommes frites." She did this continuously throughout the evening, and whatever they ate there, the lieutenants could count on having their spirits lightened a little just hearing her voice. The food was outstanding and they were tired to the bone so as soon as they had eaten and finished their beer they headed back to their room.

The mist had lifted – only to be replaced by a stiff wind, so they hurried. While they were walking the Lieutenant turned to Matthews and said, "Just so you know, in case Wilson hasn't clued you in, Ramsey is bucking for E-8, so his main mission here is to make you and me and our sergeants look bad, to get a leg up at the next board. He's after me mostly, but he'll take a shot at you and Wilson when he can."

"Hmm – I hadn't heard that, but I have to admit I never thought I'd be in a company with an all-white platoon or that I'd get a platoon that was all brothers."

"I know."

"By the way – what have you got?"

"Leftovers – but Robb is great and the troopers are too when they aren't stoned. Anyway, if we stick together, maybe we can make it out of here in one piece."

"Rog-O," said Matthews with "a little pep in his step" as he liked to say. Back in the room, they dressed in PT gear and made their beds. This consisted of unzipping a sleeping bag, laying it out on the mattress, and then throwing a poncho liner over everything. Too late, the Lieutenant realized he hadn't packed any kind of a pillow, so he rolled up his second poncho liner, tied it, and used that. He figured he would be tired enough not to care.

Both lieutenants put their wet fatigues and field jackets on hangers and placed them near the heaters in their room to dry, and after assessing the situation, set out their clothing for the next day. They decided they would need the heavy field shirts and pants for warmth, field jackets, liners, and rubber ducks and they would carry their B-bags, with overshoes and their ponchos and poncho liners, since every day promised to be, if anything, more cold and more miserable than the one they had just been through. The Lieutenant set his travel alarm for 0530 and turned out the lights. They were both asleep immediately.

COLD AS ICE

The next morning, as soon as the alarm went off, the Lieutenant shaved, and dressed in his fatigues, field jacket, and rubber duck to go to the mess hall for breakfast. When Matthews was ready, they walked over together and saw that the Mess Sergeant and the cooks from the battalion had set up in the mess hall the battalion would rotate through for gunnery. Both Lieutenants had decided that they would need a hearty breakfast since they never knew when they might get a chance to eat on the ranges. When they were finished, they stopped in at the First Sergeant's office to see if Borman had made any changes in the schedule. Since he hadn't, they went back to their room and dressed in all of their cold weather gear, shouldered their B bags, and walked together to the motor pool.

It was a little before 0700 so they threw their bags up on the back decks of their tanks and carefully climbed up to put them in the bustle racks. By then, they could see their crew members walking toward them after being released from the morning formation. Sergeant Robb ordered the members of First Platoon to get to work checking all the systems on their tanks and their tracks, before coming over to the Lieutenant. The Lieutenant greeted him and asked if there was anything he needed to do other than check out the systems on his tank. Sergeant Robb told him, "No, sir. I checked the tanks out yesterday and they were all operating properly. Due to the changes in temperature, and the ever-present dampness, you never know when something might short out or fail to connect so once we have rechecked everything on our respective vehicles, we'll walk the line to see if any of the tank commanders have problems."

Shortly before 0900, Captain Borman pulled into the motor pool. He called the platoon leaders to him and when they had assembled, he made sure that they had their tanks on the proper frequencies according to the CEOIs. He also gave them the frequency for range control, that they would have to monitor when they were on the ranges. He asked if any of the platoons had any problems with their vehicles. When all the platoon leaders confirmed that they had no issues. Captain Borman directed the platoon leaders to do an internal commo check and then check with him on his radio. The Lieutenant and his tank commanders had already done an internal check but did a second one anyway. The Lieutenant and Sergeant Robb checked in with the Company Commander on his frequency. They also preset the range control frequency so they could shift to that frequency whenever they wanted to.

The platoon leaders completed the company commo check and Borman signaled for Ramsey to have Third Platoon fall in behind his Jeep with Second and Third Platoons to advance in their normal order and follow him to the zero range. The company maintenance vehicles brought up the rear. The weather was even worse than the day before. Heavy fog and light rain had left a thin coat of black ice on the road so the tanks could only move at six miles per hour. Sergeant Robb came over the platoon net and reminded the drivers to try to ease off to the right if they lost traction because brakes were worthless and they were in sixty-ton toboggans if they were sliding.

First Platoon finally arrived at a little after noon. Third Platoon had picked up some practice rounds for their main guns and ammo to check fire their machine guns and fifty-caliber for the M-2 machine guns borrowed from Headquarters Company to be installed on firing simulators attached to the main guns. Each tank took up a position on the firing line and oriented on a large square boresight target down range. The crews taped twine over the gun barrel of the tank and sited it at the boresight target. Borman and Jackson, the master gunner had taken up positions in the tower that overlooked the range. The company frequency was used as the firing frequency and Borman monitored range control. The firing tanks were all on the company frequency and the platoon leader and platoon sergeant monitored the range control frequency too.

When the crews had mounted the M-2 machine guns, the tank commanders came on the firing net and announced that they were ready. Borman had each tank fire at his command and adjust their sights based upon the impact of the main gun round in the boresight target. When that was done, each tank fired the M2 simulator and adjusted the machine gun mount so that the round fired from the machine gun struck the target at the same place. By this time, the weather conditions had worsened, and the fog made it impossible to see the markers for the range fans so Borman had to shut the range down until they had better weather.

Guerra pulled up and issued warm C-rations to the crews, who drank hot coffee and cocoa with them. It took the balance of the daylight that was left to finish getting Third Platoon zeroed and used to firing both the main gun and the M-2 simulator and check fire their own machine guns. With dusk approaching, and ice forming everywhere in the face of a brisk wind, Borman gave the order to follow him back to the motor pool. The tank platoons slid into the motor pool just as darkness fell hard behind the mountain. The crews quickly checked the tanks, and when they were done, the lieutenants made their way back to their room to get out of their field gear, grab a fast hot shower, and go to dinner.

Since their platoons had not made it to the firing line, both lieutenants had spent the afternoon listening on the radio and trying to stay out of the wind by crouching in their tank turrets with their rubber ducks and ponchos on to retain warmth from the heater around them. But no matter how hard they tried, it seemed that water always leaked into the turret at the back of the cupola, right where their neckline was, so they got colder and wetter as the day progressed.

STUMBLIN' IN

Matthews was practically frozen, so he took the first hot shower and headed for dinner. The Lieutenant told him he would catch up, so after he was cleaned up and dressed, he walked to the club close behind him. When he came through the door, he saw Sergeant Robb seated at a table on the far wall. He went over and sat down beside him noting that Matthews had identified Wilson and was setting down with him. The Lieutenant asked Robb, "Do you think we'll get the company through bore siting and zeroing tomorrow?"

Robb thought for a minute and then said, "I'm not optimistic, sir. We have heavier weather coming in so imagine that we'll get a change of plan on the training schedule to slip us to Monday for another day of zeroing and maintenance. Third Platoon could go on ahead and start firing Monday morning but generally battalion likes to keep the companies together. That's so tanks are not running around to the training areas getting lost in the fog or slipping off the road if there is ice. As you can see, a big part of being a Baumholder involves sitting and waiting for the weather to clear long enough to shoot for a while before it socks in again. Most tank battalions go to gunnery twice a year at Grafenwöhr which is farther to the east. The weather there is better for firing long range tables. Several years ago, the Eighth ID stopped sending battalions to Graf but instead elected to do gunnery here where four of its battalions are located. We try to do gunnery at least three, and if possible four, times a year so you will see a lot of Baumholder."

The Lieutenant thought for a minute and then asked, "Is the Jaeger Schnitzel any good?"

Robb looked at him and said, "Yes, sir, of course. Germans manage this facility; they know how to cook and eat."

"Great. What's the First Sergeant's schedule for tomorrow?"

"Platoon leaders and sergeants need to be in the First Sergeant's orderly room at 0600, for further instructions, to include when the company will march back to the gunnery range. He won't have a firm time until he had spoken with the CO and the S-2 to see what the weather will be like."

"Thanks, I'll see you in the morning. I'm going to go over and meet the new lieutenant in Alpha, I think we were in basic together with my best buddy from the Third ACR."

REMINISCING

The Lieutenant rose from his seat, told Sergeant Robb that he would see him in the ordinary room in the morning. Then he went over to the counter and ordered his dinner and hot chocolate, before sitting down with the new platoon leader of Alpha. His name was McCrae. He was a West Pointer in the same year group as the Lieutenant. He knew McCrae a little because he had been assigned to the 4th Training Brigade but to a different battalion. Consequently, other than brigade social gatherings, he had never met the man. He was a nice guy, from Pennsylvania and was a good friend of Scotty, his best friend, who was from Pittsburg. They had been in basic together and the same infantry battalion at Knox. They had immediately taken a liking to each other and got along well. He and McCrae swapped a few stories about Knox and Scotty who they agreed was a real character. Everyone has a guy that gets them in and out of trouble on a regular basis – Scotty was that guy.

They had arrived in Germany the same week. The Lieutenant asked him how he was enjoying the tour so far, since they had barely seen each other. McCrae said, "When I got here, I was assigned to Alpha Company and I stayed with Harry Coyle in a room reserved for German maids in the basement of his quarter's building. I kept my uniforms and other stuff there when I went on Reforger as an observer, with Lieutenant McCallister, who is now the Scout Platoon leader, and Captain Coyle. I spent most of my time with battalions of the Third Armor Division, although I spent a day as liaison to a German tank battalion to get a better understanding of how they operate. When I got back from Reforger, I moved into an apartment on the economy, and my platoon went into Q-service. I've been training with my platoon as well as getting up to speed on Alpha's duties when we are cross attached to 2/13 Infantry. I also spent quite a bit of time at the Infantry Battalion getting to know the Charlie Company Commander and the members of the staff there as well."

The Lieutenant compared notes with him, recounting his own adventures over the last three months. When he had finished, McCrae followed up, "When I came in country, Shea sat me down and explained that I would spend a few months in a tank platoon until I'm qualified in that position. Then I should expect to join operations as the S-3 air or assistant S-3. Then, after about a year on staff, or a little less, I will get assigned as XO in one of the line companies."

The Lieutenant shook his head and thanked him for the insight. Then he wondered to himself what Shea had in store for him. But at that moment, he just needed to eat and get some sleep. He heard the lady call out his order, so we went to the counter and picked it up on a tray before walking back to his seat. He saw that Lieutenant Matthews had also been called to pick up his dinner, so the Lieutenant motioned to him to have him join them.

The Lieutenant introduced Matthews to McCrae, who asked him where he had come in from and how long he had been in. Once Matthews had answered those questions he asked the same questions of McCrae. When they were acquainted, Matthews asked McCrae how he liked Alpha. He said, "I have a great platoon and I'm getting along well there, but since I'm a first lieutenant, I'll probably transferred to the staff after gunnery."

As Lieutenant McCrae finished talking, the Lieutenant and Matthews finished their dinner, expressed their happiness at meeting him but that they needed to get some sleep. They excused themselves and headed back to their room where they relaxed a bit. Matthews was excited because his wife was supposed to arrive in country in time for Christmas and that it looked like he would be in Second Platoon for at least a year. The Lieutenant then chimed in saying, "I heard through the grapevine that I'll moving to a new position when we get back from gunnery. I have no idea what it's going to be, but then, I have never known where I'd go next. Counting orientation training in the summer of '76 when I was still a cadet, so far, I've been in and out of two infantry battalions, a cavalry regiment, and a tank battalion in a separate armored brigade, before getting sent to this tank battalion in an infantry division overseas. All those changes have made me pretty flexible when it comes to duty but my primary duty is just getting back to the States, getting married, and getting my wife in country."

There was no response, so the Lieutenant looked over at Matthews to see that he had already passed out, fast asleep, so the Lieutenant turned over and went to sleep himself because morning would come soon enough.

Chapter 14
I Can't Stand the Rain

Both lieutenants woke up early and went to the orderly room for coffee and to catch up with Guerra, Borman, and the platoon sergeants. It was Sunday, so Third Platoon would do maintenance on their vehicles, and individual training, while Second and First Platoons zeroed their main guns and test fired their machine guns. Consequently, the Lieutenant found himself slowly following Matthews's platoon back up to the firing range through light rain and a strong breeze.

When they arrived, he was happy to see the sun at last. He laagered his tanks behind the firing line while Matthew's platoon zeroed their main guns, test fired the machine guns and practiced with the M-2 machine guns mounted on their 105mm tubes. When Second Platoon was through, Matthews's tanks pulled off the firing line and laagered back well behind the range so that First Platoon was finally able to pull forward into the firing positions. Specialist Blaylock and the driver, with the help of the loader, were able to get the main gun bore sited quickly despite the wind that was blowing stronger and forming ice on the front slope of 1-1. The other crews, inspired by the cold weather, completed fairly quickly as well. The only holdup was 1-3, where his Guamanians were learning by doing and arguing amongst themselves in their own language.

Sergeant Robb finally went over and helped them. After about half an hour, all of his tanks had fired their main gun and were bore sited. They quickly test fired the coax and M-85 machine guns. The last one was a real pain to load and charge, but fun to fire, since it was his own personal weapon in the tank commander's cupola. Then each crew attached a Ma Deuce, a M-2 machine gun, to the main gun and adjusted the platform to bring the rounds fired from it in line with the strike of the main gun round on the bore site target.

So far, while cold and windy, it had been a fairly clear day and both platoons were able to complete their tasks in the space of less than four hours. By midafternoon, however, the drizzle and fog had returned, so Borman shut the range down. The crews took the M-2s off and passed them to the supply sergeant to take back to the orderly room. The CO then pulled his Jeep out onto the main road and the platoons followed him back to the motor pool. On the way back, low clouds rolled in and heavy rain poured down, which slowed the convoy to five miles per hour. As they arrived, Borman radioed the platoon leaders and told them that the First Sergeant would have meeting following the company formation at 1700. He also reminded them that they needed to be up early the following morning to begin firing the first gunnery tables.

Afternoon schedule set, the Lieutenant and Blaylock removed the machine guns and handed them down to the driver and loader. Then he slid off the tank's front slope with his B-bag and headed for his room. He was soaked, despite the rubber duck, and he was fast figuring out why this area was a gunnery range. With the weather, no one would choose to live there.

TEQUILA SUNRISE

The next morning, following a meeting in the First Sergeant's office, the Lieutenant enjoyed a hearty breakfast in the mess hall. By the time he was back in the motor pool, the sun was trying to break through the clouds with a sliver of red and yellow that initially lifted his spirits. However, the mist settled in almost immediately and the normal dismal gray took over again. Borman called for a commo check with the platoons and when that was done, he summoned the leaders to his Jeep for an update on their situation. The company would road march to a firing range where the platoons would go online. Each individual tank would fire a series of pop-up silhouettes beginning with targets at the farthest range and closing to five hundred meters, using the M-2 machine gun simulators. Due to visibility problems, the farthest range of targets would be at eighteen hundred meters, which was fairly close compared to the Lieutenant's experience doing gunnery in the desert at Fort Bliss where targets started at twenty-five to three thousand meters.

Borman then explained that due to fog, rain, and sometimes snow, he expected that it would take the better part of two days to get the company through this range. After that, the platoons would go to the firing ranges to complete individual tank gunnery on the move, using the M-2s. Platoon leaders would rotate through tactical exercises as well. From what he'd experienced so far, the Lieutenant guessed that for most of the exercises, just getting there through the fog in the morning and returning in the rain or sleet in the afternoon would be the most arduous part.

Once Borman came over the radio, the platoons pulled up behind his Jeep and followed him to the firing range. After they arrived, the crews loaded .50 caliber ammunition for the M-2 simulation devices they would be used on the range. The Support Platoon and details from Headquarters Company assisted in moving the ammunition onto the tanks. When the crews of Third Platoon had their ammo, they moved their tanks to the firing line and waited for the fog to clear. Second and First Platoons laagered in a large parking area behind the control tower and waited for their turns on the line.

By the time they could see the range markers, it was late morning when Third Platoon was allowed to fire at targets individually. The

Lieutenant told Sergeant Robb that he wanted to observe and walked up to the tower to watch the firing. He stayed to watch Matthew's platoon as well. By shooting as quickly as Third Platoon had, Second Platoon was able to complete firing and pull off the line. Since First Platoon couldn't start until after 1400, he and Robb would have to come back the next day with their crews and finish up. At that point, the Lieutenant was catching on to the greatest advantage there was to being the best platoon in the tank company: you marched and fired first and thereby received several days back in garrison to maintain your tanks and enjoy your free time as opposed to sitting on range after range, wet and freezing.

When the Lieutenant's platoon was halfway through, heavy fog rolled into the range, followed by sleet and driving rain so the two platoons and Borman drove slowly back to the motor pool. When they arrived, the Lieutenant slid off his tank and went over to Sergeant Robb and asked, "Do you think we'll be able to get through gunnery tomorrow?"

Sergeant Robb replied, "The weather looks bad but if we can get at least three hours, we should be able to do it. The main holdup will be the Guamanian crew on 1-3, who have never done tank gunnery before, so we're on the fast track of OJT. The tank commander was an artilleryman for his first enlistment in the Nam. In his head, he continually wanted to fire low and high and then bracket the target, which is simply too slow for a tank."

The Lieutenant nodded his head and then followed Robb and his troops back to the company area for evening formation, where Guerra informed them that First and Second Platoons would be staying back to do maintenance and individual training the next day, while First Platoon would finish individual gunnery on the range where they had been today. They would stand to at 0700 and pull at 0730.

LIONS

When he got back to his room, the Lieutenant was surprised to see his roommate packing up. Matthews told him that additional rooms had opened up since another battalion from the division, that had to hold over for a few days due to the weather, had finally completed platoon gunnery and pulled out that morning. Apparently, officers were usually billeted to individual rooms unless there was some overlap. The Lieutenant told Matthews that he would miss him, but he was also sure that he would appreciate the chance to sleep a little later, since he was getting up early to make sure everything was ready to drive to the range and get through individual gunnery with the six tanks under his command.

When Matthews was gone, he went over to the club to have a chicken sandwich and some hot peppermint tea to warm up. Then he went back to his room to try and get to sleep. There was a small field desk and chair in the room that neither of the lieutenants had had occasion to use. The Lieutenant took out his notepad and sat down to try and figure out how many days he had been in country and how many days it would be before he left to get married. Having been determined that he was almost halfway through the seven months before he flew home, he climbed into his sleeping bag, pulled a poncho liner over himself, and fell asleep.

The next morning, he went down to the First Sergeant's office for a cup of coffee and to meet Sergeant Robb. Robb told him that the word was they would try to leave thirty minutes earlier than announced the day before, so he had the platoon down for breakfast already. Taking the hint, the Lieutenant hurried to the mess hall and ate quickly before running back to his room to put on field gear and hastily pack his A-bag for the day. He was surprised to see the First Sergeant in the motor pool but then realized he was issuing C rations for lunch, so he didn't have to drive out to the range for one platoon.

At 0800, Captain Borman pulled into the motor pool. Sergeant Robb already had the platoon finish with an internal commo check, so the Lieutenant did a quick commo check with Borman, who then gave the order to follow his Jeep as it pulled out of the motor pool to the firing range.

When they got to the firing range, his tanks pulled into position on the firing line with 1-2, 1-3, 1-1, 1-5, 1-4, and 6-6 lined up from left to right. The master gunner, and a detail from HHC, moved from tank to tank with the M-2 machine guns. When the Ma Deuces were attached to the main guns, Borman let each tank fire individually to adjust the mechanism for the machine gun to strike the center of a target at 1200 meters. Once each of the tanks had gone through that exercise, the crews took a short break to get some hot coffee at the tower where it had been delivered from the mess hall by the XO's driver.

While the crews were doing that, Sergeant Robb and the Lieutenant met with Borman and gave him the order of fire for the tanks in the Lieutenant's platoon. He and Robb had decided to have the Lieutenant's tank fire first so that his crew would be complete, as the word was that when the company moved to individual gunnery tomorrow, he would be tasked to go on a firing exercise with one platoon leader and crew from each tank and infantry platoon to simulate combat on the West German border with an M-16 rifle affixed to the tube and small metal targets. That exercise would also involve the fire support officer since the platoon leaders would have to call for artillery from miniature mortars to place little smoke rounds on the miniature range made to look like a series of German villages.

ENJOY YOURSELF

The Lieutenant agreed that that was a good plan and directed his crew to man 1-1 and prepare to fire the range. The crew members for his other tanks and 6-6 took their positions to watch as 1-1 began to engage. Fortunately, the weather was better that morning so the Lieutenant received approval to fire at will as soon as he announced ready.

When he radioed Borman on the firing frequency, the farthest target popped up immediately. The Lieutenant used his pistol grip to lay the gun, announced the target type, range and ammunition selection for the gunner who identified the target and fired on command. Since they were using the fifty-caliber training device, the loader and driver spotted and announced if they saw a hit. This kept both of those crewmen engaged and talking. Based on his experience on other ranges, the crews that were most successful kept up communication and usually the driver was in a better position to see many of the targets and the loader was in a better position to look to the left flank. The Lieutenant had a better vantage point from the top of the tank so usually he would be the first to announce the target and move the gun so the gunner could pick it up and shoot.

Specialist Blaylock was an excellent gunner, and the crew fired the targets from farthest to nearest almost as quickly as they appeared. In the span of less than twenty minutes, they fired multiple target types at various ranges. The Lieutenant enjoyed tank gunnery, so he was sorry to see the time go by but happy that he was able to slew the turret with the tank commander's override and fire from his position to hit one target at five hundred meters. When he received the ceasefire order from Borman, he relayed that to the crew and had Blaylock and the loader clear the machine gun, disconnect the ammunition, and give the "all clear." When they were through, he requested permission from the tower to come up and observe his other tank crews fire and to have Sergeant Robb join him.

When he reached the base of the tower, he met Robb who asked how he'd done. Robb congratulated him on doing very well and the Lieutenant smiled a little to himself before climbing the stairs to the tower. They then went up to see Captain Borman, who was just giving the order to 1-2 to begin firing. Sergeant Webb, on 1-2, had about two years in his current position and a gunner that he had worked with him for over a year. They fired their table fairly quickly and only missed one target. Sergeant Robb was pleased with their performance and prepared for 1-3 to begin. Since they had never engaged in tank gunnery before, the Guamanian tank commander and gunner were nervous, and he could tell that there was quite a bit of hesitation in the firing commands as the targets came up.

The tank commander, as usual, was trying to fire at the base of a target which was proper but then he would fire over the target and try to bracket the difference. This not only required more time but used much more ammunition as well. Since they were using fifty caliber machine gun ammunition, this was not so much of a problem, but when they got to table nine, and used the main gun, it would be a significant hindrance.

It was only mid-morning so Robb asked the Lieutenant if he could get with the Company Commander and see if they could have 1-3 fire the table again because they were just learning. Captain Borman agreed and relayed that order to the Guamanian tank commander. He then told them to stand by and clear the gun so Sergeant Robb could go down and talk to them a little before they re-fired.

Robb had the crew dismount the vehicle and come to its rear, where he spoke with them briefly about the basics of tank gunnery and how to improve their skills. When he was done, they remounted the vehicle and the tank commander called the tower to announce that he was ready. Borman gave the order to begin firing and this time the crew of 1-3 did a much better job. For targets past a thousand meters, they generally got a hit after two shots, and they were able to hit the closer targets with one round.

When the tank commander on 1-3 called in that his machine gun was clear, Borman gave the crew on 1-5 the order to commence firing. The tank commander on 1-5 had only been through gunnery as a tank commander once before, so the coordination between him and his gunner needed some work. He also had the most inexperienced gunner, who was only on his second gunnery. Nevertheless, they fired their table in less than twenty-five minutes and hit all of their targets except for two. Robb was pleased with their performance, so the tower ordered that crew to clear their weapons and when that was done, Borman ordered 6-6 to engage.

Sergeant Fields, "The Buffalo," had about as much experience as Webb on 1-2. The gunner, however, had less experience so they went through the table in a little over twenty minutes and missed two targets. As he was watching the crew of 6-6, the Lieutenant thought to himself it would be a good idea if Borman would crew his tank for at least part of gunnery because the perspective you had in moving from the top of a tank was totally different from that you got in the passenger seat of a Jeep. As a practical matter, however, company and battalion commanders did not shoot tank gunnery themselves. The thinking was that they would probably be in the loaders side of the tank anyway to have room to lay out the maps and deal with the radios if the time came when they would have to traverse an actual battlefield in an armored vehicle.

LIDO SHUFFLE

Sergeant Robb was happy with 6-6 and he immediately left the tower to mount 1-4. Once he was in position, he announced to the tower that he was ready and put his crew through their paces. It was an awesome spectacle and a good training exercise for the Lieutenant to see that crew engage targets rapidly and effectively so that they cleared their entire table with no misses in fifteen minutes. Afternoon had arrived, so Borman gave the order for a thirty-minute break to eat their C- rations before they road marched back. He pulled the Lieutenant aside and told him that the following day he and his crew would be going to a battalion exercise involving a lieutenant platoon leader from the mechanized infantry battalion and him on 1-1 simulating his platoon. He would use an M-16 rifle attached to his main gun to simulate fire on a miniature range. He sat down with Borman and the map sheet for Baumholder. The CO gave him the range number and pointed to it on the map. The Lieutenant used his grease pencil to circle that range on the outside of the map case that he had covering his map of the area. Borman told him that since he would be putting the other platoons through individual gunnery, he would not be accompanying him and that he needed to move out before 0700 in order to get to the range by 0800.

The Lieutenant walked over and told Robb, "I'll be on a platoon leader fire exercise tomorrow, so I won't be with the platoon on the firing range with the Old Man."

"I understand, sir; I've fired that range before myself."

The Lieutenant then assembled his crew behind their tank and gave them instructions to meet him in the motor pool at 0630 the next morning so they could crank up the tank and road march out to the range by themselves. He told them to be sure that they got with the First Sergeant first thing in the morning and pick up C-rations to take with them. As he was finishing, Borman stepped out of his Jeep and gave the hand signal for everyone to mount and prepare to move out. He then came over the radio on the company frequency and gave the order to march back to the motor pool. The Lieutenant passed the order along to his crews and one by one, the tanks backed up and then moved forward to fall in line behind the CO to march back to garrison.

By the time they got back, it was late in the afternoon, with darkness falling fast, so the Lieutenant turned the platoon over to Sergeant Robb to check the vehicles and do any necessary maintenance. He and Blaylock, with the help of the gunner and the loader, all checked their portions of the of the track and looked to see if there were any oil, fuel, or hydraulic leaks. 1-1 was in good shape so when the other crews had finished visually inspecting their tanks, Sergeant Robb released the crews to go to their rooms and the Lieutenant trudged back to his.

After a quick shower to wash off the diesel, dust, and mud that had accumulated on him during the day, he changed into dry clothes before going over to the club for dinner. By now he had developed a taste for chicken sandwiches and French fries as well as hot peppermint tea to warm up. When he looked over into the far corner, he saw

Lieutenant McCallister, so he took his tray over and joined him. McCallister was excited because he was already transitioning to the Scout Platoon and was training with the outgoing platoon leader during gunnery. His tank platoon was now commanded by his platoon sergeant, but the word was that they would be getting in a new lieutenant from the Minnesota National Guard when they returned. McCallister also told him that he had stopped by the Adjutant's office and learned that a new platoon leader, a second lieutenant, was coming into B Company when they returned to Mannheim.

McCallister was also excited because of his new girl. He liked her a lot and her father was the one-star commanding the MP detachment in Mannheim, so that couldn't hurt his career. The Lieutenant congratulated him and wished him the best as he himself had been fortunate enough to find his girl when he was still a second lieutenant. Then he asked McCallister, "How's the new Company Commander in Alpha doing?"

"Okay, but I miss Harry Coyle, mostly because he was quieter and a little more easygoing. Pallas was the Headquarters Company Commander for the last year. That unit has so many problems that it has made him a little paranoid. A line company had fewer soldiers, and the CO has more control, and he's great guy, so I expected that he will settle in before long."

The Lieutenant said, "I'm going to miniature range tomorrow using M-16 rifles."

McCallister nodded and said, "I've fired it several times. You'll have a good time but be sure and leave early because the weather forecast for tomorrow calls for low clouds, and freezing drizzle in the morning, and the range is in a low area where the fog will be bad."

The Lieutenant thanked him for the head's up and went back to his room to relax for a few minutes, before setting his alarm clock for 0500 and turning in. As he lay on the bed, he thought about his girl back in Kansas. It was a little after 8:00 pm in Germany so it was early afternoon there and he imagined her getting up and getting ready to go to work in a couple of hours, to check IV sets. Then, he pulled out his wallet to look at the small copy of their engagement picture that always cheered him up. He put his wallet back on the small table beside his bunk and crashed.

JOURNEY OF THE SORCERER

He heard an alarm go off in another room and came to immediately from a sound sleep. He turned his clock off, took a deep breath, and decided that he would shave and take a shower to make sure that he was awake. Once dressed, he went down to the orderly room for a cup of coffee and to visit briefly with Sergeant Robb. Robb was an early riser, so within five minutes he came through the door, saluted the Lieutenant, and greeted the First Sergeant. Robb told him that his tank crew was up, and that he had sent them over to eat breakfast early and draw C- rations. The Lieutenant thanked him and told him, "I'll be eating breakfast and taking off with the crew shortly. I expect I won't see you again until late afternoon when the company is back from the range and I'm back from the mini range."

"I'll see you then, sir."

The Lieutenant thanked him, went over to the mess hall, and ate a larger than usual breakfast. Then he went back to his room and changed into his field gear, making sure he had plenty of licorice in his field jacket pockets to tide him and the crew through the day.

When everything was ready, he tossed his A-bag over his shoulder and walked to the motor pool. Specialist Blaylock had the tank cranked up and the heater on high. Everything was in working order, but the fog was horrific. He could barely see fifteen feet in front of the tank. He threw his bag up to Blaylock and slowly climbed up on the 1-1. Blaylock looked at him and said, "Sir, I'm afraid you're going to have to ground guide us to wherever we're going. You have the only map, and I can't recall how I got there since the last time I fired the range as gunner, and you can't see shit from the gunner's seat."

"Fair enough. All right," said the Lieutenant as he slid down from the tank commander's cupola. The entire tank was slick from the diesel and dust that coated it and the freezing rain that was starting to accumulate. He crawled down the tank and then slid off the front slope to the ground. Having been warned about the fog the night before, he had his L-shaped field flashlight with him, but it was in his A-bag. He asked Blaylock to get in the bag and toss it to him. Blaylock pitched it down and tossed him his map case with the map of Baumholder as well.

Finally prepared for the march, the Lieutenant began walking out of the motor pool with the flashlight over his shoulder so the driver could see him and not run over him in the fog. Fortunately, it was early in the morning and there were no other vehicles on the road. He stopped at the first intersection and tried as hard as he could to see in every direction, but the fog had every vantage point completely whited out. Taking a look at the map, he determined that he needed to take a left to get on to the road that would lead to the range they would be firing. Having walked down that road for five hundred meters, he came into an intersection that at first looked like the road he needed to take. About one hundred meters in, however, he could see a large building looming in front of him that wasn't on his map. He signaled to the driver and Blaylock who could hardly see him from the turret, with his flashlight to stop. He then walked up to the tank and told them he needed them to stay put while he walked over to see what this building was and try to figure out where he had gone wrong.

As he walked toward the building, he saw that there was a sign next to the door in a metal frame. When he reached it, he shook his head in amazement and chagrin. It was the Baumholder officers club and his tank was in the middle of the parking lot. He sat down with his flashlight and looking at the map realized that the turn that he thought would take him to the range was, in fact, a long driveway to the club. He went back out into the parking lot to his tank and looked at the surface. It was concrete but not of the strength you would find in a tank motor pool. Consequently, he told the driver to follow him slowly, directly forward, and not to turn the tank using the neutral steer method at all. Instead, they would take a long loop back out of the parking lot to the main road to avoid tearing up the surface. Blaylock, the loader, and the driver all shook their heads in agreement and followed the Lieutenant as he walked carefully back to what now appeared to be the main road to the firing range.

As he reached the road, he could see that the sun was just beginning to burn through the worst of the fog and the drizzle was lifting. He walked down the main road another five hundred meters and visibility was now up to fifty feet. From the map, he could tell that he needed to make better time in order to arrive at the range by 0800. He signaled again for the driver to stop and walked up to talk to his crew. He told them that based upon the small amount of traffic and better visibility he believed they could make about eight miles an hour to get to the range unless they ran into denser fog. They looked at him quizzically, but he said, "Don't worry – officers are magic." Then he stored the map case and crawled slowly up onto the tank and then pulled himself up on the cupola like he was trying to climb an ice cube. When he finally got to the cupola, he slid into the tank commander's seat and made sure that his helmet was connected and he could communicate by intercom with all crew members. He also checked the frequency for range control to make sure that he could contact them by radio if necessary and then finally told the driver to move out slowly; first going five miles per hour and then when it looked like they could move faster, up to ten. Within five minutes, he realized that they were climbing a little and that the range they were headed to was on higher ground than garrison and the officers club.

After ten minutes on the road, the visibility improved until the sun broke through, so the Lieutenant ordered the driver to increase speed to fifteen miles an hour, and twenty minutes later, they reached the entrance to the firing range. He ordered the driver to stop so that he could dismount and lead the tank into the range complex.

As he came through the gate, he met the NCO in charge of the range who directed him over to a captain who was the range officer.

The range officer led the Lieutenant, followed by 1-1, to the firing position where the tank pulled in and parked. The range officer congratulated the Lieutenant on having the first firing vehicle there and being almost half an hour early. He told him that he had another tank coming and two infantry tracks for a total of four platoons on the firing line. One of the infantry tracks would be from 2-13 infantry, which was part of the Lieutenant's brigade and another track with a new lieutenant from one of the mechanized battalions stationed at Baumholder would be firing as well. He also told him that when those vehicles had arrived, he would have all of the officers come to the tower for a briefing. The range control officer then gave the Lieutenant the firing radio frequency he would use to control everyone on the range and ordered them to the tower for the briefing.

The Lieutenant looked over toward the tower and saw what he thought were thermite containers for hot beverages, so he asked the range officer if his crew could warm up with coffee or whatever else was available. The range officer told him that that was not a problem; the range had mess service from one of the local units and coffee and hot water for soup and cocoa was delivered in the morning and at noon, so he could have this crewman help themselves whenever they needed it. The Lieutenant thought to himself this was the first real piece of luck he had had so far in gunnery. So, he thanked the range officer and told him that he looked forward to firing.

As the range officer walked back to the tower the first infantry track came through the gate and pulled up next to him. It was a track from one of the battalions of the 87th Regiment, so the Lieutenant waved hello and waited for the rest of the vehicles. The next vehicle in was another tank from Charlie Company that was commanded by an E-7. The last track to come through was from 2-13 infantry. The Lieutenant shook his head in amazement because he recognized the track commander. It was Lieutenant Bird, known as T-Bird, who he knew from ROTC Advanced Camp, over three years earlier. In many ways, it seemed like forever and yet it also seemed like only yesterday because the Lieutenant had met his future wife for the first time in May of that year. She was just nineteen, and he was twenty with a new crew cut, getting ready to go to Fort Bragg for six weeks of infantry training, before shipping out to Fort Knox, Kentucky, where he was to be assigned to a tank battalion for three weeks.

As the other vehicles came in, the range officer spoke with each vehicle commander. When he had spoken with Lieutenant Bird, he walked back with him to the tower and, having climbed up to the top, radioed the other vehicle commanders to his location for a briefing. All of the platoon leaders came to the base of the tower and climbed to meet the range officer, who showed them a diagram of the firing range that identified the firing positions for the vehicles and the targets. They were laid out in rows at various spots in the terrain, set up to look like a series of German villages, small roads, and bridges similar to the type you found on the border. The range officer then explained that when the range was open for live fire, they would see targets approaching closer and closer to them as the rows of silhouettes flipped up from the bands farthest away to the closest.

The range officer also explained that they could also call for fire on a separate frequency. He would give each platoon leader an Infantry PRC 77 radio and attach an M-16 to the main gun with the device that would allow the gunner to activate a solenoid and fire the rifle. Indirect fire with simulated by the use of miniature mortar-like devices that could fire projectiles into the range. The purpose of the range was to give the lieutenants some feel for what it would be like in the Fulda Gap under attack by two regiments of motorized infantry with tanks in support. Each officer was given a map of the range with preset registration points for the artillery so they could call in for indirect fire as well as use their tanks. In the case of the infantry vehicles, the M-16 was affixed to the barrel of the M-2 machine gun and it could be fired to simulate a Tow missile at longer ranges and the shorter-range infantry weapons closer in.

DEVIL'S GUN

When the range officer was finished, he told the platoon leaders to go back to their vehicles and check the M-16 devices. After they were satisfied that the devices were properly installed, they were to signal with their right hands held up. When all the vehicles were prepared to fire, the range officer would then radio them on the firing frequency and give them instructions to test fire their systems. Going out the door of the tower, the Lieutenant walked over to Lieutenant Bird and asked him, "When did you get in country?"

Bird answered, "In August. I spent most of my time in the field as an observer for Reforger and then in platoon and company training in the maneuver area near 2-13."

"Where were you before?"

"The 197th Infantry Brigade at Benning. Then I was transferred to Europe to strengthen the units here. How about you?"

"I was at Knox and Bliss before I was sent here in August myself. Anyway, it's great to see you. We need to get together back in Mannheim." The lieutenants parted company and hurried to their vehicles.

As he headed toward 1-1, the Lieutenant thought to himself, *T-Bird is one of the funniest people I've ever known.* When he got back to his tank, he walked around to the left front and climbed up the front slope, carefully, practically crawling to the turret and the cupola. Then he stood up and holding on to the bustle rack, made his way around to the main gun where a rifle was mounted in the simulation device attached to it. He inspected the device and saw that it was aligned with the tube and securely mounted. He raised his right hand to signal that he was ready and climbed into the turret.

The range officer came over the firing net and directed each of the firing tanks to identify a small square target at a simulated range of twelve hundred meters and test fire the M-16 to allow their gunners to adjust the mechanisms so that they matched the site for the gun tube. Unfortunately, when the Lieutenant tried to fire the rifle, it did not function. The Lieutenant raised his hand for assistance from the range NCO who came over and told him that the solenoid for the device was out and they had no spares. The Lieutenant then told the NCO that he would simply remove the M-16 from its mount and fire from the turret beside the gun, in the seated position, while having the gunner move the turret back and forth to try to identify the targets. The range NCO acknowledged the adjustment and informed the range officer.

All of the platoon leaders signaled that they were ready, and the range officer gave the order to prepare to fire, and then to fire at will. As the last of that command echoed through the radios, the Lieutenant could see in the far simulated distance large numbers of targets appearing on the horizon. These targets were too far to engage accurately with the rifle, so he looked at the map sheet that had been provided and identified the pre-registered targets for the artillery. As he was prepared to call for artillery at a road intersection over three thousand meters away, he heard T-Bird come on the net requesting a battery, followed by "map sheet," and then the series of numbers on that map sheet that reflected which one he was on. This was followed by, "Fire at any bridge or intersection, you're bound to hit something." He looked over at T-Bird and laughed a little to himself, then he asked for fire on registered target, "A11."

He looked back over his left shoulder and saw one of the miniature mortars fire small projectiles toward that target area. They landed short so he called for fire again and to add one hundred meters. More projectiles landed effectively at the target point and the Lieutenant could hear the other officers calling for indirect fire as well. The targets were now closer in at approximately the two-five-hundred-meter range. At this point, he inserted a fresh magazine into the M-16, sent the bolt home, and began firing at any unusual vehicles he could identify and tanks. Initially, his rate of fire was fairly slow as he adjusted to the range, but as the targets began to roll in like a wave he was soon firing more and more, slapping in a new magazine. As he was firing, he yelled over his shoulder, "I'll show you what a tank platoon online can do!"

He fired roughly one round every second until he had exhausted three more magazines. At that point, the targets were close to five hundred meters, and the range officer gave the order to ceasefire and clear. Once all of the vehicles had stopped firing, the range officer directed the platoon leaders and their crews to the base of the tower where the diagram was set up. The crews formed a semicircle around the diagram, and he critiqued each crew in turn. The Lieutenant had been effective with his rifle at shorter ranges but had not made effective use of the artillery at longer ranges.

The Lieutenant noted this and told himself that he would do better on the next round. The range officer then told them that they would take a short break for lunch before firing the range again in the afternoon. After that, they would have a final critique and be released back to their companies.

As the members of his crew ate their C-rations, the Lieutenant pulled out a couple of pieces of licorice and ate them with a cup of black coffee, while he studied the pre-registered artillery targets. When the crewmen had finished lunch, he asked them to step over so he could talk to them. He told Blaylock, "I want you to fire the M-16 from beside the gun like I did earlier. Start engaging when I tell you to. I'll lay the gun and slew the turret in the general direction of the targets that I want you to fire on. I'll call for artillery continuously to try and destroy the maximum number of enemy vehicles; that will allow you to better engage those that are close in."

The two range NCOs saw that the crews had been fed and came and told them to remount their vehicles and prepare for commands from the tower. The Lieutenant remounted 1-1, but this time took his seat inside the turret while Blaylock held the M-16 with the barrel up and down range and a fresh magazine in it, with more beside him. The range officer addressed the crews on the firing net and asked each platoon leader to raise his right hand if he was ready. When all hands were raised, the platoon leaders again got the order to prepare to fire and then fire at will.

This time, the Lieutenant called for artillery fire at the farthest bank of targets by using the registered target number nearest them. When the vehicle targets appeared at the simulated two-five-hundred-meter mark, he slewed the gun slightly and ordered Blaylock to begin engaging any unusual vehicles and tanks. The Lieutenant then continued to call for artillery as the banks of targets approached. Blaylock performed admirably, firing round after round and hitting target after target, until the final bank of targets at the modified five-hundred-meter mark came up.

When they received the order to cease-fire, they also received the order to clear all weapons and waited while the range NCOs checked to see that the rifles were cleared, and that the magazines and unspent ammunition were returned to them. The devices were detached and returned as well. When all of those tasks were done and the NCOs waved to the tower, the range officer ordered the crews back to the same debriefing area. The range officer was quite happy to inform all of the platoon leaders and their crews that they had done a much better job the second time. It was now midafternoon, so the range officer gave the order for the firing vehicles to return to their companies. His tank had been the first to arrive, so the Lieutenant was allowed to leave first and walked in front of his tank, back to the edge of the road, where he remounted and the driver took off for the motor pool, in sunshine for a change. It felt really good.

Chapter 15
Knock on Wood

Back at the motor pool, the Lieutenant and his crew checked the tank for any issues following its range time. While they were doing this, Borman's Jeep pulled into the motor pool, followed by the tanks of the Third Platoon, Second Platoon, and the remainder of First Platoon. When all of the company's tanks were parked, the crews began doing their maintenance checks and the Lieutenant went over to talk to Sergeant Robb.

Robb told him, "With the good weather, all of Third Platoon's tanks and half of the Second Platoon's tanks completed individual gunnery. If it's good tomorrow, it looks like we'll complete individual gunnery. After that, the companies will be moving on to night fire as a battalion, with each company firing online through one evening, if we have good visibility. I expect we'll also have a day of maneuver training with an infantry company pitted against our company. Both of the companies are cross attached to one platoon so that there will be a tank-heavy team versus an infantry-heavy team in some sort of scenario. Following those field exercises, each company would rotate through to Table 9, full platoon gunnery shooting our main guns. After that, with any luck, we'll get a day or two in garrison for maintenance to allow Battalion to fix any broken equipment before the road march back to Mannheim on Monday. And, sir, the CO is having a meeting of all platoon leaders and sergeants after the afternoon formation." As usual, the Lieutenant thanked Robb for the information.

He then waited for Robb to form the platoon in two ranks, saluted and turned the platoon over to the Lieutenant so he could address the crews briefly. He explained, "I've been away all day, firing the platoon leader mini gunnery range, but tomorrow I'll be shooting the individual range along with you. I know Sergeant Robb has trained you well, so I'm sure you'll do a good job, and I look forward to it." He called the platoon to attention and turned the men over to Sergeant Robb, who gave them instructions to be prepared for stand-to at 0700 and then dismissed them for the day.

The Lieutenant turned to walk back toward the company area and slowed his pace so that Robb could catch up with him. He asked, "How are the men were doing?"

"Sir, mostly they're just tired of being cold, wet, and bored waiting for the weather to clear so they can shoot."

"I understand. I feel the same way, and with any luck, we'll be done with the ranges and ready to get back to Mannheim in the next few days."

"Knock on wood, sir."

The Lieutenant stepped into the orderly room to find Borman, Marley, as well as the other platoon leaders and sergeants, waiting for them. As they took their place in the semi-circle around the CO, he gave them their instructions for the upcoming days. First, the remainder of Second Platoon and First Platoon would have to complete individual gunnery the next day if at all possible. After that, Bravo and the other two companies would rotate through a company-on-company exercise with an infantry company, a day of Table 9, and depending upon the weather, night fire.

Borman had hoped to do night fire with seventeen tanks online the following evening, but the weather forecast was terrible, so

Battalion had decided to delay it for a day. That meant that the company that had been engaged in company-on-company maneuver would go from that to night fire. If they couldn't get the night fire range completed on the second night, they would go on to Table 9 the next day. If at all possible, given the weather, they would try to execute night fire after Table 9 for the whole company, but it appeared that a large storm was coming in off of the Atlantic, rolling up to their part of Germany, bringing with it snow and freezing rain. So, it didn't look like they would get to night fire before returning to Mannheim, but they would try. Borman asked if there were any questions. There were none, so he turned the meeting over to the First Sergeant, who asked the platoon sergeants to meet with him separately. The Lieutenant caught Robb's eye and knocked on the wooden frame as he walked out the door.

Having been on the range all day with little to eat, he hurried back to his room and took a long hot shower to warm up. Then he went to the mess hall for a quick dinner. As he was walking back to his quarters, he ran into an Armor second lieutenant that he had not seen before. He introduced himself and found out that the new lieutenant, Borovich, was assigned to Charlie Company. He had just arrived in country two days earlier. After getting his TA-50 issued and moving into the BOQ building across the street from the Lieutenant's, a sergeant from the rear detachment had driven him to Baumholder to take over as platoon leader to fire the tables that remained and especially Table 9.

Borovich had just left Fort Knox with a couple of days leave in New York before coming to the battalion. He was beyond tired and a little disoriented from jet lag. The Lieutenant told him that he understood since he had been placed on alert and put in the field for two days the day after he arrived in country. Since they were both exhausted, they went to their respective rooms, but the Lieutenant gave him his BOQ room number and told him that if he needed anything to stop by and see him. And, when they were back in

Mannheim, if he would like an evening meal with other officers, he was sure that Buzz wouldn't mind an extra face at the dinner table.

THE LONG RUN

The Lieutenant had settled into a routine: sleep and quick preparation for the next day. The monotony of fog watch and constant road marches had blended all the days into each other until he had no idea what day of the week it was any more, nor did he really care. The one thing that he did know was that he planned to finish gunnery so he could road march his platoon back to Mannheim as soon as possible.

The next morning, he headed to the orderly room for a cup of coffee. Robb was already there. He told him that the men were up and would be through with breakfast in the next hour, ready to move out at 0700. The Lieutenant thanked him and went to the mess hall for breakfast before dressing in his field gear, shouldering his A-bag, and trudging back to the ready line for yet another day. At 0645, Borman's Jeep appeared in the motor pool. He summoned the platoon leaders to him and they compared radio frequencies. Then they boarded their tanks, did a commo check with him, and followed him to the range.

Due to heavy fog, they had to sit and wait for well over an hour before the range officer could identify the panels of the range fans well enough to allow firing. When he gave permission, Sergeant Wilson, Matthews, and the tank commander for 5-5 pulled up, one behind the other, to begin the individual gunnery range. Lieutenant Matthews had elected to go first, so once he received the "all clear" and permission to engage, 2-1 moved forward at about five miles an hour down the series of trails that made up the range. Every hundred meters or so, he could hear the fifty caliber Ma Deuce machine gun fire one shot followed by the sound of the tank rapidly accelerating to continue down the course. It took almost thirty minutes to complete the full course. When he got back, Lieutenant Matthews was beaming because his crew had done well, and like most armor lieutenants, he liked to fire tanks. Once he pulled in and cleared his machine gun, Wilson entered the range on 2-4 and began to fire. In slightly less than thirty minutes, that tank completed the course and cleared its guns, so that 5-5, the final tank from Second Platoon, could begin firing.

With 5-5 downrange, the Lieutenant dismounted to go talk to Sergeant Robb. He told Sergeant Robb, "Since we'll be in a company-on-company exercise tomorrow, and I'll undoubtedly be busy, I'll go ahead and fire first, so I can get my crew some rest before night fire. I also want to go up to the tower and observe the rest of my tanks fire the course." Sergeant Robb agreed, so while the bulldozer tank, 5-5, cleared the range, the Lieutenant climbed aboard 1-1 and ordered the driver to move forward to the entrance to the range.

Borman radioed and ordered the Lieutenant to move forward and engage. He told the driver to move out at five miles an hour. As they moved down the tank trail, to his left, he saw an infantry fighting vehicle target at about a thousand meters. He ordered the driver to stop, and through the intercom system, told Blaylock where the target was and what type. He also called for high explosive anti-tank ammunition to be shoved in by the loader. Blaylock identified the target immediately and the loader came on the intercom screaming "up!" meaning that he had simulated inserting a 105, and it was fully loaded in the gun. The Lieutenant then ordered the gunner to fire. He did, and the fifty-caliber round hit the plywood target, causing it to fall back. The Lieutenant then said, "Target – ceasefire," over the intercom and ordered the driver to move forward again.

He and his crew continued in this fashion through the remainder of the range, knocking down targets from two thousand, down to two hundred and fifty meters. The Lieutenant and his crew members picked up a rhythm for halting, firing, and moving out again. Since the loader did not have to load actual ammunition, he could spot targets that the Lieutenant missed and also relay information as to whether or not a target had been hit. The driver was also able to identify targets, and because of his vantage point could sometimes pick out a target on the horizon even before the Lieutenant could, since the Lieutenant was trying to observe a 180-degree fan around his vehicle. They continued on for about thirty minutes, making a couple of large curves so that they could fire directly and off the flanks of their tank. At the end of the course, he told the driver to turn hard to the right, and he notified the tower that they were complete. Then he reversed the turret so that the gun was over the back deck. He turned himself around in the tank commander's cupola and the driver went back to the firing line at over fifteen miles an hour.

When he came to the intersection of the firing line and the exit from the range the Lieutenant stopped and gave the tower the "all clear," after making sure that the machine gun was unloaded completely. Then, when given permission, he moved his tank back behind the tower, where Second Platoons tanks that had already fired, were parked. As soon as his driver parked, the Lieutenant heard Borman order 1-2 to begin the course. The Lieutenant requested permission to come to the tower and see the rest of his platoon fire. He soon received permission from the tower and walked to the base. Then he climbed up several flights of stairs to the observation deck.

As expected, the crew of 1-2 performed very well, hitting all but one target. Borman congratulated the Lieutenant on his crew successfully completing the course, missing only two targets. When 1-2 was done, 1-3 began firing. Due to limited training, the Guamanian tank commander and gunner were still working on getting their rhythm down so that they could halt, fire, and move out again, trusting the range finder on the tank to allow them to hit the target, as opposed to firing multiple rounds to bracket the target like an artillery piece would.

Still, they did a much better job than they had previously and were making progress, so the Lieutenant was happy. 1-5 did well, missing two targets, while 6-6 negotiated the course quickly missing one more target than 1-5 had, mostly due to the younger gunner on board that tank. Finally, Sergeant Robb pulled up to the entrance and when given permission, entered the course a little faster than the other tanks. He and his gunner identified and fired more rapidly than any other crews, and finished the course five minutes faster, having hit every single target.

When Robb was done with the course, Borman ordered all of the crews to the base of the tower where he and Sergeant Jackson, the company master gunner from Third Platoon, critiqued each crew. Overall, Matthews' platoon had done a good job, hitting only three targets less than Third Platoon, and First Platoon was Second Platoon with two fewer targets. The Lieutenant could tell that he had his work cut out for him but overall was pleased with the performance of his crew and the crews on the other tanks in his platoon.

As the commander was concluding, a vicious wind started to whip in from the northwest, with blowing sleet and reducing visibility. On top of that, fog rolled in and freezing rain came down onto the range as well. Borman radioed back to Battalion to see what the weather report was for the evening. He was told that conditions would be terrible so night gunnery would have to be scheduled for the next evening. Borman passed that information along to the crews. Then he told them to go ahead and finish lunch if they hadn't eaten, and to get some hot coffee or cocoa and prepare to follow him back to the motor pool in fifteen minutes.

Once the lunch break ended; Borman signaled to the platoon leaders to have their crews mount. Then the tanks lined up and followed him back to garrison. The road march was painfully slow because visibility was severely limited and the road was slick due to Baumholder providing its worst weather: freezing rain, with sleet and snow as well. The wind was bitter cold and drove the sleet into the tank commanders' faces, as they tried desperately to give the drivers instructions to keep them from going off the road. After nearly two hours, the tanks creaked back into the motor pool, where the crews parked them, and checked the tracks and undercarriage for any maintenance issues, before scurrying to the company area for the 1700 formation, which was over almost before it began. Before releasing the soldiers, Guerra told the platoon leaders and sergeants to meet with Borman after the formation.

As the troops left, the platoon sergeants and leaders assembled in the corner of the orderly room where the CO was standing. On the Baumholder map, he showed them the maneuver area where they would be working the next day. He estimated that it would take at least two hours to get there, so they needed to pull by 0630. Once they reached the maneuver area, First Platoon would be detached and come under the command of Alpha Company 2-13 Infantry. The First Platoon of the other company would be cross attached to Bravo. The infantry company would operate under the call sign "Team Mech," and Bravo use the call sign, "Team Tank," during exercises they would take up the whole day.

Borman finished his briefing and left for Battalion. The Lieutenant approached Robb and asked, "Have you been through these training scenarios before?"

"Yes, sir. If we have a good infantry Company Commander, we should do well. I'll have the troops up by 0530 so they can get breakfast, since they will be out in the woods all day with only C-rations." The Lieutenant thanked him and decided that he would eat a full meal in the mess hall that evening, as there was little guarantee that after breakfast, he would have an opportunity to drink very much of anything, and he was running low on licorice.

The rhythm of the days became second nature to him, so he ate dinner quickly and in solitude, after taking a shower. When he got back to his room, he did the best job he could of cleaning his field gear for the next day and setting it out to dry before going to bed. He only hoped that the weather would be better the following day.

HOLD THE LINE

The next morning came too soon. He was worn out and sore everywhere from clambering on and off his tank and being banged around in the turret. Luckily, the snow, sleet, and rain had stopped, and the wind had died down to a gentle breeze. As he walked to the orderly room to get a cup of coffee, he could see the moon fairly clearly through the haze and fog. After eating breakfast, he caught up to Lieutenant Matthews on the way back to their quarters and told him that it looked like he would be attacking him and defending against him, so it ought to be a lot of fun. Lieutenant Matthews shook his head "yes" with his usual broad grin and a cocky attitude and promised to give him a fight.

The lieutenants went to their respective rooms and put on their field gear before shouldering their A-bags and walking together down to the motor pool to meet their platoons. When he arrived, Robb had his platoon's crews mounted on their tanks. Surprisingly, Borman was already there. He had arrived early, probably because he was nervous, since in this exercise, he was the one who would be graded. The Lieutenant hurried over and checked with Robb to make sure that all of the tank commanders had the company frequency that would be used on the march. Robb affirmed that so the Lieutenant crept carefully up the right side of his tank to the turret where he pulled on his CVC, pushed the toggle switch for the radio, and asked for a commo check from the tank commanders. When that was through, he called Borman on the radio and let him know that his platoon was ready.

When Borman heard from the other platoon leaders, he ordered the platoons to follow him. Third Platoon pulled up behind him, followed by Second Platoon which was trailed by First Platoon. It was still foggy and misty, so the company took almost an hour and a half to get to the maneuver area. When there, Bravo's tanks laagered up and waited for further instructions. Behind them, they could hear the tracks of a mechanized infantry company. When the APCs arrived, they formed up in a small clearing next to the tanks. Borman ordered his platoon leaders to his position so he could lead them to the control officer for the exercise.

The control officer was in an M-577 command track. He had a large map of the training area inside it and welcomed the officers to stand in a semi-circle around it. He explained that the First Platoon of Bravo Company would cross-attach to the infantry company and that the First Platoon of the infantry company would cross-attached to Bravo. He then gave the frequencies for each of the teams and the platoons. The team leaders would command the platoons on the team frequencies, while the platoon leaders and sergeants would use their platoon frequencies within their platoons. The control officer would monitor range control, as well as the company frequencies and the platoon frequencies, to evaluate communication in the elements.

The control officer next directed the officers to the map of the engagement area beside him. Team Mech, in the morning, was tasked to defend against an attack by Team Tank that would come from the south. This would pit two infantry platoons and one tank platoon against two tank platoons and one infantry platoon. The control officer then turned command of the platoons over to the team commanders, so Sergeant Robb and the Lieutenant went to visit with Captain Collier, the infantry Company Commander. He was a pretty senior captain and was easy to talk to.

The captain laid out the basic defense which was two infantry platoons in the woods around his headquarters and the tank platoon guarding the firebreak road, which was on the left flank before branching into the objective area. Looking at the map, the Lieutenant asked Robb what he thought would be the best thing to do. Robb suggested that most of Second and Third Platoons would probably attack up the road and try to overwhelm them. For that reason, they needed to slow them down a little.

His suggestion was to take 6-6 and put it on the right flank of the company where it could make a lot of noise moving around and keep the infantry platoon that had been attached to Bravo Company busy. The infantry there could be counted on to fight the other infantry in detail. He assumed that Ramsey would convince Lieutenant Matthews to lead the assault, and take heavy casualties, so that Third Platoon could come in and win the engagement.

Sergeant Robb then suggested to Collier that he place 6-6 on the right flank with the infantry platoon there and then split First Platoon into heavy and light sections, letting them take up positions slightly past the intersection and allow Second Platoon to pass in the woods where the infantry could destroy the tanks. Once Third Platoon rushed in, First Platoon could pull out of its hiding position and allow the first two tanks of Third Platoon to turn the corner before getting shot in the flank. Robb would then shoot the last attacking tank, while the rest of First Platoon wiped out any remaining tanks in Third Platoon. After that, they could assault down the road against the rear of the remaining infantry platoon, while their infantry platoons engaged on the right flank. The Lieutenant could tell that Sergeant Robb had been in this engagement area before. Collier hesitated for a minute, but once the Lieutenant explained the plan to him in more detail, he was in full agreement, and they began to execute.

Once the platoons were in position, the team commanders notified the control officer, who gave them permission to begin the engagement. Borman was hesitant in the attack and tried to have his infantry platoon creep forward to find a weak spot for the assault. Once the infantry heard a tank in front of them, they relayed that to the Team Tank commander, who split off Bravo 5-5 and 2-5 to support the infantry. Listening to Sergeant Ramsey, the remainder of Second Platoon assaulted up the road followed by Third Platoon. Lieutenant Matthews was surprised when he didn't meet any resistance on the road and turned the corner toward the objective area. After getting past the intersection, however, he was immediately attacked by infantry with anti-tank missiles that were effective at the shorter ranges in the wood line.

The Lieutenant and Sergeant Robb waited for Sergeant Ramsey to come up and try to pass through Second Platoon. When he did, they came out of their hiding positions and killed the remaining tanks of Second Platoon, and the first and last tanks of Third Platoon. Then they annihilated the rest of that platoon in the space of less than fifteen seconds. Once that was done, the Lieutenant radio

Collier, who ordered the tanks of First Platoon to assault down the road, while his infantry platoons attacked the lone infantry platoon in the woods.

In less than ten minutes, the infantry platoons had infiltrated and destroyed all of the tanks that had been on the right flank and most of the infantry that they were attacking. At that point, the control officer called off the exercise. He ordered all of the vehicles to come back to the original lagger position and asked the platoon leaders and their crews to assemble.

They took their seats in the bleachers, back behind the original briefing area. The control officer came out and asked some questions of each of the team commanders and the platoon commanders regarding their ability to communicate effectively and understand what was happening in real time as the engagement unfolded. Each of the officers explained their maneuver scheme as they understood it and then explained what happened in actuality. Once the unit commanders finished, the control officer explained that it was noon and after a short break of thirty minutes, the teams would change positions so that Team Tank would defend and Team Mech would attack. He then turned the platoons back over to the team commanders.

THE DEVIL WENT DOWN TO GEORGIA

Captain Collier signaled for the platoon leaders and sergeants to come to his track. While walking toward the track, the Lieutenant turned to Sergeant Robb and asked him again what he thought, since he had been through this same exercise before. Robb said, "Sir, we could take 6-6 and put it over on the left flank leg where it can make a considerable amount of noise in that general area but with enough terrain and trees between it and the infantry to keep from being shot. I assume that Captain Borman will keep Sergeant Ramsey in reserve and place him on his right flank, opposite 6-6, in case of a surprise tank assault. However, speed will be everything since we only have one tank platoon. So, if we can tie up Third Platoon with infantry and 6-6, and pin down Borman's infantry with our other infantry platoon, we can attack as facts as possible up the road. I bet that Second Platoon will be back in the wood line guarding the road into the headquarters on the objective."

The Lieutenant gave that some thought and listened to Collier as he explained his plan. As anticipated, he intended to keep two infantry platoons attacking through the woods and used the armor platoon to come up the road. When asked what he thought, the Lieutenant explained that it was good to have at least one tank on their left flank to keep a tank platoon busy and some of the infantry as well. Since a tank platoon was probably going to be guarding the intersection leading to the headquarters, they would attack as quickly as possible up the road to shoot the first tank they saw at that intersection.

The Lieutenant and Robb thought the first tank would probably be the platoon leader for Second Platoon, because he was new to the country and nervous. When that tank was killed, they needed the infantry to overwhelm the rear of that tank platoon and engage them on the road while his tank platoon continued to attack down the road. If they were lucky, his platoon could breakthrough, attack, and take the headquarters. If the second infantry platoon could overwhelm the headquarters with what was left of the third infantry platoon, they could probably kill most of Third Platoon's tanks in the woods as they tried to come to the aid of the commander and Second Platoon.

Collier nodded his head, signaling that he was in agreement with the plan. Then he told the Lieutenant to brief his tank commanders and execute the maneuver he had described. Collier called his infantry platoon leaders to the side and briefed them on their responsibilities. When that was done, they released the platoons to go to their positions.

When they were ready, the Lieutenant notified the range officer. After two minutes, Collier came over the team frequency and gave the order for his tank platoon to attack up the road. At the same time, both infantry platoons began probing the defense in front of them. As soon as the Lieutenant hit the main road, he ordered his platoon to attack as quickly as possible. They did so and got to the intersection before the first tank from Matthews' platoon reached it. B-12, his lead tank, engaged 2-1 and destroyed it. 1-3 passed to the right and engaged the second tank from Second Platoon and destroyed it. The Lieutenant could see that the infantry was coming onto his left flank to attack the remaining thanks from Second Platoon.

As the turrets of those tanks turned to face the infantry, the Lieutenant saw that he had a gap and passed to the right with Sergeant Robb and 1-5 behind him. He told the driver to run flat out down the road and they came up on Batson's vehicle almost immediately. As 1-1 pulled up beside Batson's trac, the Lieutenant slewed his machine gun over and called out to his Company Commander, "You, sir, are my prisoner."

Almost simultaneously, the Lieutenant radioed Collier to inform him that he was on the objective and that he expected to have it cleared within one minute. Sergeant Robb and 1-5 swept around to his right again and blunted the attack from what remained of Third Platoon, which had been attacked by the infantry, and knocked out Sergeant Ramsey's tank and his platoon sergeant's tank as well. When the firing had stopped and the smoke from the Hoffman devices had cleared, and while the Miles beacons continued to flash and blare, Borman called the range commander to let him know he had been captured.

Having received the message from Borman, the range officer halted the exercise and ordered all members of both teams back to the briefing area. When they arrived there, he told the teams that he had made the decision to start the march back to garrison because bad weather was coming in. Although they had light fog and mist, most of the day, freezing rain and sleet would soon be starting. This would also preclude Bravo Company from doing a night fire exercise. Consequently, Borman ordered them to meet him back in garrison and gave them a building number where he would debrief and critique them.

The road march back to the motor pool was the most miserable one that the Lieutenant had endured so far. Along the entire way, a sharp wind blew stinging ice pellets into his face and the driver's as well. The loader and gunner took shelter in the turret, and even though they had the heater in the tank turned up as high as possible, it couldn't keep pace with the howling wind that was whistling through the openings. When they got back to garrison, Robb directed the crew members, other than tank commanders, to do maintenance on the vehicles and check them thoroughly. He also told them that they had a 1700 company meeting, where they would get instructions on what to do for next day's Table 9 range.

Once the crews were working on the tanks, Sergeant Robb and the other tank commanders walked to the building with the number given to them by the range commander for the debriefing. They were soaked to the skin, and almost frozen in some places, but when they walked into the room, they were happy to find that the heater was working and there was hot coffee and cocoa to help them to warm up. They each took a beverage and sat down in chairs that had been placed in a semi-circle at the end of the room. When they were in place, the range commander came in and reviewed his comments from the earlier exercise before adding his additional comments with regard to the exercise that had just taken place.

He noted that communication between the platoons and the team leaders and between the platoon leaders and the members of their platoons had improved. He also explained how the infantry team had overwhelmed the tank team when one of the tank platoons was separated from supporting infantry. It was therefore vulnerable to an initial armor shock attack, followed by infiltration of the attacking infantry on the flank. The speed at which the tank platoon moved also precluded rapid reinforcement from Team Tank's Third Platoon, as they were also under attack by an infantry platoon on the right flank.

When the range officer concluded his briefing, he asked for comments from the team commanders. Borman thanked him and acknowledged that he would do better at the next gunnery. Collier stood up and said, "That Lieutenant is the best tank platoon leader I've ever seen in my life," pointing to the Lieutenant.

The Lieutenant then stood up and replied, "No, I'm the lieutenant with the best tank platoon sergeant you've ever seen in your life." Sergeant Robb showed no emotion, but the Lieutenant could swear he saw a slight grin that came and went immediately.

The range officer thanked everyone and released the platoon leaders to their companies. Sergeant Robb and the Lieutenant walked out together. The Lieutenant saw Lieutenant Matthews coming up behind him and told Sergeant Robb to go ahead. The Lieutenant and Matthews walked together to the company and the Lieutenant congratulated Lieutenant Matthews on doing a fine job. He just hadn't had the advantage of the experience that Sergeant Robb had since it was his third tour in Mannheim. Matthews looked up and smiled at that, and then said, "I'll get you next time."

They walked into the crowded orderly room together. The heat was turned on high and they could almost see steam coming off of the wet uniforms of the soldiers as they massed together. The smell was like a wall. Guerra called the company to attention and then put them at ease. He explained that the next day was the capstone day of gunnery for Bravo Company. They would be going to the Table 9 range, and each platoon would be firing 105 training ammunition from the main guns, plus firing the coaxial and M-85 machine guns as well.

Consequently, the company would stand-to at 0600, after an early breakfast at 0500, because they needed to get all of their machine guns mounted and make sure the tanks were ready to roll by 0700 in order to get to the Table 9 range on time. They would have to rotate all three platoons through in the morning since Charlie Company was scheduled for the afternoon. Alpha Company and Headquarters' tanks had already fired. Also, once they were done with Table 9, the entire battalion was going to try to make one last attempt to do some type of night fire exercise, since it was the last night, but visibility was expected to be poor. They would be using M-2 machine guns attached to the tubes against targets at closer ranges, using their passive night vision systems.

Guerra finished and Borman called the platoon leaders and sergeants to his corner of the room and then he released the rest of the company. Once platoon leaders and sergeants were assembled facing Borman with Marley beside him, Borman relayed to them that doing well on Table 9 was absolutely imperative for both him and the battalion. He continued, "I know they will all try to do a good job, but every crewman needs to be involved in sighting targets and communicating their location so they can fire fast and hit the maximum number of targets in the small amount of time allotted. I also want all of your machine guns loaded, charged with ammunition, and function checked as soon as we get to the range. Third Platoon will fire first, followed by First and then Second Platoons. I will observe from the tower and look forward to seeing how all of you do."

Chapter 16
Got to be Real

After Borman finished, he released them, and they steeled themselves against the sleet to walk back to their quarters. The Lieutenant stopped Sergeant Robb and asked, "What do we need to do to help Sergeant Aguinaldo hit enough targets to qualify?"

Robb told him, "Sir, the best thing would be to put 1-2 on the right flank, with 1-3 to the left of it, and your tank in the middle, with 1-5 to the left of you, then 6-6 further left and my tank on the far-left flank. That way, my gunner can shoot any targets that pop on the left flank. Sergeant Webb, on 1-2 is very good and he can handle the right flank; if you watch closely, and make sure that Blaylock picks up any targets that 1-3 misses, I figure that 1-5 and 6-6 will do a pretty good job of clearing the targets directly in front of them and if I can hold the left flank and clean up all the targets on that side that are missed by the tanks to my right, we should have a shot at qualifying."

The Lieutenant again thanked him and Robb replied, "I'll be down at the motor pool early to install the machine guns, and we'll have a platoon meeting at that time. We'll go over what they need to do in order to qualify and reinforce that once they get to the range.

Since Third Platoon is firing ahead of us, they will have a chance to see what the range is like before they have to shoot it."

The Lieutenant and Sergeant Robb steeled themselves against the sleet and parted company. Robb was going to the club for dinner. The Lieutenant planned to eat quickly and get to bed early. Having spent almost a year on day and night tank gunnery ranges in New Mexico, he planned to eat a large meal in the mess hall and eat a large breakfast as well, because with the weather, they would probably be stuck on the range for most of the day followed by night fire, and the chances of getting hot food out there at night was essentially zero.

As the Lieutenant walked into the mess hall, he looked up at a calendar, and for the first time realized it was Friday night. He thought, *The last few months have been so strange compared to how I used to look forward to Friday evenings back in the States when I would try to catch a movie or go out for a drink.* Looking at the chow line, he decided that he would make do with hot cocoa and the spaghetti that was being served.

When he looked across the room, he saw Lieutenant McCrae, so he sat down beside him to visit. Alpha company had fired Table 9 along with Headquarters' tanks. The Lieutenant asked him, "How was Table 9?"

"It was pretty intense. There were a lot of targets popping up quickly in front of every tank's lane. My platoon did well. We hit one more target than necessary to qualify." The Lieutenant congratulated him, and for the first time noticed that he was still in field gear. He asked why, and McCrae told him, "Alpha and Headquarters tanks are going to the range tonight, with Colonel Shea and my new CO, to try and do night fire. However, the weather is so awful that it looks like it's going to be cancelled, and Alpha and Headquarters tanks will spend tomorrow in the motor pool. We'll try to do night fire tomorrow, just as soon as the sun goes down." The Lieutenant wished him well on getting night fire done, stood up and made his way back to his room, where he set his alarm, climbed into the sleeping bag, and drifted off, thinking there was finally light at the end of the tunnel.

MISTY BLUE

His alarm shattered the morning. The Lieutenant opened his eyes wide, looked around the room, and took a deep breath before standing up. He had been in so many places in the last few months he wanted to make sure that he knew where he was when he came out of the soundest sleep he had in the last few weeks. Almost immediately he realized that he was still in Baumholder, and he was up an hour earlier than usual because his platoon was going to fire Table 9. He got dressed quickly and ran down to the orderly room for a large cup of strong coffee and to see Sergeant Robb.

Robb was already there, visiting with the First Sergeant and coordinating for a hot meal to be brought out for lunch, since the company should be done firing by then, and the weather was still bitterly cold with drizzle and freezing rain. Sergeant Robb said, "Sir, I have the platoon up and they are eating breakfast. As soon as you're finished eating meet me in the motor pool. I'll have the crews install and function check the machine guns with the Supply Sergeant and the armorer in case any of them need to be repaired. We can't have any problems on the firing range that could slow the battalion down."

The Lieutenant thanked Robb and excused himself to the mess hall where he loaded up with as much hot food and coffee as he could take. Then he went back to his room, put on his field gear, shouldered his A-bag, and trudged to the motor pool. When he arrived, Sergeant Robb had already been there for five minutes. The crewmen had the M-85s and M-240s on the back decks of the tanks, and the loaders, gunners and tank commanders were wrestling them into the turrets. When all the machine guns were installed, the gunners and tank commanders checked them to make sure that they would function. When that was done, the tank commanders and loaders checked the tanks again, as the drivers warmed them up.

The Lieutenant finished checking 1-1, climbed into the cupola, and called on the radio for a commo check with his tank commanders. When the commo check was complete, he called Borman on the company net to let him know that his tanks were ready. Ramsey had already radioed that he was ready, so they waited about five minutes for Second Platoon to finish installing machine guns and checking them. Matthews radioed the Company Commander to let him know that his platoon was ready, and Borman gave the order to march to the Table 9 range. Third Platoon moved forward and fell in behind Borman's Jeep, followed by First Platoon and then Second Platoon. Lieutenant Marley followed a Second Platoon with the maintenance vehicles, Supply Sergeant, and armorer, so they would be ready if anything broke down.

It was eerily silent as the company moved out in the dark with heavy fog swirling and a light mist freezing on the tanks. After only ten minutes, the Lieutenant was already chilled to the bone despite the fact that he had on fatigues, a field jacket, field shirt, field pants, and a rubber duck. He also had to keep stamping his feet to force circulation to keep them warm as well. What a way to spend a Saturday, he thought, as the tracks creaked uphill towards the Table 9 firing range.

When they arrived at the firing range, the sun was just peeking through the gloom with a dark blue glimmer on the horizon, but it was still quite foggy. The Support Platoon was there with their GOERs, giant four-wheel trucks, full of tank ammunition, and five-ton trucks with M85 machine gun and M240, thirty caliber coaxial machine gun ammunition. The tank ammunition was training ammunition. The machine gun ammunition was cap and ball with tracer every five rounds.

Third Platoon was lead platoon so Sergeant Ramsey and his crews loaded the tank ammunition in the turrets and lowered the machine gun ammunition into the turrets as well, so they could load and charge the machine guns once they were on the range. After almost half an hour, Third Platoon was ready to fire the range. Second and Third Platoons lined up and loaded their ammunition as Third Platoon pulled into place on the firing line and waited for the fog to lift enough for the platoon to shoot.

By 0900, the wind had come up and the fog cleared away. The range officer, who was the assistant S-3, immediately ordered Third Platoon's tank commanders to practice fire their machine guns and one round of main gun ammo, before taking on the range. The Lieutenant was sitting up in the cupola of his tank with the CVC on, listening to the company net. Sergeant Ramsey was a seasoned professional and did a good job of directing his weaker crews to targets that they might not have identified, and also having his gunner, a Filipino Specialist Five the Lieutenant knew from Fort Bliss, to take out targets that have been missed by those same crews.

As usual with Table 9, the firing stopped faster than you would think as the last targets closed in to five hundred and then two-hundred and fifty meters, where the crews exercised their machine guns against them and personnel targets. When the last targets had been presented and time expired, the range officer called for a cease-fire. All the crews cleared their machine guns, and the main gun, and slewed the barrels to the rear where they locked them in place to signify that they were completely clear. Third Platoon then left the range and laagered behind the tower, as First Platoon moved forward to get online.

SULTANS OF SWING

Once all of his tanks were online, the Lieutenant requested permission from the tower to load their machine guns and test fire them and the main gun. The sun had started to come out and it was actually beginning to be, for Baumholder, a nice morning, so the tower gave the Lieutenant permission that he passed on to his crews. They loaded their machine guns and test fired them and then loaded one round of man gun sabot anti-tank ammunition and fired at a target that was twelve hundred meters away. When that was done, the Lieutenant signaled to the tower that his tanks had completed their function checks and were ready to fire the range.

A tanker is never as happy as when he is firing a range, and a tank platoon leader is never as alive and focused as when he was firing Table 9 platoon fire. The first thing that the platoon leader sees is a row of targets almost twenty-five hundred meters away with smoke from the attached Hoffman devices simulating that the tank targets are firing at his tanks. At that time, the platoon leader must order his tank commanders to engage, while at the same time directing his gunner over the intercom to pick up targets in front of his tank as he lays the gun to get the unity window in a position where the gunner can see them.

This calls for considerable coordination between the Lieutenant with his left hand on the radio or intercom, while using his right hand and the tank commander's override to slew the tanks turret back and forth to put the gun in position to fire the next target. The tank commander also has to select the ammunition type for the target type and relay that to the loader so the loader could put the right type of ammunition in the main gun tube. Additionally, the Lieutenant also had to listen for his gunner who would let him know when he had identified the target and the loader to know when he had the 105-millimeter tank round shoved into the gun and his hands out of the way. There is simply nothing quite like it as the crew members of the Lieutenant's tank and the crews of the other tanks in his platoon picked up the rhythm of loading and firing and identifying targets. It was almost like ballet or synchronized swimming as the tanks rapidly engaged and fired and then engaged and fired again.

Out of the corner of his eye, he could also tell that 1-3 was having trouble. He ordered Sergeant Webb, on 1-2, to try to pick up targets in 1-3's lane and slewed his gun over to the right so that Blaylock could try to pick up targets that 1-3 missed, in addition to keeping up with the targets in the Lieutenant's lane. It was unbelievable how fast the time went, with the firing of the guns and the smell of cordite all around him, followed by his engaging a truck target with his M-85 machine gun at five hundred meters as Blaylock used the coaxial machine gun to engage troop targets at two hundred meters. Far too soon, the command came from the tower for the platoon to ceasefire and clear all weapons. The Lieutenant ordered his tank commanders to do that and when they were all clear, slewed their tubes back over the back decks and locked them in place. Just as soon as that was done, he requested permission to move behind the tower. The range officer granted it and the Lieutenant ordered his platoon to laager in the parking area behind the tower next to Third Platoon.

1-1 swung into position and the Lieutenant took off his CVC for the first time, to let the breeze dry his hair that was drenched in sweat. The sun was shining, warming him up in is rubber duck, and it felt wonderful. He crawled out of the cupola and sat on the mantlet, hollering down to his crew to turn off the tank and take a break. He passed down some licorice to his crew. They were all laughing and congratulating each other on completing the table. As he was resting, out of the corner of his eye, he could see a member of Third Platoon running over to him. It was Specialist Segar, the gunner on Ramsey's tank. He yelled up to the Lieutenant over the noise of the tanks around him that were not yet switched off. He said, "Hey, lieutenant, just like old times, isn't it?"

"Yes, just like Donna Anna. The smell of cordite and the crashing sound of guns. I'm just glad that I'm not on the detail to replace any shot out targets." Segar laughed, because he had served on the targeted detail with the Lieutenant at Fort Bliss for several months in the desert by White Sands. Then he yelled up, and the Lieutenant could not hear over the sound of Second Platoon's tanks that were moving up to test fire their guns. The Lieutenant asked, "What?"

"It's Lieutenant Matthews, sir. His eyes were wide the entire time you were firing, and he asked his platoon sergeant what the hell was going on. Wilson just told him, 'That's Lieutenant Manteuffel dealing with it.'" With that, Segar turned to run back to Ramsey, who was calling for him.

Lieutenant chuckled to himself and thought, *That's the truth, back on another tank range, in some hell hole, and dealing with it.* He could hear the tanks of Second Platoon practice firing machine guns and test firing one round from their main guns. Then he heard Lieutenant Matthews on the company net, dealing with it himself. He was doing a good job for a first timer and, as had been the case with the Lieutenant earlier in his career, it takes a while to settle down and develop the rhythm of the tanks and see the whole battlefield to identify the targets that are not being dealt with, but pose the greatest threat.

In less than twenty minutes, the firing ceased and Second Platoon came around the tower to laager next to First Platoon. They had good weather and the tanks had all fired well with no breakdowns, so they were done a little after 1100. The First Sergeant pulled up with his Jeep and a trailer full of hot food from the mess hall. It looked to be ham and mashed potatoes. Personally, the Lieutenant had been expecting chili mac: elbow macaroni with hamburger, spaghetti sauce, and chili powder, which was a mainstay of units eating in the field. When he brought this up with Guerra, the First Sergeant informed the Lieutenant that the battalion commander could not stand chili mac, and it would never be served in the battalion as long as he was in command.

The Lieutenant thought about that a little but then resigned it to the idiosyncrasies that commanders had sometimes, and you just had to be accustomed to them. In any event, he was too tired to care and he just wanted a hot meal, so he climbed carefully down to the front slope of the tank, slid to the ground, put on his helmet, and walked over to the makeshift serving line. When he had a plate of food, he went back to his tank and leaned back against the glacis plate to eat. He had a cup of hot coffee that he placed beside him next to a headlight and drained it immediately.

Once the crews finished eating, Borman stepped out of his Jeep and gave the signal for the platoons to mount up. The tank commanders relayed that to their crews and everyone climbed aboard, prepared to road march out behind the commander's Jeep. Borman radioed the platoon leaders that when they got back to garrison, they would debrief with the operations officer and the Alpha Company master gunner, who had served as a range NCO for them.

TWO OUT OF THREE AIN'T BAD

The weather held and the march back to garrison was a pleasant drive that took a little more than an hour. When they reached the motor pool, Borman ordered all of the tank commanders to park on the ready line and then assemble in the briefing room that was previously used for the critique of the company-on-company exercise. Borman ordered the platoons to take seats and asked Third Platoon to come to the front of the room to be debriefed. As suspected, Third Platoon did a very good job and the platoon qualified as the crews hit the required number of targets in the time allotted. Since Third Platoon had qualified, there wasn't a lot to critique given that Borman was completely new to Europe. The S-3 did note that the gunner on 3-1 was particularly good, and that the crew on 3-5, which was the youngest crew, needed more practice in order to hit more targets and bring up the platoon's score.

Borman called First Platoon up next and the Lieutenant took a seat in the center of the first row. Sergeant Robb sat beside him, to his right, and the other crews fanned out to the left and right in the same order that they had fired on the range. Borman said that the platoon did very well and only two targets shy of qualifying. The crew on 1-5 was short one target, and 1-3 was short three targets. The platoon could miss two and still qualify next gunnery. The Lieutenant acknowledged that, and Robb shook his head in agreement. They just needed to work more with the crewmen on 1-3, who were still trying to bracket targets as opposed to using the unity window and the gunner's and tank commander's sites.

The Lieutenant, despite not qualifying, was still happy and proud of his crews, as the crew on 1-3 had improved immensely since they first arrived at Baumholder. And, given Sergeant Robb's expertise, there was no doubt that the Lieutenant's platoon would qualify at the next gunner gunnery and, in fact, he looked forward to qualifying with them.

Borman excused First Platoon to the back of the room and Second Platoon came forward. Lieutenant Matthews was still excited from firing his first gunnery and he was also very enthusiastic as Borman began his critique. Second Platoon was only one target short of qualification and the tanks that could bring up the platoon's score were Matthews' and 2-5, that had a new E-5 tank commander and a Specialist Four gunner. The Lieutenant smiled to himself, again because he knew that Matthews' crew would be able to get enough targets to qualify at their next gunnery, once they got used to working with him and he settled down in the difficult tasks of trying to direct fire from the other tanks while identifying and laying the gun for his own gunner to shoot the targets in front of him.

Second Platoon's debriefing was over, so Borman ordered the platoons back to the motor pool to check all the systems on the tanks and clean the machine guns, to make sure they were ready for night fire. Alpha had not been able to complete the company's night fire range as scheduled, so those crews were already on the night range waiting for darkness to fall. To be sure they were there on time and not slow the battalion down, Bravo would stand-to at 1700 and march to the night fire range immediately after. The battalion commander was hoping that the weather would hold long enough to get all three companies through night fire.

However, as usual there was a snowstorm with a mix of freezing rain coming in from the west, so everyone had to move as quickly as possible. Guerra had already ordered the platoon sergeants to have the crew members draw C-rations from the mess hall before stand-to. They would eat whenever they had the chance on the range.

As the crews made their way to the motor pool, the Lieutenant caught up with Matthews and congratulated him on a good shoot. Matthews was grinning from ear to ear and already excited about how much better he would do when they came back in February.

The Lieutenant agreed and then excused himself to go back to his room and replenish his supply of licorice in the pockets of his field jacket. He knew from experience on night ranges; it would be a long night.

I LOVE THE NIGHT LIFE

At 1700, the crews were ready and mounted. After a quick commo check, they followed Borman as his Jeep pulled out into the dusk. An hour later, they arrived at the night fire range, where Alpha Company, all seventeen tanks, plus three tanks from Battalion, were parked in line, installing machine guns, and function checking them. The weather was already starting to turn gloomy and the red lights on the range fans were barely visible through the fog. Borman came over the radio and ordered all three platoons to laager behind the tower. All the crews did a quick ammunition check and notified Guerra if they needed machine gun ammunition. The First Sergeant had a truck with him from Headquarters Platoon that he used to distribute ammunition to the crewmen, who loaded it on to their tanks.

With the machine gun ammunition stowed in the turrets of their tanks, the tank commanders of Bravo had nothing to do but wait. They huddled in the turrets as the wind came up and the sky darkened. Borman was in the tower with the Shea, observing Alpha and waiting for his turn to fire. Occasionally, one of the NCOs who was observing would come down and talk to the First Sergeant and let him know how it was going.

The wind picked up with a heavy mist that stung like sleet. The word was that the weather would probably get much worse throughout the evening, so Shea decided to have Alpha's crews fire as best they could right away, as long as they could see the red lights on the range fans. Visibility was so poor that they had a hard time seeing anything, even with the tanks' passive night vision systems they had for the drivers, gunners, and tank commanders.

The weather finally worsened to the point that it was difficult if not impossible to make out targets coming up in banks like they had fired during the daylight Table 9. Shea ordered the Alpha Company Commander to allow all the tanks that were online to engage in what tankers called a "mad minute." That meant that all the targets would be raised and the crews would fire every last bit of ammunition that they could get through the guns as quickly as possible. When the command was given to fire at will, the sound of twenty main guns booming in the darkness was deafening. Main guns rocked the tanks, and M-85 machine guns were clattering with an occasional burst of .30 caliber bullets from the coaxial machine guns adding to the frenzy. In the end, it couldn't have taken more than ten minutes for Alpha to blast all of its ordinance into the night. In fact, they could have been done in less than five minutes, but a couple of the M-85 machine guns jammed and had to be cleared, after which the tank commanders expended all of their ammunition against targets at the two-hundred-and-fifty-meter band.

When the Alpha tank commanders radioed their CO that they were "all clear," they raised the M-85 machine guns to forty-five to fifty degrees and slewed their main guns back over the rear deck. Then they backed up and drove behind the tower to link up with Captain Pallas, who had just come in from Headquarters Company before gunnery, and followed him back to garrison. You could feel the sigh of relief in the air as those soldiers knew that their adventure in Baumholder was finally over.

With Alpha marching into the darkness, Borman came on the company frequency, which was the firing frequency, and ordered his platoons to march around the tower into firing position. It was so dark, and the fog was so bad now that the loaders dismounted and used flashlights to ground guide the tanks to the firing line. Once the tanks were next to the firing line, they neutral steered to the left and pulled up to the line, so all of them were facing down range. The loaders mounted the vehicles and the crews loaded the machine guns. The commanders reported to Borman that they were ready to fire, Borman radioed them to stand fast and await further instructions, since sleet was beginning to fall. Between the sleet and the mist, the red lights glimmering in the distance that marked to area that was safe to fire were barely visible.

After half an hour, when they were nearly frozen, the moon peaked through the clouds well enough to illuminate the first five hundred meters of the range. The wind had come up as well and blown some of the mist away, so the red lights were visible. Borman quickly ordered all tank commanders to fire at will at any target in their lane. Not needing a second invitation, the Lieutenant immediately looked through his night vision sights and identified what looked like an infantry fighting vehicle target to his left front at about five hundred meters. He slewed the gun and using the tank commander's override, identified the target and when the gunner identified it in his unity window, the Lieutenant gave the command to fire. The loader already had placed a round of training HEAT ammo in the gun so Blaylock could fire as soon as he was ready.

By now, all the other tanks on the range were firing and the noise was so overwhelming that the Lieutenant could hardly hear himself think. As soon as he gave the command to fire, he felt the tank rock backward violently and then forward again as it settled back down to its original position resting on its shock absorbers. From what he could see, it looked like they had hit the target, so he looked to his right front and slewed over in that direction to what appeared to be a tank target at four hundred meters. He screamed into the intercom that there was a tank at three hundred meters to their right front. Blaylock looked for the target, while he ordered the loader to slam one round of SABOT anti-tank ammunition into the gun.

As Blaylock shouted that he identified the target, the Lieutenant gave the order to fire, and again, 1-1 rocked violently back on its rear haunches, almost like a horse, as the round left the tube and then rocked violently forward again as he peered through the darkness and his night vision sights to try to see if they had hit the target. It looked to the Lieutenant like they had, so he tried mightily to identify another target in his lane. He thought he saw one at about five hundred meters, but couldn't tell what type it was, so he ordered anti-tank ammunition; when the gunner identified what he thought the Lieutenant saw, Blaylock fired.

This time, where the target had been there was now only darkness, so the Lieutenant identified another target that was much closer and appeared to be a truck. He screamed into the intercom, "M-85 from my position," and fired his machine gun until the target dropped. Then he saw a row of troop targets pop up quite close, less than a hundred meters away, so he ordered Blaylock to expend his coaxial machine gun ammunition on those targets.

After that, he and Blaylock searched as well as they could in the darkness and the mist to find any additional targets. They could not find any, but Borman came over the radio and ordered all the tank crews to fire all of their remaining ammunition at anything they could see and if not to expend it all down range right away. With that order, it seemed like literally thunder, lightning, and all hell broke loose as all of the crews fired their main guns. The machine guns chattered as well.

Finally, the Lieutenant heard from the loader over the intercom, "Ammo," that meant they had used all their tank ammunition. Blaylock and the loader checked to see if there was any more co-axial machine gun ammunition. The loader confirmed that there was none. The Lieutenant then waited for an opportunity and told Borman that his tank was clear.

The Company Commander then conducted a roll call of all the tanks by bumper number after the Lieutenant's tank, 1-1. The roll call continued through the rest of First Platoon, Second Platoon, Third Platoon, and then 5-5 and 6-6. When all were clear, Bravo's tank commanders elevated their M-85 machine guns and slewed the main guns to the back deck to be locked down by the loaders. When that was done, Borman asked for a roll call by bumper number to make sure that all the tanks were cleared and ready to leave the range. When all of the tank commanders answered in the affirmative, he ordered the platoons to march around the tower to the entrance where they had laagered previously, to await his Jeep.

Standing in the tank commander's cupola, with head down to keep it out of the sleet, the Lieutenant silently wished Charlie Company well, trying to fire after them, and hoped that they would finish up so that the battalion would complete gunnery in full. In any event, at that time, Borman came over the radio and gave the order for Third Platoon to follow him. The platoons left the range in the order that they had arrived and marched very slowly back through the driving sleet and intermittent snow that swirled around them, as they tried to get back to the motor pool. An hour later, the Lieutenant thought he could barely make out the lights in the motor pool four hundred meters ahead.

Borman's Jeep made a last long turn, then drove up to the ready line. When the tanks were online, he radioed that the First Sergeant and Supply Sergeant would be picking up the machine guns and taking them to the supply room. Since the next day was Sunday, they would spend that day in the motor pool, preparing the vehicles to march and cleaning all of the machine guns before reinstalling them for the road march back to Mannheim. As those orders came over the radio, the Lieutenant thought to himself, *I'm finally done with my first gunnery.* In the back of his mind he continued, *I'm looking forward to the next one because I really want to qualify my platoon.*

He and Blaylock wrestled the M-85 machine gun to the side of the tank. The loader pulled the coaxial machine gun out and handed it up to the gunner, who sat it down beside the M-85. They then sat in the turret, out of the weather, shared some licorice, and waited for the Supply Sergeant and Guerra to come by and pick them up. When that was done, Sergeant Robb came over and motioned to the Lieutenant. He slid off 1-1, walked over to Sergeant Robb and asked him what the plan was for the next day. Robb told him, "Sir, I have spoken with the tank commanders and the company will stand-to at 0700, after breakfast at 0600.

The Lieutenant thanked him for the update, and told him, "I

expect that I'll see you in the orderly room a little after 0500. I really want to warm up before coming out and spending another day in the endless winter of Baumholder."

"Understood, sir. I'll see you then."

DUST IN THE WIND

It was after 2100 when he reached his room, but the Lieutenant was both cold and hungry, and it was Saturday night, so he decided to go to the club and have a beer and a chicken sandwich before going to bed. When he walked into the club, he saw Lieutenant McCrae, so he ordered his dinner, and walked over and sat down beside him. He took a drink of his beer and asked, "How did night gunnery go?"

"There was nothing out of the ordinary, we just shot everything in sight and got rid of all our ammo and finally got off the range." He was extremely glad about that, and continued, "I'm looking forward to moving up to staff when we get back."

The Lieutenant responded, "I appreciate that, but I personally hope to stay in Bravo Company for at least one more gunnery so I can get my platoon qualified."

McCrae told him, "I understand that, but probably due to the influx of new lieutenants and rotation of first lieutenants in the battalion, that's unlikely."

When Lieutenant Matthews came in, the Lieutenant motioned for him to join them. He sat down with them, still grinning from ear to ear; he was exuberant at the thought of getting away from Baumholder and pleased with how he and his platoon had performed. He was also excited, because his wife was due to join him about two weeks after Thanksgiving. The Lieutenant told him that he looked forward to meeting her, because she sounded like she was a lot of fun. The Lieutenant then rose, and said goodnight, as he was ready for bed and looking forward to the last day on the Rock.

When he got back to his room, the Lieutenant took a shower, set the alarm for 0400, and promptly fell asleep. When the alarm went off, he sat up in bed and looked around the room realizing that he was still in gunnery. He cleaned up his field uniforms as best he could, then shaved and took another shower to wake up before dressing and heading down to the orderly room for some strong black coffee. When he got there, Guerra did not disappoint. He could smell the coffee all the way out onto the gravel walkway in front of the building. When he got inside, he took a cup, and stood in the corner, drinking and warming up, waiting for Sergeant Robb. After five minutes, Robb came in. He poured a cup and joined the Lieutenant. They compared notes on what needed to be done that day before they made their escape from the mountain.

Robb told him, "From the reports I have from the tank commanders, 1-2 and 1-5 have a couple of pieces of dead track. All the tanks need to have their wedge nut tightened and the tension on the tracks adjusted before we can begin road marching back to Mannheim. Doing all the required maintenance, and getting the machine guns cleaned to the armorer's satisfaction, will take the better part of the day. And, sir, make sure you're ready to have lunch in the mess hall at noon. Traditionally, the battalion cooks make a Thanksgiving dinner that's also a celebration of our last day on the Rock."

"Very well, I'll see you in the motor pool at stand-to after the platoon sergeants take roll."

As he walked into the mess hall for breakfast, he saw Lieutenant McCallister sitting with Lieutenant Pavel, across the room. He hadn't seen Pavel during gunnery and had seen little of Lieutenant McCallister too. Pavel had been shadowing the outgoing Combat Support Company Commander and when he was not joining his platoon in gunnery, McCallister had traveled with the outgoing Scout Platoon leader to begin assimilating into his duties which would require extensive field work. The Lieutenant also knew that Pavel always had the latest intel from the staff as to what was going on since he was highly attuned to the gossip grapevine in the unit.

He went through the line and took his tray over to them and sat down. Pavel nodded toward him and asked if he knew McCallister.

He said, "I do. We were together at VMI and I saw him at Captain Batson's farewell."

Pavel told him, "From what I hear, all first lieutenants who have qualified in their positions will be rotating to new jobs in the battalion when they get back to Mannheim."

The Lieutenant told him, "I hope that's not true for me. I've just been with my platoon a little over ninety days, and I really want to go to gunnery with them again to get them qualified."

Lieutenant Pavic said, "I understand that, but I bet you'll be moving somewhere in the battalion when we get back."

The Lieutenant then looked over at McCallister and asked him, "How are things going with your new girlfriend?"

"Things are going well, and I hope that it works out."

"I hope things work out for the two of you. I'm just counting down the days to March, when I go back to Kansas, marry my girl, and bring her back to Germany." As he said that, the Lieutenant could hear the troops rustling out of the mess hall, preparing to go to the motor pool, so we quickly finished his breakfast, excused himself, and hustled back to his room to put on as much warm weather gear as he could, before going down to the ready line for stand-to.

He ran out to fall in behind the troops as they walked down to the motor pool and looked up at the sky. As expected, it was cold, gray, and foggy. He blew into his hands to warm them up, slipped on his gloves, and continued to the ready line where Sergeant Robb was already at work. Robb had taken roll and reported to Guerra, as had the other platoon sergeants. The Lieutenant asked Robb to go ahead and form the platoon in front of 1-1, so he could talk to them briefly.

When they were assembled in two ranks, he stood them at ease and told them how proud he was of them and that he hoped he would bring them back to Baumholder for the next gunnery to qualify. He also told him that he would be working with them that day to get the tanks and machine guns ready to get out of there the next morning. He then turned the platoon over to Sergeant Robb who ordered the crews and their tank commanders to work until 1700, when the First Sergeant would have a formation in front of the orderly room.

Along with the other tank commanders, the Lieutenant climbed up into the turret of his tank and made sure that all of the accumulated trash, pieces of machine gun link, and any spare ammunition was policed up. Along with his crewmen, he inventoried the basic load of tank ammunition on board and reloaded the combat load ammunition that had been removed to make room for the training rounds they had fired. A detail from Headquarters Company and Support Platoon, with their cargo trucks, assisted in that effort. Any loose machine gun ammunition was turned over to them as well.

NEW YORK GROOVE

A little before noon, Guerra walked up to the ready line and announced to the platoon sergeant that they could release their crews for an hour to eat lunch in the mess hall. As his men began walking slowly toward the mess hall, bent against a cold wind, the Lieutenant fell in behind them. As he walked, he saw that there was a new officer coming over from Charlie Company. He stopped and waited for the officer to catch up with him. He was a brand-new second lieutenant, so the Lieutenant introduced himself and learned that his new compatriot was named Borovich. He had just arrived from New York a few days earlier, so the individual tank gunnery, Table 9, and night fire were his introduction to the battalion. The Lieutenant asked him, "How long it had been since you graduated from Armor Basic?"

"About two weeks, most of which I've spent on leave and en route to Germany before arriving and linking up with Charlie to shoot."

"How did you do?"

"I did pretty well on Table 7 and on Table 9, my platoon was just three targets short of qualifying. I missed two since I didn't have any time to train with my gunner."

The Lieutenant then told him, "Well, I'll be going back to the States in March to get married and bring my wife back to Germany."

Borovich nodded and said, "I have a girlfriend back in New York, but I'm not interested in getting married anytime soon."

As they entered the mess hall together, the Lieutenant looked up and saw that Matthews had beat him into the building and was trying to warm up as he waited in the chow line. As the Lieutenant looked around, he saw that there was milk, coffee, cocoa, and turkey, lots of turkey. The odd thing was that all of the food in the serving line was C-rations. It appeared that the Battalion had been saving turkey loaf C-rations for the entire last year, to have enough to serve them this C-ration Thanksgiving dinner on their last full day at Baumholder. They made their way through the line; both lieutenants picked up C-ration cans full of turkey loaf, and fruit. At the end of the line, they grabbed cans of John Wayne bars and crackers. They then picked up mugs of coffee and joined Lieutenant Matthews in the far corner of the room. They were soon joined by McCallister and McCrae, from Alpha Company, and Bernie, the FSO, who was in from a live-fire artillery exercise. The Lieutenant hadn't seen him since he'd been in Baumholder, so it was good to catch up. Bernie was doing well, business was good, and he was looking forward to getting back to Mannheim so he could make a trip to Amsterdam. He was low on inventory.

The cooks had placed extra P-38 can openers on the tables but the Lieutenant had one hanging around his neck, with his dog tags that that were still wrapped in black electrical tape from Ranger school, along with a small charm he picked up in a pawn shop in Juarez, a miniature of the Third ACR's crest, to remind him of the prisoners that made up his first platoon. The P-38 was one of the most useful things in the Army, an emergency screwdriver, and he used it to open the mail if it made it to him in the field. They cut open the cans and dumped their contents onto their trays. While eating, they had a good time talking about incidents that had happened in the last couple of weeks and how they looked forward to getting back to Mannheim, to enjoy some sort of real Thanksgiving dinner in the next few days.

When they were through eating, the lieutenants took a deep breath, finished their coffee, and bundled up to go back to the ready line. When he arrived at his tank, the Lieutenant saw Sergeant Robb and walked up to him. He asked him, "What do you need me to do to help out?"

"Sir, if you would, you could go to the arms room and supervise some of the younger crewmen who are attempting to clean the machine guns." Since the Lieutenant had been trained on them earlier in his career, this did not present a problem and, he thought to himself, *At least it will be a little drier and warmer*.

When he arrived at the arms room, he found Blaylock wrestling the M-85 machine gun, so he helped him disassemble it. They cleaned the barrels that were lying beside it and then took apart the coaxial machine gun and cleaned it as well. When that was done, they reassembled them and the Lieutenant walked around the room to supervise his other crews who were doing the same. When all of the crews were ready, he asked the armorer to inspect them. They passed and the armorer had all of his platoon's machine guns stacked in one corner of the arms room where they would be available in the morning to pick up and install on the tanks for the trip back.

With that task complete, the Lieutenant ordered the crewman from his platoon to head back to the ready line. When he approached 1-1, he saw Robb and walked up to him. He asked how things were going and Robb told him that he had one piece of dead track to finish replacing and then in about twenty minutes all of the maintenance on the platoon's tanks should be done. The Lieutenant asked, "What do you expect will happen tomorrow?"

"We'll probably pull at around 0900, to allow the weather to lift, so we won't have to drive down the mountain roads and get on the Autobahn in the darkness or fog. Also, sir, you need to coordinate with the CO for each tank to divert for about five minutes on the march to pick up Kentucky Fried Chicken for the long haul back to

Mannheim. I expect that we will stop at the same rest area where we laagered on the way here, and all of our soldiers will be cold and hungry, so a lunch of chicken with coffee will go a long way toward making what's left of their gunnery experience enjoyable." The Lieutenant nodded and told him that he would take it up with the old man after evening roll call.

The Lieutenant left Robb to his work, supervising the replacement of dead track and he walked the ready line to see how his tanks were doing. With the exception of some mud and grit that was still stuck to portions of them, they appeared to be in good shape, and he was looking forward to getting back to Mannheim and his own bed. He was also anxious to get to the bank to try and give his girl a call to let her know that he had made it back from gunnery. As he was thinking about that, he saw Guerra appear out of the mist on the edge of the motor pool. He ordered the platoon sergeants to march back and assemble their platoons in front of the orderly room. All the troops followed him and formed up. The Lieutenant stood in the back since the company was under the control of the First Sergeant and the other noncommissioned officers.

Once the company was formed, Guerra informed them that they needed to be packed up and ready to go by 0600. They would have breakfast in the mess hall after which they would draw the machine guns and install them on the tanks, check the radios, crank the vehicles, and make sure that all of them would move and were ready to march. Their gear needed to be as clean as possible and stowed in their A and B bags in the bustle racks with everything lashed down for the trip. The weather was supposed to be bad in the morning, but with any luck, by the time they got down off the Rock, and at least halfway to the rest area, the overcast should break and the last half of the march would be in sunshine. This raised everyone's morale. The First Sergeant then called the company to attention and turned the soldiers over to Borman, who stepped forward to address them.

Captain Borman congratulated the troops on a successful gunnery. He also appreciated their hard work and looked forward to the next gunnery, where he hoped that all of the platoons would qualify. He also told them to have a good time at dinner, but not to drink too much so that they would be ready to roll by 0800. Having said that, he turned the company over to the First Sergeant, who dismissed the platoons to the individual platoon sergeants.

Since Robb had the platoon well in hand, the Lieutenant walked rapidly around the edge of the company and caught Captain Borman in the orderly room. As he came up beside him, he said, "Sir, it's traditional to let the tanks and their crews pull over briefly at the Kentucky Fried Chicken at the bottom of the mountain, to pick up food for lunch, similar to what they did when they came up to Baumholder."

Borman was a good officer, attuned to his soldiers, so he agreed, "That's fine. We'll just maintain radio contact as the tanks pull in next to the restaurant and then come back onto the road."

LOOKS LIKE WE MADE IT

Captain Borman then turned his attention to Sergeant Ramsey, who was coming up to speak with him, so the Lieutenant saluted, did an about face, and walked out of the orderly room to his quarters. As soon as he walked through the door, he climbed out of his field gear and placed the pieces of that uniform on hangers so they would dry. Then he took a shower and put on the cleanest pair of fatigues he had left to go to dinner on his last night in the club, where he ordered Jaeger Schnitzel and a beer. Once he had them in hand, he joined Pallas, the new Alpha Company Commander, and Lieutenants McCrae and McCallister, who were already eating. He sat down, took a long drink of his pilsner, and asked Captain Pallas, "How's Sieglinde doing?"

Pallis told him, "I just got a letter from her and she's doing just fine but misses me a lot. She's been living in the BOQ alone, and as usual, the place is crazy." The Lieutenant sympathized with her, and he had only lived there a little over a month. But that was Mannheim.

As the others started talking about gunnery and an upcoming winter exercise, his thoughts drifted back to one of his drill sergeants back at Fort Knox. When he found out that the Lieutenant was assigned to Third Brigade in Mannheim, he sat down over a cup of coffee with him in the mess hall after breakfast one spring morning and told him that he would tell him a story that would explain Mannheim to him. He said that bachelor officer quarters and officer's married quarters were all three-story buildings. Enlisted housing was four stories high.

In Mannheim, it was not uncommon for wives to have affairs with soldiers who stayed back while their husbands were in the field, so one day as he was shining his combat boots while watching his two little boys play, a male body came shrieking down onto the sidewalk and screamed even louder after impact. It turned out to be a male soldier in fatigues and boots who had severely broken both ankles and was writhing in agony. The drill sergeant explained that he looked back up over his shoulder to a fourth-floor window, where a young German woman was leaning out of the window crying out "Don't worry, Liebchen, it's not my husband, he's only my other boyfriend."

At that time, the Lieutenant looked at the NCO in disbelief, but the drill sergeant told him, straight faced, "Mannheim is just like that, sir, so don't be surprised if weird shit happens." The Lieutenant had thanked him for the heads-up, even though, at the time, he didn't believe it. He believed it now.

Turning his attention back to the officers at the table, he could hardly wait to get back to his own BOQ room, despite the craziness in the building. Even Sieglinde, who was a German national, was surprised at the goings on and heaven only knew what he had to look forward to when he got back after weeks in the field. It just reaffirmed his conviction that he was not now, and never would be, the sex police. He was wholly focused on getting married, getting his wife in country, and getting out of the BOQ. As he returned his attention to Pallas again, the Lieutenant realized that Pallas had just said that he and Sieglinde were planning to get married next summer. The

Lieutenant congratulated him and said, "I know you two are looking forward to that date."

Pallas nodded in agreement and replied, "There are several new officers coming into the battalion and Alpha is due to receive a new lieutenant who was just activated from the Minnesota National Guard. Due to the shortage of combat arms officers in Germany, this is becoming more and more common."

"At least," the Lieutenant said, "he's arriving during something of a lull in the schedule, so maybe he'll have a chance to get used to the country and the battalion before we have to go to the field again."

Pallas agreed and said, "I'm glad to get him in. With both McCrae and McCallister leaving, I'm short-handed lieutenants so I've requested the next new one coming in from Knox, too."

The Lieutenant finished his meal and took what was left of his beer and the tray with him. When he turned in the tray, he saw Lieutenant Borovich coming in, so he decided to finish his beer with him when he sat down. Borovich was beyond exhausted. Between jet lag, bad weather and gunnery, he looked like one tired puppy. As he ordered, the Lieutenant came up beside him and asked if he could join him for a beer while he ate. Borovich said yes, so the Lieutenant ordered another Pilsner and walked over to an empty table by the wall that separated the officer's side from the enlisted side. Borovich picked up his dinner, came over, and joined him. After he had a couple of bites of food, Borovich came back to life, at least a little. The Lieutenant told him, "It will be a lot better once we're back in Mannheim, where the pace of life will slow, at least a little, and you'll have a chance to explore the country a bit and enjoy the tour."

This made Borovich's eyes perk up considerably and he said, "I intend to go out on the town as soon as when we get back. I've heard a lot of stories about the red-light district."

The Lieutenant looked at him, shook his head, and said, "Well, good luck with that. I've been in enough bad places to last me a lifetime, and I'm engaged, so I'm just looking forward to getting as much sleep as I can, to try to get out of here, and get married in March."

Having said that, the Lieutenant finished his second beer and excused himself to go back to his room to sleep, as the next day would be a long one. Before he left, however, he told Borovich, "Try to get some rest when you return, because undoubtedly you will be on the duty roster within a few short days."

"I'm already on to pull battalion duty after we get back on Wednesday night." Borovich thanked him for the advice, and the Lieutenant went back to his room, set the alarm for 0500, hung up as clothes to dry, and passed out, dead to the world.

Chapter 17
Running on Empty

He sensed motion down the hall and came to before his alarm went off. He shot out of bed, shaved and took a long, hot shower to warm up as much as possible, since the road march would probably be brutally cold. Once he was dressed, he went to the orderly room and drank two cups of hot coffee. Halfway through the second cup, Sergeant Robb came in and told the Lieutenant, "The troops are up and packing. As soon as they have had something to eat, they'll start transporting their gear down to the tanks. Then they'll come back to the arms room to get the machine guns."

"Good, I'll eat breakfast and pack my equipment down to the ready line, so I can help with drawing and installing the machine guns."

The Lieutenant hurried down to the mess hall and ate as much cream beef, scrambled eggs, and black coffee as he could, before returning to his room. Once there, he pulled on several layers of field gear and packed everything else into his A and B bags. On the way to the ready line, he stopped by the orderly room and turned the keys to his quarters over to the First Sergeant, before walking down to 1-1. Blaylock and the driver were already there. They had drawn their sidearms and the grease guns for their tank. The driver was checking the vehicle, and Blaylock was stowing his gear when the Lieutenant arrived. He threw his A and B bags up on the side of the tank, and Blaylock slipped them into the bustle rack. The driver's gear was already stowed, so the only person they were missing was the loader.

He arrived five minutes later after being waylaid by Guerra to become part of a cleanup detail to police the area and clean out the troop billets. Once his gear was stowed, the Lieutenant left the driver to continue working on checking out the tank, so that the Lieutenant and the other two crew members could go to the arms room and pick up the machine guns, and he and the loader could sign for their forty-fives.

The Lieutenant, Blaylock, and the loader trudged slowly through the early morning fog as they ascended from the motor pool. As they walked, they could tell that the wind was starting to blow harder, which meant that they would probably be experiencing sleet and snow, before much longer. They made their way up the steps into the orderly room, and then to the adjoining room that served as the company arms room for gunnery. The Lieutenant took his crew and walked over into the corner where he saw Sergeant Robb.

The machine guns were marked with masking tape to match the bumper numbers of the tanks. The Lieutenant slipped over to Sergeant Robb and asked, "Any problems that need my attention?"

"No, sir. The word is we'll stand-to on the ready line at 0800, make commo checks, and crank up all the vehicles so we can give headquarters maintenance and company maintenance the opportunity fix anything that can't move on its own. That way they will have recovery vehicles to spare to drag anything that breaks down along the way."

The Lieutenant thanked him again, before going over and shouldering the M-85 machine gun for 1-1. He told Blaylock to pick up the coaxial, and the loader picked up two spare M-85 barrels. He told the loader that he would need to change with him about halfway to the motor pool since the M-85 weighed so much. They both drew their pistols. The armorer checked off that they had picked up the machine guns and the Lieutenant signed for them. They then walked slowly, through the ever-thickening mist, to get to their tank. When they were about halfway to. 1-1, the Lieutenant stopped and told the loader to hold up, so he could shoulder the M-85, and carry it the rest of the way to the tank. The Lieutenant, in exchange, took the two M-85 barrels.

When they got to the tank, the driver helped the loader put the machine gun up on the sponson box on the right side of the tank. Blaylock hoisted the M-240 coaxial machine gun on top of the side of the tank as well. When that was done, the loader, Blaylock, and the Lieutenant climbed up to the turret and picked up the machine guns, so Blaylock could install the coaxial on the loader's side. Then he helped the Lieutenant install the M-85 in the tank commander's cupola and attach the barrel. The Lieutenant then stowed the extra machine gun barrel in the ready rack behind him. He was completely out of licorice, so it promised to be a long ride. He did, however, have a tin of John Wayne bars and crackers, leftover from one of the C-rations he had eaten previously in case he had to make do.

With the machine guns installed, the Lieutenant now changed frequencies on the radios that were in the CEOI on the lanyard around his neck. He slid down off the front of 1-1 and called the tank commanders to him. He ran over the frequencies with them and ordered them to get their radios ready and be prepared for a commo check in five minutes. Once the platoon check was done, he came up on the company net and called Borman to let him know that his platoon was ready to march. The Lieutenant was a little bit early, as Borman was just pulling into the motor pool in his Jeep when he answered his call. It took another fifteen minutes for the rest of the company vehicles to be up and running and in communication. This was because the First Sergeant had to pack up the orderly room, and what was left of in the arms room, with the armorer, Supply Sergeant, and his clerk. Once the company assembled in the motor pool,

Borman gave a warning order that they should be prepared to pull at 0900.

At exactly 0900, Borman gave the order for Bravo to move out. The command group and Alpha had already departed, so Third Platoon pulled into line behind the CO's Jeep, which was following the maintenance section of Alpha Company. As the last tank in Third Platoon neared the exit gate, Second Platoon followed. The Lieutenant looked for Matthews and saw that he was crouched down so low that he could hardly see him. The wind, which had picked up considerably, fiercely whipped sleet all around the turrets.

The Lieutenant, who also crouched down trying to keep himself warm, slid up onto the tank commander's seat, where he steadied himself with one hand, and kept low on one side of the machine gun mount. He thought to himself, *at least there aren't going to be any trees, so we don't have to worry about piano wire on the march.* As Second Platoon left the motor pool, he ordered 1-2 to follow close behind, trailed by the rest of his tanks in road march order. He looked back and could see 6-6 following 1-5, so he radioed Borman, to let him know that his platoon had just left the motor pool. The company was moving slowly, less than eight miles per hour, due to the fog and sleet. Visibility was terrible. The Lieutenant only hoped that things would improve when they came down off the Rock.

JUNK FOOD JUNKIE

As the tanks made their way slowly down the mountain, the intensity of the storm increased, with the wind blowing at least thirty miles per hour. Over an hour later, the sky began to lighten, and the sleet turned to snow, swirling around them, falling heavily on the road, making it slippery and more difficult to see. 1-1's driver was doing his very best, but the Lieutenant from time to time had to call him on the intercom to have him make a correction, to stay closer to the middle of the road, since he didn't want to go into the ditch and throw a track.

The Lieutenant was freezing. Snow was sticking to his face and accumulating on his rubber duck. Finally, after another twenty minutes or so, the Lieutenant could see where other tanks had pulled over to the right and he could make out the unmistakable logo of Kentucky Fried Chicken. He went on the platoon frequency and told the commander of 1-2 to pull over to the right when he came even with the restaurant. The other tank commanders were to follow suit, have their loaders dismount to get the chicken, and bring it back. The Lieutenant had two ten-mark pieces in the pocket of his field jacket that he fished out and gave to his loader to pay for his share. As 1-1 pulled up next to the restaurant, the loader jumped out of the turret, slid off the front of the tank, and ran in. He came out with two buckets of chicken that he handed up to the Lieutenant, who was squatting on the side of the tank next to the turret. He set them down beside him, slid around to get on his knees, and handed them up into the turret to Blaylock, who placed them near the heater so they would stay warm for lunch.

It was funny how the smell of chicken, a little memory of home, made everyone feel so much better. He looked out and ordered 1-2 to move out, followed by 1-3. He then told his driver to follow 1-3, and radioed Robb with instructions to bring up the Light Section and 6-6 as soon as they were out of the way. As his tank moved down the road toward the Autobahn, the wind died down, and the snow began to fall in heavy flakes; it was almost like they were in a snow globe.

As they got up onto the highway, the Lieutenant looked behind him and saw 1-4, 1-5, and 6-6 slowly coming up behind him. Once the entire platoon was on the Autobahn in the far-right lane, the Lieutenant informed Borman that all of his tanks were on the road. The Lieutenant and all of the other tank commanders then settled into the Autobahn seats in their turrets, hunkered down as best they could to keep warm, and enjoyed what they could of the scenery that they could make out through the snow.

While the weather was better and the snow had died down, the sheer monotony of the march caught up with the crewmen. To counter the monotony, the Lieutenant traded stories with them on the intercom, and they told him about things that had happened to friends of theirs at gunnery to pass the time. *It was interesting,* the Lieutenant thought, *how you got to really know people after being on a tank with them for a month or so.* It was almost like his roommates back at VMI, who, after four years, were closer to him than many of his own relatives, because they lived in close quarters, and visited so much.

It was almost noon and the sky was clearing; he could see that Alpha Company had left the road to the rest area. He watched as Third Platoon and Second Platoon pulled in as well. He came over the platoon net and ordered Sergeant Webb to follow close behind Second Platoon and pull straight in, so they wouldn't have to neutral steer the tanks and tear up the cobblestones.

Once all his tanks were stopped, he ordered his platoon to dismount their vehicles in shifts, to use the facilities, and get hot drinks. They would eat the chicken on the tanks, policing up the trash and throwing it away, as they exited to the Autobahn. Once they were done with lunch, Borman came over the company net and gave the order for Third Platoon to move out, followed by the others in due order. The Lieutenant relayed the order to his tank commanders, and as soon as 5-5, the last tank in Second Platoon pulled out, First Platoon's tanks pulled in behind it, and they began the last leg of the march to Mannheim. The sun was finally out, and the Lieutenant was getting warm, so he took off his rubber duck and field shirt to enjoy the sunshine. The rest of the trip was uneventful and almost pleasant. The Lieutenant was also becoming increasingly excited at the prospect of getting to a phone to call his girl back in Kansas.

ROCKIN' DOWN THE HIGHWAY

Two hours later, he saw the tanks from the two platoons ahead of him pulling to the right, as they approached a road sign. When he saw the sign, he realized it was to exit for Mannheim, so he relayed that to the crew, and they all cheered. The whole battalion was anxious to get back, so they did not march back through the maneuver area from Coleman Barracks, but continued on the road, to Käfertal, the suburb that contained Sullivan Barracks, which was the home of the Pagan battalion. The vehicles made their way carefully through the narrow streets of the town, finally driving slowly around the traffic circle, through the front gate of their Kaserne, and to the motor pool.

As his platoon pulled through the gate, the Lieutenant breathed a sigh of relief and stamped his feet to wake them up so he would be able to dismount 1-1. When all of his tanks were on the ready line, he slid down the front slope to meet with Sergeant Robb, who came over as soon as he was off 1-4. Both of them then summoned the tank commanders over. Robb directed Staff Sergeant Aguinaldo from 1-3 to supervise the drivers and gunners who would be cleaning the tanks and doing maintenance, while the tank commanders and loaders took the machine guns down to the arms room to sign them in. When the other five tanks' machine guns had been turned in, Sergeant Aguinaldo and his loader would take theirs down. That way there would always be a trustworthy NCO in the motor pool with the crewmen to make sure that no one took off early to go downtown.

The Lieutenant had plenty of adrenaline pumping to get everything squared away so he could make his call to his girl. He left the barrel in the M-85 and placed it on the side of his tank. Then he slid down the front, pulled it off, and slung it over his shoulder. He carried the spare barrel in the other hand while the loader grabbed the coaxial machine gun and followed him to the arms room. When they arrived downstairs in the barracks, the Supply Sergeant and armorer inventoried the weapons as they came in. The crewmen turned in their sidearms and grease guns as well. Guerra was also in the arms room. He let them know that the following day they would be cleaning all their weapons; but the first order of business today was to make sure that the tanks were checked out and squared away.

YOU CAN'T CHANGE THAT

He walked up the steps to exit the barracks, feeling almost a hundred pounds lighter. He was in good spirits. Before he could cross the street, however, he saw the battalion commander coming toward him with Borman in tow. Lieutenant Colonel Shea walked up and the Lieutenant saluted. Shea returned his salute and said, "I observed you during the exercises and gunnery. You acquitted yourself well, so I have to rotate you out of Bravo Company, to a new position on staff."

Before he even thought about Shea's comments, the first words out of the Lieutenant's mouth were, "Sir, I don't want to leave Bravo. I can qualify my platoon next time and I want to do that."

The lieutenant colonel looked at him sternly and said, "I just don't have the time to do that. I have new lieutenants coming in and I need to get you branch qualified. That means you have to serve as a tank platoon leader, a junior staff officer, and a company executive officer. To get that done in the next two years, you have to move."

The Lieutenant said, "Very well. Where am I going?"

"You will be going to the Support Platoon. It's currently commanded by Lieutenant Kowalski. Do you know him?"

"Yes. He was a year ahead of me at VMI."

Shea said, "That's what I thought. So, you should be able to sign out of Bravo and into Headquarters Company tomorrow. Coordinate with Lieutenant Kowalski and sign for the equipment in Support Platoon on Wednesday morning."

Captain Borman chimed in, "You need to get with Sergeant Robb and have a full layout of the TO&E equipment tomorrow at 1000 to inventory and sign it over to a new lieutenant who had just arrived. Name is Marr."

With that, Lieutenant Colonel Shea informed the Lieutenant that he needed to get up to battalion to address other officers who were changing positions. The Lieutenant saluted sharply and Shea returned it sharply as well, before doing an about face and hurrying to headquarters. The Lieutenant looked to Captain Borman and asked, "Has the new Lieutenant arrived yet?"

The captain told him, "He's getting settled in since he took quarters on the economy. He'll be in for PT in the morning and you can meet him then."

"Yes, I'll coordinate with Sergeant Robb to have the equipment laid out at 1000."

Captain Borman then asked, "Can you make yourself available to introduce him to the battalion and brief him on the status of the platoon? You could sign out of Bravo Company at the end of the day tomorrow and into Headquarters Company at PT on Wednesday."

"Yes, but I will also need to catch Lieutenant Kowalski tomorrow afternoon to get some kind of sense of what a Support Platoon Leader does before I sign for the platoon."

Borman told him that it would be fine, but he needed to have him back for evening formation on Tuesday, because he intended to have a brief informal farewell for the Lieutenant and a hail for Lieutenant Marr.

The Lieutenant saluted Borman and walked as quickly as he could back to the ready line. When he arrived at his tank, he motioned for Sergeant Robb to come to him. He quickly brought Robb up to speed on the need for a full equipment layout next morning at 1000 and his departure to Support Platoon, effective Wednesday morning. Then he asked him what the status of the tanks were with regard to maintenance, cleaning them, and getting them ready for the layout. Robb told him, "Sir, all the vehicles have been checked. There are no mechanical issues. I'll have the crews wash down the tanks and make sure they are squared away tomorrow morning, after PT. You and I can make sure that all of the machine guns are clean and in good order, as well as the sidearms and grease guns for the layout."

"That's fine. I just need to get my A and B bags back to the BOQ and stop by to see the Headquarters Company Commander enroute, to let him know that I'll make PT formation, on Wednesday morning, and sign in shortly before."

With Blaylock's help, the Lieutenant wrestled his A and B bags to the ground. He strapped on the A bag like a pack and slung the strap for the B bag over his shoulder. Then he walked to Sergeant Robb and told him that he regretted having to leave the platoon before he could get them qualified, but that the colonel was adamant, so he would have to go. Sergeant Robb told him that he knew that already, as the NCO grapevine always kept him up to date. The Lieutenant saluted Robb and took off for Headquarters Company to see if he could catch Captain Boris, the Company Commander, and his First Sergeant, to find out where the Support Platoon Leader's office was.

He humped his bags slowly up the steps and through the big wooden door into Headquarters Company. He passed the CQ desk where he had asked directions when he first arrived in Germany. The walls were still black, and the building was as depressing as ever. He carefully walked down the stairs to the First Sergeant's office. When he walked in, he set his A and B bag down on the floor in the corner and approached him. The First Sergeant already knew who he was, so the Lieutenant told him that he anticipated signing into the company before PT on Wednesday, and asked if the Company Commander was in. The First Sergeant told him, "No. He's still on the road because we have some straggler vehicles, but I'll pass on your intentions to him. I look forward to seeing you on Wednesday when you sign in officially."

The Lieutenant thanked him but then remembered the question he had meant to ask. "First Sergeant, where exactly is the Support Platoon Leader's office?"

The First Sergeant looked at him quizzically, and then told him, "It's in the S-4's supply officer, in the building between Bravo and Alpha Company's barracks, in the line company courtyard."

The Lieutenant nodded and asked, "Is the Support Platoon Leader back from gunnery?"

"No, sir, he's not. He probably won't get in until late tonight. He's responsible for returning unused ammunition. He also has to transport the battalion's equipment and provide fuel for the march. He probably will roll in at 2000 or 2100."

The Lieutenant thanked him, shouldered his bags, and hauled them back up the stairs. Once he was at ground level in the courtyard, he took a hard left through the gates and headed around the traffic circle to the BOQ.

LOVE TO LOVE YOU BABY

As he neared the BOQ, he saw Lieutenant Borovich in civilian clothes headed back toward the battalion. He greeted Borovich and asked him where he was going. He said, "My platoon is squared away. I just dropped all my gear in my quarters, and I have a pocket full of marks, so I'm taking the streetcar downtown, to 'Penis' Strasse to partake of some German culture."

The Lieutenant asked him, "Wouldn't you rather come up to my room and have communal dinner in the kitchen with my friends? Our chef, the dentist, will surely have something worth eating prepared."

"Thanks, but I'm not interested. From what I've heard, my platoon has a tradition of taking care of business on P Street as soon as they get back from gunnery."

The Lieutenant wished him well and added, "Since I'm older, wiser, and engaged, I won't be going with you."

PEACEFUL EASY FEELING

As Borovich headed for the streetcar, the Lieutenant watched him over his shoulder, shook his head, and thought to himself, *There is nothing more dangerous nor endangered than a twenty-something second lieutenant loose in a foreign country.* But he would never be the sex police, so he humped his bags up the stairs to his room, unlocked it, and threw them in. Then he walked down to the kitchen to see if anyone was there. The room was vacant, but he could tell that Buzz had a roast in the oven, so that was one less thing the Lieutenant had to worry about.

The building was quiet. At that point he went back to his room, washed his face to wake up a little, and then ran down the stairs and over to the pay phone by the streetcar station to see if he could call his girl. It was only about 1800 in Germany, so it was before noon in Kansas. He fed some large five-mark pieces into the phone, dialed her number, and prayed that she would pick it up. When he heard her voice on the other end, a wave of relief washed over him. As soon as she realized it was him, she told him how much she missed him and loved him. She also added that she had not received a letter from him in about a week. He told her that he had been away at a place called Baumholder, firing the tanks, and that kept him busy. He quickly promised that he would get a long letter out in the next couple of days. He had staff duty on Wednesday night, so he should have quite a bit of time then. He also told her that he was being transferred to Headquarters Company to become something called a "Support Platoon Leader." From what he could figure out, it meant that he would be in charge of the trucks and drivers who hauled ammunition, fuel, and supplies, for the battalion.

Anyway, it really didn't make any difference to him. He just wanted to hear her voice and see how things were back in Kansas. She told him that, as usual, they had light snow with driving wind and freezing temperatures and that she missed having him beside her to keep her warm. He told her the same, before he asked her how the wedding plans were going. She told him that their wedding announcement was out in her hometown paper, arrangements were made with her church, and she had ordered invitations.

She went on to tell him that her uncle, the Monsignor of the Cathedral in Kansas City, Missouri, had agreed to marry them and that he had made arrangements to travel to Abilene the day before the wedding. He would go back on the day of the wedding, so he could say Mass in Kansas City on Sunday. He told her to be sure and thank her uncle and her parents for everything they were doing. As he finished, he could see from the meter that he was running out of time on the call and he was out of marks, so he told her that he would have to catch her later and he surely would get a letter out to her by Friday. They then both said, "I love you." And the phone went dead.

LOWDOWN

Depressed, he walked slowly up the stairs to the second floor. He thought to himself, *It is surreal that I am getting ready to take my fifth job, between a regiment, and three different battalions, in two continents, in the space of two years.* In the words of the Grateful Dead, "What a long, strange trip it's been." The only question he had was how long it would be before he rotated out to yet another assignment as an executive officer. But he had no control over that, so he peeked into the communal kitchen. Bigfoot and Buzz were sitting in the corner. He asked if he could join them for dinner later that night. They told him that he certainly could, and to grab a beer out of the refrigerator. He did that, opened it, took a drink, and told them he'd have to excuse himself to take a shower because he had been on and off tanks for the better part of two and a half weeks, and he was coated in diesel. He would, however, join them shortly and bring them all up to speed on the joys of tank gunnery on the Rock.

When he got back to his room, he turned the water on as hot as it would go. Then he made sure he had a poncho liner draped over his bed before he put his A and B bags up on it. While the water warmed up, he laid out all of his field gear and clothing. He took the cleanest set of field pants and field shirt that he had and placed them in the closet in case he needed them for an alert. He hung his cleanest field jacket beside them. He had some clean fatigues left in his dresser, so he took all of his dirty fatigues and other laundry and jammed them all in his B bag, so he could take it to the laundry in the morning. When he was done with that, he jumped in the shower and let it run over him for five minutes. The water was running black, from the diesel smoke and oil that completely covered him. Once he felt fairly clean, he dried off, slipped into an old VMI T-shirt and jeans, grabbed what was left of his beer, and headed back to the communal kitchen.

He stepped into the kitchen to find that Lieutenant Pavel had joined Buzz and Bigfoot. Pavel greeted him and said, "I heard that you're going to be the Support Platoon Leader, effective Wednesday."

"Yes, and I expect my life to get a lot busier when that happens."

"It sure will. I'd like to visit with you a little longer, but I need to get cleaned up, to go out on the town with the outgoing and incoming scout platoon leaders, and Armstrong, the outgoing Combat Support Company Commander."

He stood up to leave, and as he walked out the door, he looked over his shoulder and told the Lieutenant, "Good luck, you're going to need it. And you'll know why as soon as you have a chance to talk with Lieutenant Kowalski." The Lieutenant thanked him, then having drained his first beer, went to the refrigerator, pulled out another one and opened it.

BLUES FOR ALLAH

As he dropped the top from the beer bottle into the trash can, he told the two men who were still seated there listening, "While gunnery itself was fun, firing the tanks and the machine guns was exhilarating, most of the time at Baumholder is spent driving through terrible weather, and sitting for hours, on fog watch, waiting for the weather to clear, so we could shoot." Both Buzz and Bigfoot nodded in assent, as that is what they had heard from others who had made the trek.

The Lieutenant added, "The only bright spot was the food; it's outstanding, in the German clubs. And the barracks were pretty good, by Army standards. Other than that, as Mike Hoare said of incarceration, 'With the exception of the salubrious effect it has one's liver, I cannot recommend it.'" Buzz laughed and turned his gaze to the door. The Lieutenant looked over to see that Tom Storch had just turned the corner into the room. The Lieutenant asked Buzz and Bigfoot, "How are things going?"

Buzz looked at him, and asked, "Have you heard about the Iranian Hostage crisis?"

The Lieutenant looked at him quizzically, before he said, "I haven't heard any news in weeks. I've been more than busy, just trying to stay warm and dry." Buzz quickly brought him up to speed on what little was known. He didn't say anything but thought to himself that the advent of more terrorists sure didn't bode well for anyone, including him. Both Buzz and Bigfoot followed up letting him know that other than that, it had been pretty much business as usual while he was away. Elizabeth was working late, preparing students for semester exams that had to be completed before she could go back to England for the holidays.

The Lieutenant thought he saw a look of disappointment in Storch's eyes when he realized that Elizabeth wouldn't be joining them that evening. However, everyone perked up when the timer Buzz set for dinner went off. He rose to pull the roast out of the oven and set up the makeshift serving line. All the officers there made plates for themselves and sat down to focus on eating. The Lieutenant, in particular, was starving. He had only eaten a piece of chicken at the noon break, having decided to leave the rest of it for his crew since they had worked so hard and enjoyed it so much. As he finished, Buzz remarked, "Bob, you look completely whipped. It looks like your ninety-mile tour of Germany took a toll." The Lieutenant agreed that he was and excused himself to get some sleep, since he had to be in the company early for morning PT, and then prepare for a mid-morning inventory of his six tanks.

GOT TO GIVE IT UP

When he was back in his room, he changed into PT gear, slipped under the poncho liner on his bed, turned to set his alarm for 0500 and passed out. He heard the click of the alarm and turned off the clock before it rattled the next morning. He crawled out of bed and washed his face to wake up. Then he pulled on his dark blue jogging suit and tennis shoes. He sat and thought a bit, while he waited for the water in his hot pot to heat up so we could make a cup of instant coffee, before facing the day. Then, remembering what he really needed to do, he pulled out a page of stationery and a pen and started writing a letter to his girl. He explained what had happened in gunnery and how he was being reassigned within the battalion. He knew it wouldn't mean much to her, but she was always fascinated by what he was doing and liked to keep up.

After he had written a couple of paragraphs, he realized that the water was hot, so he made a double strong cup of instant coffee and drank it quickly. He finished the third paragraph, before turning the hot pot off, locking the door and leaving to go to the orderly room.

As he walked through the front gate of Sullivan Barracks to Bravo Company, it struck him that this would be the last morning he would go down to Bravo's orderly room to see Sergeant Robb and catch up on the news of the day. He would certainly miss his platoon and Robb in particular, as he kept him abreast of everything happening and looked out for him on a regular basis.

When he got to the orderly room, he saw that Robb had arrived ahead of him. He was standing next to Guerra, drinking a cup of coffee slowly. When he walked in, Sergeant Robb came over to him, to let him know that he had the platoon under control. All the equipment was ready to be laid out by 0830, the machine guns had been cleaned once, the evening before, after the Lieutenant left, and that as soon as they were inspected at 0900 by him, they would be ready as well. The Lieutenant smiled and said, "Thanks for everything you've done for me."

Sergeant Robb replied, "I tried to get ahead of the schedule because I know you need to catch up with Lieutenant Kowalski and coordinate transferring over the Support Platoon tomorrow." The Lieutenant thanked him again as they walked up the stairs and joined the formation for physical training. Since they had been in the field for almost three weeks, the company only ran one mile, and did some mild calisthenics, to warm the soldiers up before breakfast and a day in garrison, maintaining their vehicles. When physical training was done, the Lieutenant, who had brought some American money with him, decided to eat breakfast in the mess hall, because it might be a long time before he saw another meal that day.

When he walked into the mess hall, he saw that Lieutenant Matthews had decided to eat breakfast there as well, so he joined him in the room reserved for officers and senior noncommissioned officers. As he took a chair across from Matthews, he said, "I suppose you've heard, I'm transferring to HHC." Matthews shook his head "yes."

The Lieutenant chuckled to himself, "I can't seem to stay anyplace longer than five months. And I only really had about three and a half here. Anyway, I'll see what Headquarters Company is like, and if you need anything let me know. I'll see what I can do." Lieutenant Matthews thanked him. They shook hands as they parted company, the Lieutenant jogging back to his room and Matthews walking to the VOQ.

As soon as he got to his room, he took another hot shower to take another layer of diesel smoke and oil off him. Then he got dressed and shouldered his B bag full of field gear to take it to the cleaners. When he had dropped it off, with his laundry for a thorough washing, he made his way back to Bravo for morning formation. When Guerra released the platoons to their respective platoon leaders, the Lieutenant looked over to his right and saw a new second lieutenant, standing quietly under the trees, that gave some shade to the open side of the courtyard. The Lieutenant walked over and introduced himself to Lieutenant Marr, who expressed that he was happy to be there and glad to be taking over as platoon leader from the Lieutenant; he had heard good things about him.

The Lieutenant asked him where he had been before. Marr told him, "I'm just out of Armor Basic, but I was prior service in a tank battalion at Fort Hood. As an enlisted man, I rose from loader to tank commander, before going to OCS."

The Lieutenant told him, "You probably have more experience than I do, and I'm sure you'll do well. Sergeant Robb is great. While we're waiting for the platoon to lay out the equipment for Borman, I'll walk you through the area as best I can, to show you the layout of the battalion." They went to the set of offices adjacent to the line companies where the supply officer, his noncommissioned officer, the transportation coordination officer, the assistant battalion Supply Sergeant, the S-4's driver, and, as luck would have it, the Support Platoon Leader, had their desks.

The Lieutenant asked the S-4 NCO if he had seen Kowalski yet. The NCO told him, "No, but he's expected by 1000, so he should be available, after that."

The Lieutenant told him, "I have a full field layout with Lieutenant Marr here at 10, but once my platoon is signed over to him, I'll come back here to meet up with Lieutenant Kowalski. Leaving there, the lieutenants walked up to Battalion Headquarters, since Marr had not met the S-3, assistant S-3, S-3 Air, or S-2. The Lieutenant introduced all of them to him and then took him to the backside of the line company barracks, where the personnel and company clerks were located. Leaving there, they walked down to the backside of the Headquarters Company and Combat Support Company billets, to the office and examination room. The medics for the battalion maintained their supplies there and performed routine examinations for common problems like blisters and colds.

By now it was a little after 0930, so the lieutenants made their way to the ready line where First Platoon had its equipment laid out and Sergeant Robb waited for them. Marr knew well what he was about, so he looked at the equipment. Having been in a tank platoon before, he could tell that the equipment was in good shape and more importantly, it was all there.

Captain Borman, with the Supply Sergeant, came to the platoon at exactly 1000. The Lieutenant saluted sharply and reported on behalf of the platoon. Borman casually returned the salute, then walked up and shook the Lieutenant's hand followed by Lieutenant Marr's hand. When Sergeant Robb was ready, the Supply Sergeant called out each item of equipment, and Sergeant Robb pointed to it, with a piece of aerial from a Jeep antenna.

When the inventory was complete, Borman asked the lieutenants and Sergeant Robb to accompany him to the arms room. Once there, they inventoried all of the machine guns, pistols, and grease guns by serial number. When that was done, Lieutenant Marr signed for all of the equipment from the Lieutenant, who had just seemingly signed for it himself when Batson left. With everything accounted for, the captain excused himself to go back to his office, and Robb headed back to the ready line.

The Lieutenant asked Marr, "Do you have any questions?"

"No, I'm sure that Sergeant Robb will bring me up to speed."

"He sure will. Would you like to get a bite to eat?"

"No. I'll grab something in the PX. I have to pick up my TA-50 since I just arrived in the battalion. I also have to get my uniforms tended to, so they'll be ready to go by the next week."

"Okay. I'll see you at formation later. If you have any questions, just look me up. Right now, I have to find the Support Platoon Leader and try to figure out what it was I'm supposed to be doing."

Chapter 18
Truckin'

As he climbed up the stairs to the courtyard, the Lieutenant thought to himself, *I really liked Bravo Company.* As he walked over to the S-4's offices, he could hear Major Reardon, the Battalion Executive Officer, in a heated conversation with Lieutenant Graham, who was the S-4. Apparently, they had a significant difference of opinion as to how things should have been handled during gunnery. The Lieutenant, being a non-participant, slid into the room, and looked around for Lieutenant Kowalski. He was at his desk, so the Lieutenant walked up behind him and tapped him on the shoulder. Kowalski looked over his shoulder at him and the Lieutenant asked him if he would like to join him for lunch so he could start in briefing him on the duties of a Support Platoon Leader. Since Major Reardon was still screeching, Kowalski agreed immediately and the two of them shot out the door and headed to the PX snack bar for lunch.

When they got to the snack bar, Kowalski ordered a cheeseburger, French fries, and a Coke, before sitting down at a table in the corner, farthest away from the counter. The Lieutenant wasn't that hungry, so he ordered a salad and iced tea. Then he joined Kowalski, taking a seat facing him. After he sat down, Kowalski said, "I'm moving over to become the Battalion Maintenance Officer. That means I'll still be in Headquarters Company and available if you need me for anything in the future."

They swapped some stories about their VMI days. Kowalski told him that he had been assigned to the battalion immediately after Basic Course. The Lieutenant briefly brought him up to speed, with regard to his time in the Third Cav and two battalions in the Fourth Training Brigade. Kowalski then told him, "I hope to become the executive officer of Bravo Company, after a few months, so I'll be branch qualified before I am promoted to captain and rotate back to the States for Advanced Course and company command."

The Lieutenant said, "I'm just trying to get back to the States to get married and after that, once I've served my time, I'll try to go to law school."

The Lieutenant also remarked, "Although I've been in the battalion for almost four months, I have not run into you before."

"That's because the Support Platoon Leader is always busy doing something. You're going to get seventeen vehicles and one radio, with a platoon of truck drivers, 64 Charlies or Chucks, an E-7 and an E-5 assistant platoon sergeant." As he was saying that the lady at the counter called to them to pick up their food. After they had done that and returned to the table, Kowalski asked if he had a notepad and pen on him. The Lieutenant told him that he did, and Kowalski told him to take down several telephone numbers that he read off to him from a pad he kept in his left breast pocket. They were the numbers for all of the other support platoon leaders in the brigade, including the Cav Squadron.

Kowalski continued, "This is because support platoon leaders have a special fraternity, and they had to look out for each other, in the field and in garrison, too. While other officers perform planning and training tasks, the S-4 officers, and in particular, Support Platoon Leaders, have to perform real world functions. For an S-3, if the plan goes bad, and a unit gets lost, that's one thing. But if a company runs out of fuel, or food isn't brought up, or ammunition runs low at gunnery, the Support Platoon Leader has to answer for it. And you only have one radio that's used mostly by the S-4, and the XO to constantly remind you how fucked up you are."

At the end of that last phrase, the Lieutenant looked at him a bit sideways, and said, "So this is another opportunity for a VMI boy to excel, right?"

"Exactly. Why do you think I've been trying to get to Battalion Maintenance? It's pretty awful there, but you have good mechanics, a fantastic maintenance warrant officer, and E-8 maintenance sergeant to help." The Lieutenant thought to himself, *This is starting to remind me of Fort Bliss: stuck in the desert, with fifteen convicts and a wacky spec five, on a target detail. Anyway, I survived that, so I'll just focus on getting married and make it through this as well.* Kowalski had finished his lunch, so the Lieutenant wolfed down the last of his salad, finished his tea, and rose with him, to return to the battalion.

When they reached the gate, the spec four who was Kowalski's Jeep driver ran up to him and told him that the S-4 needed to see him right away. He needed the inventory of all of the fuel before tomorrow morning. Kowalski told the Lieutenant that he had discussed the transfer with Captain Boris, who the Lieutenant knew a little from the BOQ. Boris lived downstairs by Pallas and sometimes came up with him to the kitchen. The Support Platoon would lay out all of its equipment and vehicles, and they would "stick" the tanks of the fuel trucks when the Lieutenant signed for the platoon. He then told the Lieutenant, "I'll see you in the morning at PT in HHC," and took off for the motor pool.

At that point, the Lieutenant felt a little adrift. His platoon was signed over to a new officer, and he had no duties in Bravo Company before evening roll call, when he signed out. He briefly ran over all of the outstanding tasks that he needed to tend to post-gunnery in his mind. He decided to go to the cleaners and pick up any clean uniforms that were ready. When he had done that and taken them to his quarters, he picked up his remaining dirty field gear and took it to the cleaners, so he would have all of his uniforms ready by the next afternoon. After that, he decided to go to the bank, to pick up some marks for a few German meals over the Thanksgiving weekend. With that done, he still had some time to kill, so he decided to go to Headquarters Company, and see if he could check in with Captain Boris and the First Sergeant.

As he descended the steps in Headquarters Company, he always felt a sense of foreboding, because there was so little light above in the barracks, and even worse in the basement. He made his way to the orderly room and greeted First Sergeant Ascencion, who told him that he had been waiting to see him. He asked the First Sergeant what the schedule was and was told that the company would form for PT at 0600, followed by breakfast, and morning formation at 0800. The Lieutenant asked if the commander was in and Ascension replied, "Yes," and accompanied him across the hall, where he knocked on the door and entered when the commander beckoned him. The Lieutenant walked through the door, came to the position of attention in front of the commander's desk, and saluted him sharply. The commander returned the salute and asked him to sit down in a chair across from him.

Captain Boris told him that he was happy to have him coming in and told him that he should keep his eyes open due to the large amount of drug traffic coming through Headquarters Company. If he ran into anything unusual, he should feel free to see him immediately, as he was trying hard to clean up the unit and get rid of the drugs. He had formerly been the S-2 and intelligence was the line of work that he preferred. He coordinated with Criminal Investigation Division (CID) with detectives trying to slow the drug trade and its connection to the terrorists who plagued Europe. Boris then told him, "I directed Support Platoon to do a full layout with all of its equipment in the morning at 1000. I also expect to see you, lieutenant, at 0600 for PT. Once you've signed for the platoon, I want you to stop by the office and go over the platoon roster with me." Then Captain Boris stood up and told him that he needed to take off for a meeting with CID, but he would see him the next day.

By now, it was almost 1700, so the Lieutenant walked over to Bravo Company and stood to the side. Guerra formed the company and turned it over to the Company Commander. Borman had the Lieutenant come to the front of the formation, where he thanked him for his service as First Platoon Leader. Then he had Lieutenant Marr come up and introduced him to the company. He informed the men that he was taking over First Platoon and that the Lieutenant would be going to HHC to command Support Platoon. He then released the lieutenants and turned the company back over to the First Sergeant, as the Lieutenant walked back to the rear of the formation, where the executive officer would have been, except for the fact that he was busy with maintenance problems.

WHAM BAM

Behind him, Charlie Company's First Sergeant was just finishing making announcements for the evening. When he was done, he turned the platoons over to the platoon sergeants and platoon leaders. The Lieutenant, out of the corner of his eye, saw Lieutenant Borovich step in front of his platoon. Borovich briefed his platoon on their activities and schedule for the following day, Thanksgiving Day and Friday, which would now be a training holiday. When he was done, he asked if there were any questions. One of his young E-5 tank commanders raised his hand and sought permission to come forward. Borovich told him that he could, so the young tank commander fished a pair of dog tags out of his top right fatigue pocket and handed them to the Lieutenant., saying, "Sir, I believe these are yours." Borovich looked at them in surprise. Then he realized he wasn't wearing his dog tags and they in fact were his.

The sergeant then said, "I was asked to return them to their owner by one of the girls down on 'P' Street. They were left under her bed. She told me she thought they might belong to someone in my battalion, but their owner hadn't stayed long enough for her to find out what unit he was from." At this, the whole platoon broke into laughter. When Borovich had regained his composure, he brought the platoon to attention and then turned him over to his Platoon Sergeant and moved quickly to the rear.

As soon as the platoon sergeant dismissed the troops, Borovich took off walking back to his BOQ, a little sheepishly. The Lieutenant quickly joined him and said, "LT, don't worry. A couple of months down the road, you'll probably be laughing about it yourself. Harlots and soldiers have a long-standing symbiotic relationship that goes back thousands of years, at least to the time of Rahab, so it's even Biblical. They get on well because they both sell their bodies for money, approbation, and sometimes, the sheer hell of it. Besides, you got off pretty light. I can tell you, from personal experience, in Juarez, that it's all fun and games in a Mexican whorehouse until three in the morning, when the knives come out. To quote Karen Carpenter, 'Knowing when to leave is the smartest thing that anyone can learn.' You just need to remember to keep your dog tags around your neck or in your pocket."

At that, Borovich smiled a little. The Lieutenant told him he would see him later, as he veered to the front door to Bravo Company, where he walked down the stairs to the orderly room. He signed out of the company at Guerra's desk and thanked him for all he'd done for him. He told him if he ever needed anything, to let him know. Then the Lieutenant climbed the stairs as quick as he could, and walked to the BOQ, hoping that Buzz was cooking something really good, because he was starving.

THE SAD CAFE

As he walked up the stairs to his room and rounded the corner, he could smell something really good coming from the communal kitchen. He peeked in and saw Buzz, who was busy adding some additional spices to a pork loin and potatoes that he was roasting in a Dutch oven. The Lieutenant had picked up some dollars when he got his marks, so we gave Buzz a twenty for the Communal Kitchen Fund and asked him when he thought dinner would be ready. Buzz said, "About half an hour, so you can clean up and relax a bit, before coming down for dinner." When the Lieutenant got to his room, he couldn't resist the opportunity to take another really hot shower to remove another layer of diesel smoke and engine oil. Clean at last, he stepped out and pulled on a clean OD T-shirt, jeans, and sneakers, before walking down to the kitchen.

Buzz had opened a bottle of Moselle white wine and put it on the counter with several of the short green-stem wine glasses the Germans used. The Lieutenant poured himself a glass before he sat down at the table. He and Buzz were the only ones there. Buzz asked him what his plans were for Thanksgiving and the Lieutenant told him, "My first order of business is to pull battalion staff duty tomorrow night. By Thanksgiving morning, I'll have been up for over twenty-four hours. At that point, I figure I'll just sleep in and then heat up a turkey TV dinner that I plan to buy tomorrow."

Buzz looked at him, almost in horror, and told him, "Under no circumstances will you buy that TV dinner. I am going to cook Thanksgiving dinner for Elizabeth and me, and anyone who comes by, and serve it at noon Thursday."

"Thanks – I hope to take you up on that." As he finished and took another sip of wine, Elizabeth, Tom Storch, and Bigfoot walked in, followed by Pavel.

Buzz pulled the roast out of the oven, and set it on the counter beside a small, tossed salad that he had put together, and some German bread. The room smelled wonderful. They all helped themselves and sat down around the table in their usual places. What was funny to the Lieutenant was that this building and these people were almost starting to feel like home. Once everyone was settled in, Buzz went around the horn to see what their plans were for Thanksgiving.

Tom said that he was going to eat the traditional meal at the mess all in his unit. The Lieutenant, Pavel, and Bigfoot said that they planned to join Buzz and Elizabeth for dinner. The Lieutenant asked if he could bring anything, since he didn't cook. Buzz said, "No, but if you could kick in another ten dollars toward the wine, that would be great."

The Lieutenant fished in his pocket and handed him a ten-dollar bill, and told him, "I look forward to having dinner with all of you, since the holidays are pretty hard. In fact, I haven't been home for Thanksgiving, in seven years."

When they were done, Buzz ran a sink full of warm water. They washed their plates, glasses, and cutlery, and put them back. They had eaten everything, so the Lieutenant volunteered to clean up. He scrubbed the salad bowl and the Dutch oven and put them back as well. He then excused himself to go to bed early and prepare for a night of staff duty...on Thanksgiving eve. After the battalion had just come back from gunnery. As he turned out the lamp next to his bed, the Lieutenant took a deep breath and wondered to himself, what he was getting into next, as a Support Platoon Leader? Then he drifted off to sleep.

LIFE IN THE FAST LANE

The next morning, he put on his dark blue PT gear and jogged over to Headquarters Company. He went downstairs to the orderly room and signed in with the First Sergeant. Then he went back upstairs and stood behind the formation, with the various staff officers – the Battalion Executive Officer, HHC XO, BMO, Kowalski, and the signal officer. When the company was formed, the First Sergeant turned the company over to Captain Boris, who led them on a two-mile run. The Lieutenant had taken a place at the back of the formation for the run.

When they returned to the courtyard, he asked Kowalski what he should do in preparation for the layout. He told him to come to the S-4 office so he could meet Lieutenant Graham, the battalion supply officer. He was sure Graham would want to talk with him, and after that, they would meet Captain Boris in the motor pool, where his platoon's vehicles were parked, to lay out the equipment for inventory and inspection of the vehicles.

Kowalski then told him to go ahead and get breakfast, change, and meet him at the morning formation. The Lieutenant jogged back to his quarters, took a shower, and changed, while heating him a cup of hot water for instant coffee and eating a breakfast bar. He then went back to Headquarters' Company and stood behind its support platoon, waiting for Lieutenant Kowalski. As soon as he arrived, he told the Lieutenant to follow him and they headed for the S-4 office.

When they walked through the door to the S-4 offices, Lieutenant Graham greeted them and asked them to come back and to his office. He had them take seats before welcoming the Lieutenant into the S-4 world and Support Platoon Leader job. Graham conveyed that the job would certainly be stressful, but that he would do everything he could to help him accomplish his missions. Graham then explained that the Lieutenant would be in charge of all the transportation of food, fuel, and ammunition for the battalion, to accomplish its training missions. That meant the Lieutenant would transport pyrotechnics, small arms ammunition for training in garrison, and tank ammunition for gunnery. He was also responsible for making sure that the battalion had fuel for all of its vehicles, and to account for that fuel on a daily basis.

Finally, Graham looked at the Lieutenant closely and told him that Major Reardon had complained that he did not see him in morning formation for PT. The Lieutenant looked up immediately and said, "I was there and Lieutenant Kowalski was beside me." Kowalski confirmed that. Graham told him, "I'll get back to the XO, because he carries a grudge and will continue to come after you, if he thinks you aren't conforming to Battalion SOP."

The Lieutenant thought to himself, *I'll fix this*, then he promised Graham, "I'll make sure I'm not missed at PT next Monday."

While the Lieutenant was still thinking, the telephone rang on Graham's desk. He answered, "I'll be there right away." Graham proceeded to explain that he was summoned to the XO's office for his daily meeting with Major Reardon. Kowalski took that as a cue for both of them to head for the motor pool right away.

As they walked, Kowalski told him, "We're going to inspect all the vehicles. You have a combination of GOERs and five-ton trucks, and one Jeep with a trailer. We'll go over all of the trucks first, then take a quick look at the Jeep. Once the Company Commander has left, I'll go over it in more detail with you." The Lieutenant nodded, as they walked across the street, to the section of the motor pool where Headquarters Company's vehicles were parked. In two rows, along the far corner of the crushed rock and dirt hardstand were sixteen trucks and a lone Jeep with a trailer. Standing at ease in front of the platoon was the platoon sergeant, Staff Sergeant McNabb. Upon seeing Lieutenant Kowalski, he rendered a relaxed hand salute and waited for them to come to him.

Kowalski asked if all the equipment for doing basic maintenance on the vehicles was laid out. McNabb confirmed it was. As he was saying this, the Lieutenant saw Captain Boris with the Supply Sergeant in tow, approaching them, so he came to the position of attention and rendered a salute. Lieutenant Kowalski, seeing this, turned slightly and did the same.

When he came within twenty feet, Captain Boris returned the salute, and the lieutenants dropped their arms to their sides. Boris then asked Kowalski where he wanted to start the inventory. Kowalski told him that he would like to start with the five-ton trucks, from the nearest to the farthest and then come back down the line of GOERs, and finish with the Jeep and trailer that they were standing in front of.

Boris nodded and they walked up to the first five-ton truck. The Supply Sergeant came up to the vehicle and checked the serial number on it, then the officers walked to the rear, where the tire changing tools and a few wrenches were laid out on a small piece of tarpaulin. The Supply Sergeant listed the equipment that was supposed to be there and since it was, they moved to the next vehicle in line. They continued up the line of five-ton trucks, both cargo and fuel carriers, before they came back down the line of the much larger GOERs. The Lieutenant had never really been around one of them. They were like the large Caterpillars that were used to move earth on highway projects. Once he climbed up into one for the first time, he realized that they had nothing but a solid suspension, so the only relief the driver had was the springs underneath the seat. The same held true for the passenger. They also made a hellish amount of noise since the engine was located directly behind the driver and passenger.

As they did the inventory, the Supply Sergeant and the driver of each fuel truck climbed up to the top of the truck and inserted a long stick that was marked, so they could make sure that the amount of fuel in each tank that morning was the same as the amount that was on the inventory that was taken the night before. With all of the fuel accounted for, each driver started his vehicle up, to show that it was operational, and stood by the vehicle for a uniform inspection in passing. One thing that the Lieutenant noted was that almost one hundred percent of his platoon was African American. His truck drivers were also a little more laid back than the armor soldiers in the other companies. They did remind him, however, a lot of his first platoon levied out of the Third Cavalry Regiment and assigned to him personally to do a job that no one else would do. In some respects, from what he knew so far, Support Platoon was much like that, getting tasked to do a lot of the support functions that no one else wanted to do.

THE PIRATES OF PENZANCE

With all of the trucks and truck drivers inspected, Captain Boris walked up to the Jeep and trailer. The Supply Sergeant checked the serial numbers and noted them on the inventory. The Jeep driver saluted smartly. He was the most squared away of any of the soldiers in the platoon, including Sergeant McNabb. As he stepped up to the Jeep, the Lieutenant asked the driver, "How long have you been in Headquarters Company?"

The driver, Specialist Simms, an African American, told him, "About a year, sir. I transferred in from one of the tank platoons in Alpha Company. I am supposed to rotate back to the States shortly after Christmas." The Lieutenant made a mental note at that point that he was going to need a new driver. He should be on the lookout for one immediately.

The Lieutenant turned his attention to Captain Boris and Lieutenant Kowalski. Both of them reviewed the inventory with the Supply Sergeant. Boris then beckoned the Lieutenant over and directed him to the bottom of the inventory sheet where he signed for all of the vehicles and equipment in his new platoon. When that was done, Captain Boris and the Supply Sergeant turned to walk back to the company headquarters. Then Lieutenant Kowalski told the Jeep driver to take a break while he talked to the new Lieutenant and motioned for Sergeant McNabb to come over to them. Kowalski thanked McNabb for his work with the platoon. The Lieutenant stepped up, told him that he was happy to be working with him, and that he should go ahead and have the troops do any outstanding maintenance on their vehicles and then tend to any other things that they needed to do in the barracks or with regard to their uniforms, so that he could inspect them, the following Monday, after the Thanksgiving holiday.

He also told McNabb to form the platoon in about ten minutes when he gave the signal. Set them out in two ranks, so that he could meet and visit with each platoon member individually, since he had just taken over. McNabb saluted and motioned for the drivers to come to him. The Lieutenant, meanwhile, turned his attention to the Jeep, and Lieutenant Kowalski, who said, "The main thing about being a Support Platoon Leader is that you had to be a good pirate. You need to be able to move fuel, and sometimes ammunition, between units. You will need to be able to pick up pyrotechnics for training and then smuggle them out of the ammunition supply points. To do this, your Jeep is configured with numerous small, concealed compartments underneath the rear passenger seat, and a long wooden boxlike structure, behind that seat. In the center of the front seat compartment, is a heater from a five-ton truck rigged up to provide maximum heat for the Jeep. This is important, because there are no windows on the driver's or passenger's doors, and the doors are made of plywood, with half round lumber across the top of the doors to rest your elbows and arms on while driving."

He noted that the Jeep came with a good radio tuned to the S-4's frequency that could also monitor the command net. The trailer had a plywood cover, with canvas curtains attached to it that could be propped up with a piece of two-by-four to make what Kowalski called, his "Winnebago." The trailer was wide enough for two air mattresses, and there was a hole cut in it to allow for the passage of an electric cord rigged to a tank dome light to provide light in the makeshift tent. Kowalski went on to tell him, "When smuggling larger items, you can use the trailer, or take the wooden box out of the Jeep, put it in the trailer and cover it with A and B bags to get past the Ammunition Point guards." That concluded the briefing, so Kowalski had the Lieutenant write down his new telephone number at Battalion Maintenance. He told him that most of his training would be on the job training and from time to time, he would probably need to call him over and visit with him to sort things out. Then Kowalski took off like a rabbit for his new office. The Lieutenant watched him, took a deep breath, walked up to Sergeant McNabb and told him to assemble the platoon in two ranks.

Once the platoon was assembled, Sergeant McNabb came to the

front of the platoon and announced, "Sir, the platoon is formed." The Lieutenant then stepped forward, saluted, and ordered him to accompany him on a brief inspection of the platoon. The Lieutenant ordered the platoon to open ranks, and walked to his left, to begin the inspection. The Lieutenant methodically made his way down the front rank, stopping, and executing a left face in front of each soldier. He asked every trooper what his name was, where he was from, and how long he'd been in the platoon. When he was done with the front rank, he passed to the second rank and went through that rank in the opposite direction.

When he had visited with each soldier in his command, he then took his place in front of the platoon again. He gave them the order to close ranks, then to go to parade rest, followed by stand-at-ease. He then told him that he was proud to meet each of them and take over the Support Platoon. He explained that he was coming over from Bravo Company and before coming to Europe, he had been in the Third Armored Cavalry Regiment, and the Fourth Training Brigade at Fort Knox, Kentucky. He went on to say that all of the organizations he had ever been part of had performed well for him and he looked forward to providing excellent service to the battalion. He gave the order for Sergeant McNabb to post in front of the platoon and then told him that he would have to go to organize his office and visit with the S-4, before going on battalion staff duty that evening.

He ordered McNabb to have the platoon complete their task, as he had earlier discussed with him, and to plan on meeting with him next Monday morning in the motor pool, so that he could talk with each of his soldiers in greater depth. After that, he and McNabb would meet in his office, to discuss upcoming tasks and set up training. When he was through, he turned the Platoon over to Sergeant McNabb who saluted and took charge.

LIFE'S BEEN GOOD

The Lieutenant returned the salute, did an about face, and walked as quickly as possible, back to the S-4 office. It was almost noon when he stepped through the door for the second time that day. The Battalion Supply Sergeant, Sergeant First Class Dryer, greeted him, and asked if the S-4 was in. Dryer said, "Yes, and you can go back to see him. He's expecting you."

As soon as he came into Graham's office, he asked the Lieutenant to take a seat. Then Graham began, "Before I forget, I need to let you know, you're invited to Thanksgiving dinner, with me and Dana, at our quarters."

"Thanks, I have duty all night, but as soon as I get off, I'll get a quick nap and come over. It will be great to have an actual Thanksgiving meal with the two of you."

Graham looked at him and asked, "Has Kowalski had an opportunity to brief you on your duties?"

"Yes, he did, and it looks like my Jeep driver is going to rotate out within the next month. How do I go about getting a new one?"

Graham thought a little and said, "Since 64C truck drivers are hard to come by in a tank battalion, you need to request one from the Headquarters Company First Sergeant, who will put the word out to the other first sergeants that the Support Platoon Leader is looking for a Jeep driver. Then, probably, one of the tankers who wants to get out of his current company for some reason will volunteer to come work for you."

Graham pointed to the calendar behind him, and continued, "The month of December is usually pretty quiet, so your platoon will be able to do individual SQT training, and maintenance on your vehicles. You will probably be tasked by Battalion to haul various training supplies and food to the other companies if they are training in the maneuver area. The battalion will be going on a major winter exercise in February. Before that, there will be a smaller exercise for the battalion officers to go over the exercise area and locate assembly areas and routes to be used by attacking and defending units, and the battalion's support vehicles, including your platoon. Since the battalion will need extensive maintenance after the winter exercise, the month of March is currently scheduled to be a garrison month, followed by gunnery at Baumholder in April."

"I am aware of that part of the calendar. Gunnery is keeping me from getting married in April. So, my wedding is scheduled for the eighth of March."

Lieutenant Graham's ears perked up immediately and he asked. "When will you need to be gone on leave?"

The Lieutenant told him, "It looks like I'll be flying out the last week of February. I plan to be back in Europe by March fifteenth. We'll have a one-week honeymoon in Germany after that, so I'll be reporting for duty on March twenty-second.

Graham looked at the calendar, and after reviewing it said, "That looks like the best time for you to go. Your leave won't be a problem."

Hearing that, the Lieutenant's mood brightened considerably, and he asked, "What can I do to get rolling in the near term.?"

"You can move into Kowalski's desk and see what files have been set up. There's an exercise coming up that you need to get prepared for right away. It is a mock alert, with issuance of ammunition from the preposition stock point, PSP, in the training area. You will have to load up various types of ammunition and mines to be issued in the event of war. I know Kowalski has started putting together several plans for the loadout to get everything moved out to a road on the route to the border, so the ammo and mines could be transferred to the tank crews as they drove by."

Lieutenant Graham then leaned forward, and told him, "Prepare yourself and your platoon for endless cleanup details and inspection in ranks on Monday. A new brigade commander is coming in Tuesday. He is expected to come to the battalion for a small mid-afternoon reception with the officers. There is no real schedule, so it will be here in the S-4 offices. We will have to move all of the desks and chairs to the far end of the room and bring in some tables for punch and cookies from the mess hall. Once the reception is over, it will be back to business as usual. Since the weather is starting to get more and more nasty, with fog and drizzle, there won't be a formation of the battalion on the parade ground. But the soldiers' appearance has to be exceptional throughout the entire day in case the colonel wants to walk around the area."

Since this was news to him, the Lieutenant asked, "Who's the new officer coming in?"

Graham said, "Supposedly it's an Armor colonel, by the name of Thomas."

At the sound of the name, the Lieutenant smiled a little but didn't say anything. If it was the same man, the last time he saw Colonel Thomas, he was a lieutenant colonel, in the Third Cav at Bliss. He was the Regimental Executive Officer at the time and came out to see the firing ranges where the Lieutenant's platoon of convicts, who barely had any uniforms between them, were working in the heat of the day, over 110 degrees, and didn't have the sleeves of their fatigues rolled down to prevent sunburn. Thomas stopped briefly, to bring that to the Lieutenant's attention, before driving away.

As per usual in the Cav, the conversation was forceful and one-way. Since, at that time, the Lieutenant believed he was headed back to Fort Knox and law school, he never imagined he'd run across Thomas again. However, the Lieutenant also knew that the colonel went to VMI, Class of '55. He thought that would be interesting, given his battalion's West Point proclivities.

In any event, Lieutenant Graham dismissed the Lieutenant to organize the materials in his desk, until it was time for him to report for staff duty. The Lieutenant stepped out of the S-4's office, and walked over to a small steel military desk, that looked out the window onto the line companies' courtyard. To his right was the larger desk of the battalion S-4 NCO, Sergeant First Class Dryer. At the far end of the room was an old wooden desk occupied by a Master Sergeant, with a smaller steel desk in front of it, that was presently unoccupied.

There was also a small steel desk next to the door, with a Specialist Four sitting in a chair behind it.

The Lieutenant walked up to Sergeant Dryer and asked if he could introduce the Lieutenant to the other members of the S-4 staff. Dryer immediately stood up with a big smile on his face. He took him down to the end of the room, where he introduced him to Master Sergeant Fykes, who stood and saluted. The Lieutenant returned the salute and asked him what his job was. Sergeant Fykes answered, "I coordinate the movement of armored vehicles and wheeled vehicles in convoys across Europe. I'm responsible for coordinating with the MPs and German authorities for those movements. I get the convoy numbers and vehicle numbers for them. Also, I've been in supply and transportation for many years. If there is anything the S-4 needs hunted down, rounded up, and transported with the Support Platoon, I take care of it."

The Lieutenant and Dryer walked over to the specialist. He stood and saluted. When the Lieutenant returned the salute, he could see that the soldier's name was Christopher. He asked what his job was, and Christopher told him that he was the S-4's Jeep driver. His duty station was there, with the S-4's Jeep combat parked in front of the door, so he can go to Battalion or Brigade, or any other place he needs to go, to put out fires."

The Lieutenant then looked at Sergeant Fykes, and the desk in front of him, and asked, "Does anyone sit there?"

Fykes smiled a little and said, "Staff Sergeant Brown normally sits there. He's the assistant S-4 NCO. He is responsible for coordinating with the S-3 NCO and his assistants to get orders published with map overlays. And generally, he locates anything the S-4 needs almost anywhere in Europe. He is also responsible for making sure that all of the tentage and equipment for the S-4 is properly installed when they are in the field."

Sergeant Dryer and the Lieutenant walked back to the other end of the room. He sat down at his new desk for the first time. He wrote down the Autovon number for his telephone on the notepad that he kept in his left breast pocket. He also jotted down the phone numbers for the other members of the S-4 staff. About that time, it occurred to him that he had forgotten something. He turned to Sergeant Dryer and asked, "Where is the Support Platoon Sergeant's office?"

Sergeant Dryer replied, "He usually stays in the motor pool, with the truck drivers. He has a small desk inside Headquarters Maintenance, with a telephone."

The Lieutenant turned his attention back to the telephone roster and finally found the number for Sergeant McNabb at the bottom. Sergeant Dryer also added, "Your Jeep driver is normally stationed in the motor pool with McNabb."

Having set up his communications, he went through the drawers of the desk to see what had been done so far. He was happy to find that there was a series of file folders, one for each NCO and the drivers. Each folder had the Individual's Skill Qualification Test scores, PT Test scores, information regarding the soldier's arrival in Europe, projected departure date, and a list of the vehicles they were capable of driving. He noted to himself that there were at least a few GOER drivers who had no real experience with five-ton trucks, or manual transmissions in general. He thought about this and located a pad of paper, on which he wrote a full roster of the platoon and the vehicles that they could drive, so he could identify the soldiers he needed to have trained. As he was doing this, the Bravo Company Commander's Jeep driver came running through the door and called out to him that Captain Borman needed him in his office right away.

Since it was almost the end of the day, the Lieutenant told Sergeant Dryer that he had to tend to the Bravo Company Commander, and after that he would be drawing a weapon to go on staff duty that night. He thought to himself, since it was a four-day holiday weekend and they had been in the field for several weeks, it would probably be a loud evening. He excused himself and walked quickly across the courtyard to the door to Bravo Company, down the stairs to Borman's office and saluted, upon entering. After Captain Borman returned the salute, the Lieutenant took the chair nearest the desk. Borman told him, "I forgot to bring this up earlier, but since you were the First Platoon Leader for over ninety days, you have to write an efficiency report for Sergeant Robb to cover that period of time, since you were his supervisor."

The Lieutenant smiled and said, "I'll be more than happy to do that, but I'm going on duty, so can I write it up tonight on duty and give it to the First Sergeant on Monday?" Borman told him that he surely could, and to check with First Sergeant Guerra on the way out to get a couple of forms, so he could fill one out. Sergeant Robb was the rater for the other tank commanders in the platoon, so the Lieutenant only had to write one.

Chapter 19
Wild West End

Captain Borman indicated that he could leave, so the Lieutenant saluted, left the office, and picked up two rating forms from the First Sergeant. As he turned to go, he told Guerra that he would return the completed form on Monday, just before the PT formation. He went quickly up the stairs and out the door, on his way to his new company. When he got to Headquarters Company, he asked the First Sergeant if the commander was in. Ascencion told him that he was meeting with the battalion commander. So, the Lieutenant told him that he needed to draw a sidearm for staff duty. It had never occurred to the Lieutenant before, but he asked, "What's my personal weapon?"

The First Sergeant looked up and told him, "An M-16. Everyone in your platoon is issued an M-16 rifle. Almost everybody in this company is issued an M-16 rifle, with the exception of the Headquarters Tank Section that consists of three tanks, one for the colonel, one for the XO, and one for the S-3. Primary staff officers are also issued 45s, but since your job requires you to drive long distances out into the woods, you get a rifle."

The Lieutenant looked at him and said, "For duty, I really need a shoulder holster, a pistol, three clips, and an ammo pouch." The First Sergeant called down to the armorer and told him what the Lieutenant needed. That completed, the Lieutenant walked down the dark hallway to the arms room and requested a sidearm and the other materials. The armorer did not have a spare shoulder holster, so he asked if he had a spare pistol belt and holster with a lanyard. The armorer said, "Yes, sir."

"Very well, we'll make do." After the Lieutenant signed for the ammunition and weapon, he attached the ammo pouch and holster to pistol belt. Then he loaded the pistol and attached it to the belt with the lanyard. As he headed to go up the stairs, he was met by a group of young soldiers, apparently from Battalion Maintenance, who were coming to see the First Sergeant for work on a detail before being released for the long weekend.

When he looked up into their eyes, he was struck by one person, who was quiet and aloof, but also seemed angry. Since he had come across several soldiers like that in the various units that he had been assigned to, he shrugged it off and climbed the stairs as quickly as possible. When he hit the first floor, he hurried out the front door so he could sign in as the battalion Duty Officer before 1700.

At that time of year, the days had grown short. It was almost dark and there was a heavy mist in the air that seemed to cling to the buildings and the trees. When he got to Battalion Headquarters, he stopped by to check in with the staff Duty NCO, who was already on duty.

The Lieutenant asked, "Is the Adjutant in?"

"Yes, sir, but he's in a meeting with the Old Man, the Company Commanders, and the other staff officers. It will probably be ten or fifteen minutes before he comes out."

The Lieutenant checked his watch and then told the NCO, "Go ahead and get dinner. Just report to me when you get back, so I can go to the mess hall and eat. I'll be waiting in the Adjutant's office. I can answer the phone at the desk there."

Fifteen minutes later, the Adjutant stepped out of Arnold's office, followed by the commanders and staff officers. Seeing the Lieutenant, he pulled out his duty book, and the Lieutenant signed in. He handed the book back and asked if there were any special instructions. "None other than the usual instructions, but since it's a holiday, besides making your arms rooms checks, you should walk the battalion from time to time, to make sure that there aren't any disturbances or fights that might require the Company Commander and First Sergeant of the unit involved to come in."

The Lieutenant said, "Very well." He heard the door at the far end of the building open and the Duty NCO coming in, so he told the Adjutant that he would be headed for the mess hall and after eating it would come back to take over his normal station, either in the duty office, or the Adjutant's office once the colonel was gone for the day. The Adjutant saluted him from his chair, and the Lieutenant took that as a signal to leave for dinner.

On the way down the hall, he slipped into the duty office, where he said hello to the NCO and walked over to the duty officer's desk. There, he made sure that his pistol was loaded like Sergeant Robb had taught him, with the spare two clips in the ammo pouch on the web belt around his waist. He then told the NCO he was going to dinner and walked through the darkness and fog to the mess hall.

When he arrived at the mess hall, the soldiers were louder and more boisterous than usual. They were looking forward to a long weekend, filled with various forms of entertainment, no doubt. The Lieutenant stepped through the door, signed in on the roster as the duty officer, and paid the reduced rate for dinner. Once he had his tray, he sat down in the far corner of the room. He ate quickly by himself, as he wanted to get his first round of arms rooms checks done as early as possible.

As he walked from the mess hall, back towards headquarters, the stereos were cranked up, battling across the courtyards. He could see small groups of soldiers, dressed in civilian clothes, smoking and laughing, heading for the train station to go downtown. The Lieutenant smiled to himself a little before he walked up to Battalion, where the Duty NCO met him at the door. He told him that there had been a small fight over at Alpha Company, but the First Sergeant had not left yet, so it was under control. The Lieutenant told him, "I'll go out now then to make my checks right away and go to all the arms rooms and then Alpha in particular, to get a feel for the troops before they get settled in for the evening."

If anything, it had gotten darker and a lot colder since dinner, so the Lieutenant put on his field jacket, thankful for the warmth of the liner, and slipped the holster and web belt over it, with the lanyard for the pistol attached to the belt. He then grabbed the big ring of Duty Officer's keys with his left hand and drew the pistol with his right, before he walked to the line company barracks. He went through the door and down the stairway that led to Alpha Company's basement. There was still a lot of noise in the barracks, including raised voices, and the music was deafening in the upper levels, but when he asked the CQ if he needed any help, the specialist told him, "No. If I need any, I'll call the First Sergeant who lives nearby."

The Lieutenant then proceeded to the arms room to make his check. And upon his return, checked out with the CQ. After that he went through Charlie and Bravo without incident. In the Combat Support Company, a couple of the scouts were in a heated argument just inside the front door. He neither knew nor cared what they were arguing about, but he asked them to settle down, so he wouldn't have to call in Lieutenant McCallister. They looked at him and realized that he knew their platoon leader, so they immediately shuffled off to their rooms.

After that he made his check in HHC. The bass of the stereos was echoing in the halls above and the smell of hash was everywhere, but it was basically quiet, so he returned to the headquarters, just in time to hear the telephone ringing off the hook, and the Duty NCO screaming, "Flash traffic!" He and the NCO picked up their telephones, note pads and pencils, to write down the series of codes for the message. Once he had the code and the NCO read it back, and received confirmation, the Lieutenant scooped up his copy with the ring of keys and flew down the stairs to the S-2 office. He unlocked the door and then the safe. He put the code book on the S-2's desk and concentrated. Once he had decoded the message, he breathed a sigh of relief. It was just a test message that would not require putting the battalion on alert.

After putting the code book back in the S-2's safe and locking it, he locked the door and climbed the stairs. He entered the duty office and sat back in his chair. Then he told the NCO that it was a test and filled out the log to indicate that he had received and decoded the flash traffic test message. Since it was now the middle of the evening, he signed out on the log, to patrol the battalion area. As he made his way back behind the barracks, pistol, and keys in hand, he could see that groups of soldiers were coming and going outside the aid station, and the back door to the mess hall. As he approached, they quickly disappeared around the corners of buildings, into the tree line, so he continued on his way. There were no acts of violence, or altercations, so he made his way around the barracks, where the noise was at an even higher level, to the motor pool.

When he arrived, he checked in with the NCO in charge of the guard. Fortunately, that NCO was one of the tank commanders from Second Platoon Bravo, who he already knew. He visited with him briefly and then checked the perimeter that was guarded by soldiers from that same platoon. When that was done, he walked past Battalion Maintenance to make sure that no one was trying to break in to steal parts. By now, the wind had come up and the temperature dropped, to the point where it was below freezing. He pulled up the collar on his field jacket against cold and made his way back to Battalion.

As he walked into headquarters, the Duty NCO motioned for him to come over and handed him a steaming cup of fresh coffee, saying, "Sir, I believe you could use this, since it's getting colder and colder out."

The Lieutenant thanked him and settled down in the duty officer's chair. He took a sip and enjoyed the coffee as it warmed up its body. He looked up at the clock and saw it was approaching 2300. He told the NCO, "I'm going back to the Adjutant's office, to try to catch a little nap. But get me up by 2330. I expect that several of the soldiers who had been downtown will be coming back into Garrison at that time. I'll walk down to the gate to be ready, in case of any trouble, to contact you to bring in the first sergeants and commanders if it gets bad."

At 2330, the NCO woke him up. He put on all of his cold weather gear and with his pistol on hip, stalked through the rigid night to the main gate. He stood to the side, next to the HHC courtyard to watch the men as they came back. The troops were pretty tired, after gunnery, with a little help from alcohol or whatever else they were taking. They mostly just wanted to come back and rest. If they could sleep through the noise, that had died down, at least a little. When the last few had stumbled in, he walked through the battalion. After checking in again with the motor pool guard, he hurried back to the Headquarters, at almost 0100.

Taking off all this cold weather gear, he warmed his limbs, and sat down at the duty desk, just in time for the phone to start ringing again. The NCO yelled "flash traffic," and they both picked up the telephone receivers, to receive and write down the coded message. The NCO confirmed it. Once again, the Lieutenant grabbed the message and ran down to the S-2's office. After unlocking the door and the safe, he placed the codebook down on a small table beside the sake, where he carefully decoded the message. When he finished, he closed his eyes for a few seconds and then did it again to make sure it was correct, because he was getting very sleepy. Having broken the code twice, he realized it was another test message. He locked the codebook back in the safe. Then he locked the door and slowly trudged up the steps. By the time that he had spoken with the NCO and made the required entry in the log, it was after 0200. Again, he told the NCO, "I'm going to try and catch a nap on the sofa. Get me up by three. That's when I expect the last of the soldiers who had been out on the town will come back to the billets. And my experience over the years has been if anything bad happens, it is usually at three in the morning."

RIGHT TIME OF THE NIGHT

As he lay back on the couch, he could feel the cold seeping in around the windows. He threw his field jacket over his chest and used the arm of the sofa and his scarf as a pillow. He quietly chewed himself out for forgetting to bring a poncho liner with him to use as a blanket. Since he wasn't planning on sleeping much anyway, it was no great loss for that evening, but a lesson learned for future duty. He promptly fell asleep, until he felt a hand on his right shoulder, and immediately came awake.

The NCO said, "Sorry, sir, but you told me to get you by three." The Lieutenant checked his watch. It was exactly 0300 so he thanked the sergeant, rose, and bundled up against the cold once more. He took his keys and pistol and went back out into the battalion area to stand in the shadows of the trees outside Battalion Maintenance. He watched the last of the troops come through the gate and walk to their barracks. Once they were all inside, he went back to the duty office and told the NCO to call the CQ and every company and check in with them to see if there were any problems.

After he had made all of those calls, the NCO reported that there were no problems in the battalion. The Lieutenant breathed a sigh of relief. He thought that maybe he could make it, with a nap until 0700, so he would be awake to get breakfast, make his final checks, and complete the duty log for the oncoming duty officer, who was Lieutenant Matthews.

The Lieutenant breathed another sigh of relief as he walked down the hallway to the Adjutant's office. Hoping to finally get some sleep, he made himself comfortable and relaxed. He was just drifting off, when he swore he heard the Duty NCO down the hallway, but couldn't believe it, so he continued to lie down. Then he heard the

Duty NCO screaming, "Flash traffic!" again, so he jumped up, and ran down the hallway, to the duty desk. He picked up the receiver on the phone, and with the NCO, wrote down and checked the third flash traffic message of the evening.

Again, the Lieutenant took the coded message downstairs and broke it. This time, it appeared to be an actual alert message, with the exception of one character that was out of place at the very end of the message. By now, the Lieutenant was pretty foggy, so he took a deep breath and decoded the message a second time, double checking every character. He finally came to the same conclusion: it was sent, probably because of the holiday, to make sure that all the duty officers were on their toes. The Lieutenant appreciated the sentiment, but really just wanted some sleep, so he locked everything up, went back upstairs, logged the message in on the duty log, and scuttled back to the Adjutant's office to lie down.

It was nearly 0400. The Lieutenant was having a hard time going to sleep, so he decided to make his second round of arms rooms checks. It was bitter cold, so he bundled up with his web belt and holster outside his field jacket. He tied his scarf around his head and neck and stepped out into the wind. There was now a steady drizzle mixed with sleet in the air. He jogged as quickly as possible to Charlie Company, to get indoors and out of the weather, to make his check.

He stopped by the CQ desk and asked how the night was going. Everything was calm, so he descended into the basement to the arms room, and initialed the sheet next to the armorer's window. He made his way back down the hallway and up the stairs to the door where he stepped out into the weather again. He rushed over to Alpha Company, where he made his check there as well. That being done, he crossed diagonally in front of the S-4 office to Bravo Company. He made his arms room check there, and visited for a little bit with the CQ, who was the gunner on Sergeant Robb's tank. They had a nice chat before the Lieutenant headed back out into the weather to make his way to CSC and then HHC.

When he was finished with Headquarters Company, while walking back toward Battalion, he remembered that he hadn't even started Sergeant Robb's Efficiency Report. Realizing, however, that he was simply too tired to do it, he decided he would make it through Thanksgiving and turn his attention to it on Friday. Since it was going to be a perfect report, there was not much to fret about, other than the relatively small amount of space he would have to commend his platoon sergeant for his outstanding work.

As he walked in the door, the Duty NCO told him that he would need to wake him up at 0600, because he had to deal with the oncoming duty NCO who was going to sign in one hour early, so he could make it home in time to get some sleep and spend Thanksgiving with his family. The Lieutenant said, "Very well. I'll try to catch a nap." Looking at his watch, it appeared he would have about forty-five minutes. He fell fast asleep but woke up to the sound of a different voice. It was the new Duty NCO, who had come down to wake him up and bring him to the duty office, to get briefed. When he stepped back into the duty the office, the NCO who had been on with him all night was packing up his belongings, smoking a cigarette. The incoming NCO signed in on the log. The Lieutenant signed the log, indicating that he had released the old duty NCO, who thanked him and shot out the door to get back home.

The Lieutenant had never met the new Duty NCO. He was from Combat Support Company's Air Defense Section. The Lieutenant brought him up to speed on everything that had been logged in so far and then directed him to go ahead and call all of the CQs to see if there were any problems. Once the sergeant had made all of the calls and reported back that everything was quiet, the Lieutenant asked him if he had any breakfast before coming on duty.

The NCO told him, "No, sir. Nothing but a cup of coffee and a cigarette so far."

The Lieutenant told him, "I'll release you to have breakfast at 0700. I'll go down to eat something and get back to complete my log and turn over the battalion to Lieutenant Matthews at 0800."

FRIEND OF THE DEVIL

The remainder of the morning went as the Lieutenant had planned. By 0800, he had finished his duty log and filled in some of the information for Sergeant Robb's Efficiency Report. He paced in the duty office, looking out the window for a while, hoping to see his relief. By 0815, he was concerned, and by 0830, he was worried. He asked the Duty NCO to call the Visiting Officer's Quarters and see if they could check on Lieutenant Matthews. The NCO called and confirmed that Lieutenant Matthews was staying there. But it wasn't the clerk's job to get him up out of bed, since he was not assigned to the battalion, but to the post. The bureaucracy never ceased to amaze the Lieutenant. He shook his head and told the NCO that he was going to the VOQ and get Lieutenant Matthews himself.

Once he got to the VOQ, he got the room number from the desk clerk and climbed the stairs to the second floor. He walked down to Lieutenant Matthews' room and knocked on the door. There was no answer, even though he swore he could hear someone snoring. He went down to the desk, identified himself as the duty officer for Matthews' battalion, and asked if he could have a pass key. The clerk handed him the key. He trudged back up the stairs and opened the door. He found Lieutenant Matthews, stone cold asleep. Despite several efforts, he couldn't wake him, so he took the clip out of his .45, ejected the bullet from the chamber and pressed it back in the clip that he knew was light a round as it had come from it originally.

He reached over and laid the cold steel of the barrel right next to Matthews' left cheek. His eyes came open immediately, and the Lieutenant told him, "Hey man, I've been up for over twenty-four hours. I have to go to dinner with the S-4 in about two and a half hours and I need you to show up at the battalion to relieve me right away." Matthews realized where he was and immediately started getting dressed. Out of curiosity, the Lieutenant had to ask, "What have you been up to?"

Matthews told him, "I ran into an old fraternity brother from college. He's assigned to one of the transportation battalions. We decided to go to K-Town to a disco to dance 'till after midnight and then go to a party at one of his friend's quarters until almost five. By the time I caught the train back to Mannheim from Kaiserslautern, I just crawled into bed about an hour ago."

The Lieutenant asked Matthews, "Have you pulled duty before in the battalion?"

"No."

"Alright. I'll walk back with you and on the way, give you the insight that I got from Sergeant Robb on how to stay out of trouble."

As they walked and talked, Matthews quickly woke up. The Lieutenant asked, "Have you had anything to eat lately?"

"No," so the Lieutenant took a quick right at HHC and led the way into the mess hall. When he came to the counter, he asked to see the Mess Sergeant. He wasn't in, but the Specialist Five assistant was, so the Lieutenant asked, "Do you have any food leftover from breakfast that you could put together on a plate for the oncoming duty officer?"

"Yes, sir." He turned back to the kitchen and soon returned with a plate of toast, sausage, and some scrambled eggs. The Lieutenant thanked him and handed it to Matthews, saying, "You can take this up to Battalion and eat there. I have to sign out. The Duty NCO can show you the ropes, with regard to messages and checks. Once you have some food in you, draw a sidearm and you'll be ready."

They finally got to Battalion. The Lieutenant came in and motioned for Matthews to sign in as the incoming duty officer on the log. The Lieutenant signed out and introduced Matthews to the Duty NCO. He then told Matthews, "If you have any questions, I'll either be in my BOQ room, or Lieutenant Graham's quarters for dinner."

The Lieutenant hurried over to HHC and turned in his sidearm, ammo, and holster. He was running late, and practically jogging to his room, before he realized that he didn't have Lieutenant Graham's address. He hurried into the BOQ, and shot down to the basement, to the Autovon phone, next to the washing machines. He dialed the Duty NCO's number at Battalion and asked him for the S-4's quarters address. The NCO checked the battalion roster and gave him the address.

The Lieutenant then climbed slowly up the stairs and looked into the communal kitchen. Buzz was busy making both a ham and a turkey. It smelled amazing. The Lieutenant said, "Hi Buzz, I just got off duty. Regrettably, I have to go to my new boss' house for Thanksgiving dinner. I'll be back in the afternoon, however, if there are any leftovers, for the evening meal." He turned down the hall, leaving Buzz to his good works.

Then, finally, he opened the door to his room. He undressed and took a hot shower to get the smell of fifteen hours' worth of stale cigarette smoke off of him. Then he set his alarm for an hour and a half and passed out cold.

I HAVE A DREAM

When the alarm went off, he stayed in bed, disoriented for a while, because his body did not want to get up. However, after a couple of moments, he remembered that he needed to get ready, so he stood up, shaved, and washed his face. Then he took another brief shower to wake up, and dressed in civilian clothes, or more precisely, a combination of uniform and civilian clothes: his black low quarters, gray slacks that he had picked up in El Paso, his uniform white shirt, and a maroon V-neck sweater from the PX at Fort Knox. He then walked out the door and down the stairs to the street, before crossing the street and walking over to the portion of the village where married officers were housed.

He recognized the address as being in the same row of apartments where Captain Coyle lived, so he found the address easily. As soon as he knocked on the door, it opened, and Graham welcomed him in. He had never met Graham's wife before. Dana was a small blonde, with glasses, and a tight little smile. She was even smaller than his girl, who weighed barely ninety pounds. After shaking his hand, she went back into the small kitchen. He could hear her scurrying, trying to get things ready. Graham motioned him toward the couch. He turned and took a step in that direction. However, his body was still half asleep and his foot hit something under the couch that tripped him and sent him headfirst, into the sofa. Sheepishly, the Lieutenant regained his composure, sat down, and looked down between his legs. He saw a large bag of something that had tripped him. Graham who was smiling and chuckling, asked, "Do you know what that is?"

"I have absolutely no idea."

"We are Mormons, and we are required to keep several hundred pounds of wheat in our home. It's part of our religion."

The Lieutenant then looked up and saw that throughout the apartment, in various locations, he could see sacks of wheat under the occasional chairs as well as the sofa. And when he looked into the spare bedroom, he could see rows of bags of wheat under the bed. The Lieutenant looked up and told him, "I've never met a Mormon before," and then apologized for squealing a little when he fell into the couch. Graham was a great guy. He just laughed. The Lieutenant could hear Dana giggling as well. Graham asked if he would like something to drink. The Lieutenant told him that he was past his limit on coffee, but water or iced tea would be wonderful, so Graham rose, went to the kitchen, and came back with a glass of iced tea. The Lieutenant drank almost half of it before Graham sat down. Between cigarettes and coffee, and running around the battalion all night, he was practically dehydrated. As he set the glass down, Graham asked him, "How was duty?"

"I had three flash messages, but other than that, the battalion was pretty quiet."

Graham continued, "Has anyone shown you a copy of the duty roster for the rest of the week?"

"No, I haven't seen it yet."

"Well, it was published at Staff Call, and believe it or not, you're going to be on duty again Saturday night."

The Lieutenant looked at him in amazement, and Graham explained, "Your duty last night, although it ended on a holiday, doesn't count as weekend duty. It only counts as a weekday, and since you haven't pulled a weekend duty yet, you'll go on Saturday morning and get off Sunday morning."

The Lieutenant thought to himself, *Damn*, and then realized that he needed to stay quiet, while thinking he really needed to get some sleep tonight and Friday night. "Well, thanks for the heads-up. By the way, when can I get some time to go retrieve my car from Bremerhaven?"

"Check on the bulletin board at the bank, to see if anyone needs to a car in to ship back to the United States in the next few days. You should make arrangements with that person to drive up with them and then bring them back."

This lightened the Lieutenant's mood considerably. He was desperately missing his car. He loved to drive, and he felt imprisoned without it. Looking back, he hadn't even seen it since the middle of August. So, he made a note to check in the bank the next day to see if he could find someone looking to make the trip and make the call.

By now, the Lieutenant was awake and able to take in the smell of Thanksgiving emanating from the kitchen. It was wonderful. Thinking back, the last civilian Thanksgiving dinner he had eaten was near Albany in upstate New York. He was visiting his best friend from VMI, who was descended from the Sergeant at Arms on the Mayflower. Although it had been brutally cold there, he had a great time. His buddy's mother had never learned to cook turkey, so they had roast beef with potatoes, and chocolate cake that she always purchased at her family's favorite bakery.

This meal promised to be much better. Dana stuck her head out of the kitchen and asked if Graham could help her. He we went back to the kitchen and they brought out the turkey with all of the side dishes and then set a pumpkin pie at the far end of the table. Graham asked the Lieutenant to join them. He took a seat at the table, to Graham's right, so his host was at the head of the table and his wife was to the left, nearest the kitchen. Graham and Dana said grace over the meal, and the Lieutenant bowed his head and prayed with them. Then, once they had filled their plates, the conversation began in earnest.

Like all Army wives, Dana was interested in the Lieutenant's plans with regard to settling down and getting married. He quickly told her that he was engaged, and that currently he was scheduled to go on leave the first week of March to be married on the eighth, in Abilene, Kansas, and then return with his new wife. Graham was a very quiet person, but Dana had an effervescent personality and asked a lot of questions. She wanted to know how they met, how long they knew each other, and those sorts of things.

When the Lieutenant told her that his fiancée was his former high school best friend's former girlfriend, and that, actually, he'd only known her about fifteen days out of his life before he shipped to Germany, her eyes grew a little wide. He reassured her that he had actually met her in 1976, although he had not spent a lot of time with her. He then asked Dana how she and her husband met. Apparently, they were high school sweethearts, who attended college together, got married, and moved to Fort Knox, followed by Mannheim. Graham added, "We're from South Dakota. It's our home state and we love it."

Dana said, "Our wedding bands are made of Black Hills gold that is a light rose color." They showed him their rings and the Lieutenant smiled. He told them that they were obviously in love and he was happy to be with them for Thanksgiving.

They finished the main course. The turkey started to weigh the Lieutenant down, but he made it through the pumpkin pie but then had to excuse himself so he could go to bed since he hadn't slept in over a day. Graham said he understood and that the Lieutenant looked like he really needed to get some rest since he had more duty coming up. The Lieutenant added, "Yes, for sure. I have to write the efficiency report for my platoon sergeant but don't have anything else to tend to, and with the exception of eating and sleeping, I don't have any other plans. I look forward to seeing you on Monday." He offered to help both of them clean up the dishes, but that offer was refused, so he made his way to the front door, shook hands with both of them, and headed back to his quarters.

The weather was nice for Germany that time of year: sunny and cool, but dry. He actually enjoyed his walk back to his room. The caffeine in the iced tea had kept him awake through dinner, but he was completely exhausted. He opened the door, stumbled in, undressed, crawled on top of the bed, and pulled the poncho liner over him. He looked over to make sure the alarm was not set and fell asleep.

GONNA FLY NOW

The next morning, his biological clock woke him up a little after 0500. He got up slowly and made a pot of hot water to make instant coffee later. After taking a shower and getting dressed in civilian clothes again, he looked out in the hallway and smelled the greatest smell in the world that a soldier can smell: fresh, hot coffee brewing. He immediately walked down to the communal kitchen and saw Buzz sitting in the corner, savoring his first cup of the day. The Lieutenant pulled a mug out of the cupboard, filled it, and sat down beside him. The Lieutenant told him, "I'm sorry I missed dinner, but I really needed to get to know my new boss and his wife. I had a good time with them, but I was too exhausted to do anything after I got back."

Buzz nodded, "I've pulled a lot of twenty-four shifts myself, so I know the feeling. Have you recovered?"

"Yes, for the time being, but I'm going on another twenty-four-hour duty starting Saturday morning, so I need to make the most of today."

"Well Bob, the good news is that neither Thanksgiving Day nor the Friday following are holidays in Germany, so you should be able to go out of the economy and have a pretty good time."

The Lieutenant shook his head in agreement and then said, "I'll probably catch the streetcar to Viernheim in the evening to get a meal at my favorite Greek restaurant, followed by a good night's sleep, before signing in for duty at 0700 on Saturday. By the way, how did the Thanksgiving go?"

Buzz replied, "Despite my best efforts to destroy the turkey, it turned out very well. Word had gotten out on the street that I was working my culinary magic, so Susan and Chris invited some of their friends, Bernie showed up with a couple of artillery officers who live in Mannheim, and Lieutenant Pallas and Siggy came up from the first floor to join us. The bad news is that everyone was famished, so there are no leftovers."

The Lieutenant told him that it was just fine. He was stuffed from the day before and was going out for dinner anyway. He bent down, looked at his watch, and realized that the bank would open in about an hour. He excused himself and told the colonel to enjoy his day as he left the kitchen.

He entered his room and sat down in the nearest black easy chair in his living room. He wanted some time to relax before he went over to see if he was going to have any luck getting his car. He also planned to get a few marks and try to catch his girl at home for a call to let her know that he had made it through Thanksgiving. Then he remembered that with the time change, it was still the middle of the night in Kansas, so he decided to delay the call until mid-afternoon, after getting his marks and checking the bulletin board, followed by some rest. As he sat there, it occurred to him that he really needed to send his parents a letter. He had not written them in quite a while. He wrote a quick note to his mother, telling her that he had been in the field in the early fall and had just returned from gunnery in another city. He also had a new job and was lucky enough to have Thanksgiving dinner with his new boss and his wife. He was starting to enjoy his stay in Germany and looked forward to seeing them in the spring when he came back to get married.

When he was done, he put a stamp on the envelope and sat it on his dresser. Then he washed up and got dressed to go to the bank. As he left the apartment, he could hear the first signs of life from some of the other officers who were waking up after the holiday. He decided that he needed to speed up to try to get to the bank and beat the line. Mailing his letter on the way out the door, he walked quickly across the traffic circle to the bank. He walked through the door and realized, too late, that due to the training holiday, he would be standing behind about thirty soldiers and their wives.

He took his place in line, checked his watch, and figured out it would be after mid-morning before he was able to exchange money. So, he changed his plan. Instead, he stepped out of line and went to the bulletin board in the entryway. After about a minute, he was excited to see what he had been looking for. It was a three by five card with a name, Bill Grimes, and a German telephone number. He had a pen and a piece of paper in his shirt pocket, so he wrote both down, and then proceeded upstairs to the telephone bank. He had a few mark coins in his pocket, and there were very few people using the telephones since it was early back in America.

He walked up to a phone, put in a two-mark piece, and dialed the number from the card. A gentleman answered the phone as "Bill Grimes," and when the Lieutenant explained the nature of his call, Grimes asked if he would be available to take him to Bremerhaven a week later. The Lieutenant told him that he believed he could, and asked him to hold that date, so he could confirm with his boss on Monday, and call him later that day. He also asked if the man had an Autovon number for work. He did, so the Lieutenant wrote that down and told him he would call him on the Autovon system as soon as he had confirmation. He then wished the gentleman a happy Thanksgiving weekend.

Then, having accomplished that part of his mission, he went back downstairs to stand in line to get some marks for lunch and dinner, with some more to get him through the next couple of weeks. It also occurred to the Lieutenant at that time that he was going to need some ESSO coupons for gasoline. He asked the sergeant in front of him where he needed to go to purchase those. The sergeant told him to go to the Esso station and also told him what a book of coupons would cost him. The Lieutenant thanked him. After waiting line for almost an hour, he was able to write a check, taking some in American currency with the rest in marks.

Chapter 20
Excitable Boy

He stepped into the warm sunshine and realized that he was getting hungry. He walked over to a currywurst stand near the streetcar station. Like the American hamburger, currywurst was Germany's most easily found and best snack food. After buying the wurst and French fries, he headed back to his room to enjoy lunch and a nap, before trying to call his girl.

When he got to his room, he turned on the radio to listen to the news and pop hits broadcast by AFN. He ate his lunch and relaxed. When he was done, he decided he would throw away the trash in the communal kitchen, so he headed down the hall to do that. When he reached the doorway, he saw that Bernie was there ahead of him, staring vacantly into a cup of coffee.

He said hello, threw away the trash in the can by the door, and asked Bernie how Thanksgiving had been. Bernie looked up at him and said, "The dinner was great," then added, "can I talk to you in your room about what happened afterward?"

"Sure." They walked back to the Lieutenant's room and sat down in the easy chairs. The Lieutenant asked Bernie, "What's bothering you?"

Bernie said, "I assume you know about my side business."

"Yes, I do, but, but like many other things that I'm not: I am not a narc."

Bernie nodded his head and said, "I understand, but last night, while you were sleeping, I had three visitors, soldiers in from Baumholder, the Second Battalion of the Sixty-Eighth Armor, our sister battalion. The said that they had bought some drugs from a friend of mine at the last gunnery and they wanted more. They had asked around and finally someone in HHC told them where I lived."

Bernie went on, "There were three of them, a slender black Spec Five, and two extremely skinny white Spec Fours. All three were medics. They were looking to score some hash and hoped that I also dealt in heroin. I told them that I didn't have anything. Two of them were alright with it, but the taller of the two whites kept pressing me. He repeatedly said he didn't believe me." Bernie then asked the Lieutenant. "Have you ever been in a situation, where you were in a room with someone who just gives you a really bad feeling?"

The Lieutenant told him, "Yes, my first assignment, in a basic training battalion at Fort Knox in '77. My company armorer was a black buck sergeant; his arm rooms was spotless and his uniform was always perfect. He looked like a recruiting poster, but there was something about him that always bothered me. It was like everything was too perfect, and you could feel a malevolence in the room when you were alone with him. He had a new van that he spent a lot of time washing in the company area when he was off duty."

"As it turned out, my uneasy feelings were justified. After five months in that battalion, my mustang second lieutenant Company Commander took a couple of days leave after we graduated a class of trainees and I took over as acting commander while he was gone. The first Sunday morning that I was in command, in the spring of '78, I got a call early in the morning from the battalion duty officer. He told me to report, in greens at Battalion to meet the Command Sergeant Major, a chaplain, and an investigator from CID," the Lieutenant began to detail the grisly story.

"When I pulled up at Battalion, there was an OD staff car parked in front of headquarters, and it was pouring rain. I went inside to the battalion conference room, where the chaplain, Sergeant Major, and investigator were standing. As I entered the room, the investigator stepped forward to shake my hand, as did the chaplain. The Sergeant Major saluted sharply, which was a little unusual since we didn't get along that well, and I returned it."

The Lieutenant took a breath before continuing, "The investigator then informed me that my E-5 armorer was under arrest for murder, of not only one, but several female soldiers. Apparently, unbeknownst to me in the two years before I was assigned permanently at Knox, several female soldiers had simply disappeared from the post. In fact, in the summer of '77, before I came back from Ranger School at Benning, a female ROTC cadet had disappeared without a trace. Hearing this for the first time I was surprised. I had never heard anything about a potential serial murderer loose on post during the time I was assigned there to a tank battalion in '76, as an officer in the basic course in June of '77, or later in that same year when I ended up pushing leg infantry trainees in the Fourth Brigade."

"Out of curiosity, I asked the investigator what made them believe that my sergeant was responsible. He told me that my armorer had purchased a large van with a re-enlistment bonus about two years earlier. Over about a year and a half, he had picked up female soldiers who looked like they needed a ride, and then taken them into the woods, where he brutally raped, murdered, and dismembered them. I knew the van well. Like everything else associated with the armorer, it was immaculately clean. That probably explained why his uniforms were always starched and pressed, and everything in the arms room was clean as a whistle and smelled like Clorox."

The Lieutenant continued his story: "The investigator said that CID probably wouldn't have caught up with him, but he made the big mistake. He took another soldier with him on one of his 'hunting trips' and unfortunately for the sergeant, that 'friend' could not hold his liquor or his tongue. He talked about the crime with some of his buddies and we got wind of it. After we picked that trooper up, he confessed completely. That, along with the evidence from the bodies from the shallow graves that your armorer had showed his accomplice, we were able to bring charges against him and incarcerate him pending court martial."

The Lieutenant paused and then continued, "I looked around the room and asked the question that had been on my mind all day so far, 'So, why am I here?'"

"The chaplain, a lieutenant colonel, spoke up at that point, 'Lieutenant, we are going out to inform the sergeant's wife that her husband is under arrest, and will not be coming home for the foreseeable future. As his commander, we need you to come with us and relay that message. I will comfort her, as best I can, and the Sergeant Major will assist her, regarding how she and her family will be taken care of.'"

"At the time I was thinking, *I'm not trained for this mission and the entire thing was completely surreal.* The Command Sergeant Major then said, 'Sir, we really need to go.' He sat in the front seat with the driver, while the chaplain and I sat in the back of the green Chrysler sedan. We made our way to the outskirts of Radcliffe, Kentucky, to a trailer park. Since leaving Battalion, it had been raining in sheets. It did not let up when we pulled up in front of a small trailer. I sank into ankle deep mud that came over my shoes as I walked up the steps to the front door of the trailer where the sergeant's wife lived," the Lieutenant paused before continuing.

"I knocked on the door and announced who I was. When she came to the door, I could tell that she was about six months pregnant, with a one-year-old child in her arms and a two or three-year-old little boy standing behind her. I wanted to throw up, but asked if the gentlemen with me could come in. She did not answer, just let us in. We all sat around the small kitchen table. I first told her what the situation was with her husband. The chaplain offered his help and support and told her he would make arrangements to have someone come out and see her the following day. The Sergeant Major explained what benefits she had and gave her his telephone number so she could call when she needed something, as the non-commissioned officers would take care of their own."

As the Lieutenant recounted his story, Bernie's eyes got wider and wider. The Lieutenant added, "You just never can tell," before going on, "my roommate of almost three years at VMI found the body of a murdered woman on his way to elementary school when he was a little boy. Have you ever read *The Michigan Murders*?"

"No."

"It's a book about a long series of murders of young women in Michigan, near Detroit, where my roommate hailed from. And it turned out that the murderer was the nephew of the state police investigator who was in charge of finding the killer. In fact, the murderer kidnapped young women and took them to the home of the investigator, where he raped and murdered them, cut them into pieces, and disposed of their bodies before washing everything in the investigator's own washing machine in the basement. So," the Lieutenant added, "I know what you mean when you feel like you're in the room with evil, but you can't put your finger on it."

The Lieutenant then asked Bernie, "Do you have a name for that soldier?"

Bernie thought for a minute, then said. "I can see him. He had a dull glow in his eyes, like a lot of heroin addicts. You know, the blue flame. And it seemed that he was just watching me too intently. Finally, the three of them decided to leave and see if they could find a medic friend of theirs in your company."

Then Bernie's eyes came wider as he remembered, "His name is Dahmer. That's what the black soldier used when he told him that they had to leave. But he just watched me the entire time as he walked out of my door. Creepy. So, when you're in the battalion, keep an eye out in case you run across them."

The Lieutenant said, "I doubt I will. I already have enough of my own drug dealers, stoners, and junkies to worry about, but thanks for the heads-up anyway since I'll be pulling duty Saturday night."

Finally, Bernie looked at him and said, "I need to get back to my room. But if you see this fellow looking for me, just tell him that I've gone out of town for a while." The Lieutenant agreed and said goodbye as Bernie stepped out the door and closed it behind him.

Despite having a good night's sleep, the Lieutenant was still pretty tired, so he set his alarm for 2:30 in the afternoon, climbed back on the poncho liner that covered his bed, and fell asleep, with the memories of that rainy Sunday morning in Kentucky still running through his mind. As soon as the alarm went off, he jumped up from bed, checked to make sure he had several marks in his pocket, and headed off for the post office above the bank, to try and call his girl.

He practically ran up the steps to the post office and looked around the room. He was happy to see that there was a phone available right away. He walked across the room, inserted a 5-mark piece, and dialed her number from memory. However, the phone simply rang and no one answered. He finally hung up, retrieved the coin, and decided that she must have gone home to her parents for the holiday. So, he would have to try and reach her on Monday.

As he left the bank building, disappointed, the weather was cloudy but at least warm. He decided that he would take the streetcar to Viernheim and visit the shopping center there before going to Nick's for dinner. This time he was able to purchase the correct ticket and hopped aboard the streetcar as soon as it arrived in front of him. At the Viernheim Station, he hopped out and walked over to the large department store where had previously purchased his German student shoes. After looking for half an hour but not finding anything that appealed to him, he noted several items that his girl would probably like.

THE GREEKS DON'T WANT NO FREAKS

The sun and temperature were falling rapidly, signaling late afternoon. He was famished and walked briskly to Nick's for his favorite dinner: sliced pork loin, grilled as a Shish Kebab on a steel spit with peppers and onions, served with French fries and salad. When he arrived, he ordered a Pilsner and the meal. He had a nice visit with Nick, who was in a better mood than usual, as the holiday had brought him extra business. He asked if the Lieutenant was still planning to get married in the spring. The Lieutenant said, "Yes. And I'm counting the days. By my reckoning, I'm almost halfway there." Nick smiled, turned, and then brought him his meal. He took his job time and enjoyed it thoroughly.

When he was done, Nick told him, "I need to bring out something special for you." The weather for late fall in Germany had grown foggy and cold, so he brought over a brandy snifter, a small alcohol flame, and a bottle of Greek brandy, Metaxa. He poured a glass and heated it up, before handing it to the Lieutenant. He had never really drunk anything quite like it. The brandy warmed him inside and out and gave him a very satisfied feeling. He thanked Nick and settled the tab, adding five marks for the brandy, placing the German bills and coins on the table of the booth where he had been sitting. Then he said "Good night" to Nick and took off for the streetcar.

When the streetcar arrived at Benjamin Franklin Village, he climbed down carefully and looked around before turning left to go back to his quarters. It was growing dark, but still early evening, so there were a lot of soldiers milling around. They were waiting to go downtown on the streetcar or finding rides with friends to go to the bars and other entertainment establishments in the city. The Lieutenant thought to himself that he was happy he was no longer so young and doing that himself but instead getting ready to turn in for a good night's sleep before full day and Saturday night of duty.

BAD GIRLS

He walked to the doorway to the communal kitchen, and paused to say hello to Buzz, Elizabeth, Susan and Chris. Buzz had driven downtown and brought back Chinese takeout. They were just beginning to eat and asked the Lieutenant to join them, but he told them, "I'm going on duty in the morning and I just ate any way, and drank some brandy, so I'm headed for bed." They wished him well, and he proceeded down the hallway, stopping across from Bernie's room. He could see that there were no lights on, so he assumed that Bernie was away, probably for the weekend with one of his many girlfriends, dodging Dahmer and his buddies.

Since he'd been out all day, and wanted to relax, he took a hot shower and settled in to read another couple of chapters of a book he just started, *The Great Train Robbery*. After an hour or so, he decided he would go ahead and try to go to sleep, in order to get up early and sign in for duty on time. He set his alarm for 0500 and settled back onto his bed with a poncho liner over him to go to sleep. He had just started to dose and was hoping that his girl was having a good Thanksgiving back in Abilene, when he realized that he could hear the shower and rock music above him. Meghan and her friends upstairs were settling in for an extended evening. That being the case, he rolled up his extra poncho liner to use it as a pillow, put his heavy feather pillow over his head to block the noise, and thankful for the Metaxa, went to sleep.

WELCOME BACK

He woke up shortly after 0430. It was quiet as a tomb. He got up immediately and turned off the alarm so as not to wake the other occupants of the BOQ on a Saturday morning. Since he was running early, he decided to make a hot pot of water and some instant coffee, to warm up. After shaving and getting dressed, he drank his coffee and took a long look at himself in the mirror. He was pretty tired, and in need of a haircut, but other than that, he looked ready to meet the day. He took a deep breath and put on his field jacket. This time he remembered to take his extra poncho liner in addition to his helmet, rubber duck, LBE, and a book, to report for duty.

After walking past the gate guards, he turned right into Headquarters Company. He asked the CQ if the armorer was in. The CQ told him, "Yes, sir, but he just got in from downtown and went to his room to sleep."

"When is the First Sergeant expected in since it's Saturday?"

"He'll be in by 0600, when the mess hall opens up."

"Alright, I'm going up to Battalion. Please call the Duty NCO when the First Sergeant comes in and I'll come back down so he can get the armorer or Supply Sergeant up to issue me a sidearm for staff duty."

The specialist nodded his head and said, "Yes, sir."

The Lieutenant saluted, turned, and walked out of the building toward Battalion, in a light mist and darkness. When he got to Battalion, the first thing he did was say hello to the Duty NCO and get a cup of fresh coffee. He then asked, "Where's the duty officer?"

"Sir, he just made his arms room checks and he's sleeping back in the Adjutant's office." He thanked the sergeant and sat down at the duty desk. He asked, "When is your relief due in?"

"At 0600, sir."

"Are you going to have breakfast when the mess hall opens at six?"

"Yes, sir."

"Good. When the new NCO comes on duty, I'll catch breakfast with you, if you don't mind. After that, I'll draw my sidearm at HHC and come up to relieve the duty officer and send the new Duty NCO down for breakfast."

At 0600, the Saturday Duty NCO came through the door to the building. The Lieutenant was heartened to see that it was Staff Sergeant Aguinaldo, the tank commander of 1-3 in his old platoon. He greeted Aguinaldo warmly as he came over to the desk and signed in. He told him that he would be going to breakfast early with the outgoing Duty NCO, and that he would draw a sidearm and be back before 0630 to relieve him so he could get breakfast if he hadn't eaten already. Aguinaldo thanked him and the phone rang. It was the CQ from HHC. He asked for the Lieutenant, who took the receiver. The CQ told him that the First Sergeant had just come in, so he asked the CQ to let him know that at 0630 he would be coming down to the arms room to draw a .45 for duty.

While he was on the phone, Aguinaldo was briefed by the outgoing Duty NCO, with regard to the preceding evening. It didn't take long, so shortly, he and the Lieutenant were walking to the mess hall together. The Lieutenant told him that he wanted to have breakfast with him, at least briefly, to see if there were any unusual occurrences that night.

They walked through the doors of the mess hall, took their trays through the line, and sat together, in the room reserved for senior noncommissioned officers and officers. Over cups of coffee, the NCO told him, "It was a tough night. There are a lot of soldiers visiting from Baumholder and other Kasernes, who want to enjoy the bars, discotheques, and nightlife of Mannheim. Many of them have friends in the brigade, and in Headquarters Company in particular. The duty officer stumbled across a couple of small altercations, but he and the CQs were able to settle the troops down, since they were in a good mood, for the most part, high but not completely stoned."

"Anything else?"

"No, sir. It was really pretty quiet for a holiday weekend."

The Lieutenant thanked him for his time and finished the last bit of hash browns and creamed beef. Then he washed that down with what was left of his coffee and rose. He took the tray to the window, to be washed by one of the cooks. Looking at his watch, he saw that he was behind schedule, so he walked rapidly across the courtyard into HHC and down the stairs to the First Sergeant's office.

He walked through the door and greeted the First Sergeant. Asuncion told him that the armorer had let him know that when he had pulled duty earlier, he had asked for a shoulder holster. The First Sergeant told him, "Don't worry with that. There are so many officers in the company who pull duty that the armorer keeps web belts with .45 holsters attached, to issue to the duty officers. The web belts also have an ammo pouch for two magazines. That should make it easier."

The Lieutenant agreed and said, "Thanks. That's what I did Wednesday," and he walked quickly down to the arms room. There he drew his sidearm with a lanyard, belt, holster, and magazines. Then, having attached the lanyard to the belt and the pistol, and putting it on safe, he slipped it into the holster, and walked down the long dark hall, past orderly room to the stairs. As he was walking up the steps, he could hear the early sounds of life, echoing through the stairwells above him, as the troops were waking up.

It was a cold and cloudy day but brightening in the east. He hoped for good weather and almost ran to Battalion so that he could sign in and relieve the duty officer a little early, since he had had a rough night. When he walked in, he asked Aguinaldo, "Do you want to leave for breakfast?

"No, sir. I had a light breakfast at home before coming on duty. All I really needed was coffee and a cigarette to get going."

The Lieutenant nodded and walked down the hallway to the Adjutant's office. The outgoing duty officer was the new S-2, a second lieutenant, intelligence officer, who was quiet and aloof. When the Lieutenant relieved him, he thanked him and told him that he was glad he had shown up early as he needed to get home to his new wife. The Lieutenant told him that he understood and went over the duty log with him. He noted a few incidents on it, but for the most part it was quiet, so he wished the S-2 well as he walked out the door and settled in with his Duty NCO for, what he hoped, would be a short and quiet day.

By now, he had learned the ropes and was a veteran duty officer. He had his LBE, and steel helmet, a poncho liner, and his rubber duck with him. Plus, he had his paperback copy of *The Great Train Robbery*. It was a holiday weekend, so there was no one in the

Headquarters except his NCO and him until around 0800 when the battalion commander came in to catch up on paperwork and relax a little.

The Lieutenant saluted when he came through the door and asked if he would like any coffee. When he said, "Yes, please," the Lieutenant replied, "I'll bring down a cup to you as soon as we have a fresh pot."

When the colonel had gone, the Lieutenant turned to Sergeant Aguinaldo and asked, "Can you make coffee?"

"Yes, sir."

"Excellent. I had a very bad experience making coffee the first night I ever pulled duty in my life, and I don't want to relive it." Good to his word, in about ten minutes, Sergeant Aguinaldo had a fresh pot of coffee brewed. The Lieutenant walked down to the colonel's office and retrieved his coffee mug from his desk. Shortly thereafter, he returned with it, filled with coffee, and placed it at the colonel's right side. He then left and walked quietly down the hallway to the duty officer's desk.

The sun was breaking through the clouds and the weather was nice, so the Lieutenant told Aguinaldo that he was going to make a quick check of the battalion area while it was quiet and sunny. Since it was daylight, he kept his pistol holstered and left his large ring of keys with the NCO.

First, he walked behind the battalion area, past the personnel office and the medics' aid station, to the mess hall. Then he hurried around the Headquarters Company's barracks, near the wire, up to the gate guards. Not seeing anything unusual, he proceeded across the road, to the motor pool, where he checked in with the guard force that was made up of soldiers drawn from Charlie Company. He spoke briefly with the NCO in charge, a tank commander that he did not know, and then walked past Battalion Maintenance, where he could see that Kowalski was at work on a report to Brigade. He stopped and spoke with him briefly and then made his way back to Headquarters.

KNOWING ME, KNOWING YOU

He walked through the door into the duty office, and asked the NCO, "Did anything happen while I was away?"

"No, sir. It was all quiet. But, if it's alright with you, I would like to go down to the mess hall and see if I can find something to eat, since I have not had anything to eat since five."

The Lieutenant said, "Yes. That's fine," and took the NCO's seat at the duty desk, that looked out over the walkway, to the entrance to battalion. Once he had settled in, he picked up his book and read a few pages, before movement, to his right front caught his attention. He could see an older gentleman dressed in civilian clothes. He was small, slightly over five feet, wearing a gray overcoat and a tweed hat. For all the world, he looked like an insurance salesman, or somebody's father-in-law. As he approached the walkway, the Lieutenant looked closely at the man's face, and then quickly looked back behind him, to the chain of command photographs on the wall behind him. He took another look at the man, and then a deep breath. He realized who the gentleman was and immediately rose to open the door for him. He was the Division Commander. The Lieutenant saluted and reported. The commander briskly returned his salute and asked, "Is Bill in?"

"Yes, sir, Colonel Shea is in his office. Would you like me to take you back to him?"

"No," the general said, "I know where to find him," and made his way down the hallway.

The Lieutenant could hear the colonel greeting the general and the sound of his office door closing. Deciding that he needed to stay awake, the Lieutenant poured himself a fresh cup of coffee and waited for his NCO to return. When Aguinaldo was back, he noted his return on the log and took his place at the desk next to the window. The Lieutenant told him that the division commander was in the colonel's office, so they needed to be ready in case the colonel needed coffee or anything else.

About twenty minutes later, the general came striding softly down the hallway, wished the NCO and the Lieutenant a good weekend, and walked out of the headquarters, toward their sister battalion next door for another surprise visit. When he was gone, the colonel came down the hallway and thanked the Lieutenant for recognizing the division commander. He then added, "General Livsey is known for showing up unannounced, and visiting with the commanders of his subordinate units. As it turns out, Mrs. Livsey wanted to do some shopping in Mannheim and Heidelberg. So, while she was doing that, he took the opportunity to stop in and see his battalion commanders. Keep an eye out. When he has left the battalion next door and departed Sullivan Barracks, let me know as I need to get home to take my wife Christmas shopping this weekend as well."

The colonel headed back to his office. The Lieutenant asked the Duty NCO, "Did you hear that?"

Aguinaldo responded, "Yes, sir."

"Let me know when you see a short gentleman in a gray overcoat with a tweed hat pass by. Once he is past the first set of barracks, let me know. I'll walk over to inspect the companies briefly. When he has left the Kaserne, I'll call you from Headquarters Company to let you know so you can advise the colonel."

After twenty minutes, the general passed by. Once he was out of sight, the Lieutenant walked to the courtyard of the barracks where the line companies were garrisoned. The place was practically a ghost town. There wasn't even anyone in the S-4 office. He waited for two minutes, then he walked over to the courtyard where Combat Service Support and Headquarters Company were located.

Once the general had passed the gate guards, the Lieutenant walked into HHC, to the CQ desk. He told the CQ that he needed to borrow his phone and dialed the Duty NCO and told him that he was in HHC and was finished with his check. The NCO said, "Thank you. I'll immediately inform the colonel." The Lieutenant hung up the phone and quickly made his way back to Battalion.

Seeing the commander as he walked out of the building, he saluted. Shea saluted him back and told him to have a good night.

Once he was back in the office, he asked the NCO if he wanted to eat lunch. Aguinaldo told him, "No, I'll wait until dinner."

"I'll do that as well. If you need me, I'll be in the Adjutant's office, resting a little. Come and get me at 1600. I want to make a check of the battalion, and an early check of all of the arms rooms then."

The afternoon passed uneventfully. While the Lieutenant was unable to sleep, he was able to rest and rose quickly when the NCO came to get him. The weather was still decent, and it wasn't too cold, so the Lieutenant put on his field jacket, buckled the pistol belt around his waist, took his keys in his left hand, and walked to Charlie Company.

After greeting the CQ, he descended into the dark hallway to check the arms room. There was a light above the clipboard. Not seeing anything, the Lieutenant kept his pistol holstered, but off safe and made his way to the end of the hallway where he signed the check sheet on the clipboard. He also noted the time on the notepad that he kept in his left breast pocket. After checking Alpha and Bravo Company in a similar fashion, he made his way to CSC, where he spoke briefly with the CQ, who was a buck sergeant from the Scout Platoon. He asked how he was doing with his new lieutenant. The sergeant told him everything was just fine, and that Lieutenant McCallister was doing well. The Lieutenant was heartened by that, and said, "Good to hear that. I was a corporal in his company when he was a private, a Rat or freshman, at VMI."

Leaving CSC, he went to Headquarters Company. As he walked in, he noticed that there was a lot of activity in that barracks. There were several faces that seemed unfamiliar to him. He made his check, left and went over to the mess hall, where the cooks were busy making dinner. The Mess Sergeant was gone, but his assistant was in charge, so the Lieutenant visited with him briefly, and told him that he looked forward to eating there a little later in the day.

As he walked back to Battalion Headquarters, the wind came up and the sun began its rapid descent into the west. Since Germany was so far north, night fell quickly. He felt cold and walked briskly back to the duty office, checked in with the NCO, and sat down at the duty desk. He made entries into the log for his checks. Once he was finished, he told the sergeant that he would be back in the Adjutant's office finishing up an efficiency report.

He sat down at the Adjutant's desk and pulled out the first of two copies of the form. He wrote the draft report quickly and easily, as he had nothing but good things to say about Sergeant Robb. He stood up, walked a little bit, and then came back to review the draft with fresh eyes. He made a few corrections. Then he carefully rewrote a better draft on the second copy and placed it with the book he had been reading to take back to his quarters the next morning. Having finished that, he walked down to the duty NCO and told him that he could take dinner first.

The sergeant was apparently pretty hungry. He immediately got up, signed out, and headed for the mess hall. The Lieutenant organized the materials on the sergeant's desk so he could find them readily and sat facing the window, looking out towards the battalion Maintenance Office. When the NCO came back from dinner, he signed in on the log, and the Lieutenant signed out, taking his pistol and key ring with him. As he reached the door, he looked back over his shoulder and told the sergeant, "I'll do my first evening arms room checks right after I eat, so they will be completed before we start getting evening flash traffic."

As he walked toward the mess hall, the wind moaned through the trees and the darkness. When he got to the mess hall, he again spoke with the assistant Mess Sergeant, and asked him, "How is it going?"

"Fine, sir. – It's been quiet."

"The troops seem to be enjoying their meal."

"Yes, we're having rabbit, roasted like chicken, they like it."

"Great, I look forward to it." With that, he went to get in the chow line. Since Shea had outlawed chili-mac and hot dogs due to flashbacks from his time with the Blackhorse in Nam, the rabbit was a staple of the battalion, because it was relatively inexpensive, and the troops liked it, since it was very much like chicken. The Lieutenant thought to himself, *I haven't eaten rabbit since Ranger School, and I haven't missed it that much.* Once his tray was filled, he took it and a mug of coffee back to the corner of the room, where he ate quickly, in silence, and hoped for a quiet night.

SATURDAY NIGHT

When he was finished, he exited and crossed the courtyard to HHC. Drawing his pistol with his right hand and carrying his keys in the left, he jingled them as he made his way past the CQ, down into the basement. When he reached the end of the hallway, he signed the checklist, keeping his pistol out, but in his left hand, with the keys. When he was done, he began climbing back up the stairs to the CQ desk. He noticed that the sound was markedly louder and the stereos were starting to crank up, not only in Headquarters Company, but in the other companies in the battalion as well. Based upon that, he came to the conclusion that it was going to be a long and restless night.

He went on to make his other arms room checks in the same fashion and went back to Battalion, just in time to hear Sergeant Aguinaldo yelling, "Flash traffic," out the window. He holstered his pistol, set the ring of keys down at the duty desk, and picked up the phone to listen to the message. Once he and the duty NCO had taken down the message, and the NCO had confirmed it, he took his pistol, keys, and his copy of the message down to the S-2 office. As usual, he decoded the message once. Then took a deep breath and did it a second time before concluding that it was a test. Once that was done, he locked everything up and went back up to the duty office to tell the sergeant that it was a test message and to log it accordingly.

As he sat down at the duty desk, it occurred to him that he could feel vibration in the ground, so he rose up, and stepped outside, where he could hear raised voices and stereos from not only his battalion, but the sister battalion next door. The result was what sounded like two competing rock concerts, with his body and ears in between them. He had the Duty NCO call all of the CQs to make sure that everything was okay and let them know that if there was any trouble, they should call the duty officer, the First Sergeant, and the commander immediately; and the duty officer would get there as quickly as possible. When Aguinaldo confirmed that he had contacted all of the CQs, the Lieutenant told him that he was retiring to the Adjutant's office to rest a little before checking the battalion area at around 2100.

The Lieutenant read for a while, then sat back on the sofa to rest his eyes and catch a cat nap. When he came to, he looked at his watch and realized that it was after 2030. He went ahead and got up and went down the hall to step outside for a minute. The noise remained deafening. It was still windy and very cold, but not nearly as bitter as Baumholder, so he put on his field jacket, with the pistol belt on the outside, as well as scarf wrapped around his throat and stuffed down into the jacket. Before putting on his hat and heading out into the cold, he kept the pistol drawn in his right hand, and the keys in his left.

As he walked around behind the barracks, he noticed several soldiers standing around outside the medics' offices. Once they saw him, one of them closed and locked the door, and they all scattered in different directions, some toward the parade ground and some around the corner of Headquarters Company. By the time he got there, they were long gone, so he walked around HHC's barracks, between the wall and the wire, from the gate to the post. He stopped to say hello to the gate guards, before turning into HHC. When he stepped into the entryway, the music, voices, and what sounded like pounding on wall lockers, washed over him. Down the hall, barely visible at the end, next to the far stairwell, he swore he saw a group of three, two slender white soldiers and one black, walking up the stairs. He asked the CQ, "Do you know who they are?"

"No, sir, I don't, but earlier today I saw them in the company of some of our medics, who appear to be friends of theirs."

The Lieutenant thought for a minute and took a step toward the far end of the hallway before the CQ stopped him, and said, "Sir, there's a party on the third floor. You don't want to go up there."

The Lieutenant looked at him as he thought, *I'm not a narc*, and simply said, "Understood. If for any reason you need me, call Battalion and I'll be back."

The CQ, a staff sergeant, did not respond, other than shaking his head, so the Lieutenant stepped out into the night and went down to the motor pool. He checked in with the guard there and the ready reaction force. The motor pool guard had been drawn from CSC, so he did not know either the NCO, who was a mortar sergeant, or any of the guards. After exchanging a few words with the sergeant, he made sure to tell him that if there was any trouble to call him at Battalion, and he would be there. He then made his way to Combat Support Company and through the line companies, making his checks. Although the noise was increasing, the CQs were calm. Apparently, there was nothing out of the ordinary going on for a Saturday night in garrison.

Having confirmed that the battalion area was at least relatively quiet, he made his way back to Headquarters and sat down at the duty desk with a fresh cup of coffee to visit with Sergeant Aguinaldo a little. He asked how the platoon was doing with the new lieutenant. Aguinaldo confirmed that he was a good officer, having been an enlisted tanker before, and between him and Sergeant Robb, they were in great shape to try and qualify at the next gunnery. The Lieutenant told him, "I miss the platoon, but since you're in good hands, I'm feeling better. Just let me know if my Support Platoon can do anything for you."

He then excused himself and went back to the Adjutant's office to relax and wait for the inevitable flash message he presumed would come in sometime around midnight. He tried reading, but really didn't feel like it, so he just sat back, relaxed, and let his mind wander a little. Over the years, he had pulled quite a bit of night duty in Virginia and Kentucky. The solitude didn't bother him, but the quiet took some getting used to. Fortunately, with all the revelry in the battalion, quiet was not a problem. Having spent a lot of time in the desert, on tank gunnery ranges, sitting and doing nothing, waiting to take a tank down range, he was in the habit of thinking about how he had gotten to the most recent surreal location.

TIMES OF YOUR LIFE

The break gave him an opportunity to think about the almost endless string of happenstances that made up his life. For one, he had gone to VMI to see what military life was like, and since his mother's people were from Tennessee, to get a feel for that part of the country. He also went there, because of a black musician, who, with some of his friends from Alabama, formed a band and integrated the college in his hometown in 1968. The Lieutenant spoke with him, the leader of the band, whom he liked a lot. He was a little shorter than the Lieutenant and played tenor saxophone. He also sang lead vocals and was extremely bright. He asked the Lieutenant what in the world had possessed them to come to western Kansas, of all places, to go to school. The musician had looked at the high school junior and explained, "When you go to college, you need to go someplace that's totally different from where you were raised. Where the people eat different foods and speak differently and have different ideas. You can read books and learn things anywhere, but where you go to school and what you learn there, that's your education." The Lieutenant told the musician that although he was still young and in high school, he was thinking about going to military college in the south, to VMI. The saxophonist looked at him and said, "Then that's what you need to do. You need to go back to the last bastion of the Confederacy. Go to school there, learn from those people, and maybe change it."

Taking that to heart, he had signed the book and entered the

VMI corps of cadets in August of 1973. His first night there, after a wild Hell-for-leather roller coaster ride, up and down the stairs of the barracks, followed by a screaming workout, and a cold shower in the bowels of the barracks, an older cadet had shouted in his face over the sound of drums, that were incessantly pounding throughout the building. The older cadet asked what his SAT scores were. The Lieutenant told him, and the man immediately told him his name and room number, and added, "You're my dyke." Although he didn't realize it at the time, that meant that he would be the personal servant of the man who had pulled him out of line. The dyke system was unique to VMI. It had no parallel in the other senior military colleges or the military academies. It forged a bond between the outgoing senior class and incoming freshman class that was unbreakable.

As it turned out, his dyke, Don Kilroy, was well known and generally disliked in the Corps. A draftee and Vietnam veteran who had been a combat medic, he was famous in the Corps for refusing to accept the Bronze Star Medal with V, for valor, that was presented to him on Veteran's Day of his Rat year. That was also VMI Founders Day, the second greatest holiday, at the school, after New Market Day, May 15th, which was the anniversary of the day that the Corps of Cadets fought as a unit and were victorious in a battle ninety miles north of the school.

Over the course of his first year, Kilroy, his dyke, got him in trouble on several occasions. Including a week on confinement, a trip to the Rat Disciplinary Committee, and a week at the Regimental Sergeant Major's dyke school. On balance, however, it was well worth it because Kilroy introduced him to the man who would have the greatest impact on his life.

The first weekend of spring break, since he was from Kansas, and unable to get home, Kilroy took him to Arlington, Virginia, to stay at his house. It was good to get away from the barracks and quite memorable. Kilroy had obtained tickets to the opening night of the movie, *The Exorcist*, with a question-and-answer forum with the director after the movie. It was an awesome learning experience, and also very creepy, because the theater was located in the neighborhood where the movie was shot, so you walked out of the theater effectively onto the set, after midnight. The then eighteen-year-old Cadet did not get a very good night's sleep, even with a Joan Baez poster peering down at him from the ceiling. But he enjoyed the rest of the weekend with Kilroy and his family, who appreciated having a quasi-little brother, since he was an only child.

Kilroy was in need of some quiet time to study, and he had a lady friend in Lexington, so they drove back on Monday. The Cadet set up shop in his room on the fourth floor of the barracks floor. He ventured into town to buy groceries and most particularly some grapefruit juice because he liked it. Unfortunately, he did not have a refrigerator, so it had gone bad by Thursday. Friday morning, he woke up feeling nauseous, so he was not enthusiastic, when Kilroy showed up in his doorway at noon, and told him that he needed to follow him, to meet the head of the English Department. The Cadet, with his stomach rumbling, followed close behind Kilroy who had him stand outside the office door of Colonel Herbert Nash Dillard, without a doubt the most famous professor at VMI. Kilroy entered the office theatrically, opening the door, motioning for his rat to come in. As he did, he told the colonel, "Sir, I would like to introduce you to my dyke, the Rhodes Scholarship candidate from the Class of 1977."

Kilroy then stepped back out of the room, closing the door behind him. The colonel motioned for the rat to take a seat at a chair to the left of his desk. Unfortunately, as he sat down, the Cadet could feel that the pressure on his stomach was rising, and that he could not control himself. The colonel, quick on the uptake, simply asked, "My God rat, are you ill?"

Barely keeping his wits about him, the young cadet practically whispered, "Yes, sir. I believe I have food poisoning." Nonplussed, the colonel simply reached behind him, grabbed his trash can, and placed it across his desk, so that it landed in front of the Cadet, who at that point, he simply released everything, while trying to say, "I'm sorry, I have to throw up."

The colonel simply looked at him and said, "Gentlemen vomit." When the Cadet was through, and had composed himself, the colonel asked if he cared for a glass of water. The Cadet said that he did, so the colonel poured him a glass from a pitcher that he had on the windowsill behind his desk. After he had swished out his mouth and swallowed the water, the Cadet looked up. The colonel looked deep into his eyes and asked, "Where are you from?"

The Cadet told him, "Kansas," and he smiled faintly.

The colonel then explained to the cadet, "There is a committee of faculty members dedicated to preparing cadets to compete for the Rhodes Scholarship. I am the head of that enterprise. I will be your Rhodes Scholarship advisor if you choose to undertake the task."

The colonel then told the Cadet that the Rhodes Scholarship was the most prestigious academic award opened to undergraduates, and the cadets from VMI had been awarded it in numbers far in excess of the number you would expect, based upon the Institute's small enrollment, and even smaller number of graduates. Unfortunately, no cadet had won the scholarship since 1971, due, primarily, to the Vietnam War. With the end of the war in sight, the mission of the committee was to try and reverse that. He went on to say, "I was heartened to hear that you are from Kansas, because as far back as I know, students who won in the state of Kansas, historically, win the Rhodes Scholarship. So, if you can win in Kansas, in all probability, you will win the scholarship, and go to Oxford, to study politics, philosophy, and economics at Magdalen College."

By now, the colonel could see that the Cadet was practically green and not feeling much better. The colonel told the cadet that he would have him summoned following week to discuss things in greater detail. But in the meantime, he wanted him to consider two things. One was the mission, that in his words, "Is easy to state, and almost impossible to do. First, you will study, based on a curriculum designed by me, because you have to not only engage in but excel at every form of human endeavor. Second, based upon a recent bad experience that I suffered through with a fantastic candidate who discovered women and became sidetracked: you will have to be willing to swear off women for the next two and a half years. That's because a Rhodes Scholar cannot be married, and more importantly, has to maintain his focus, up to and through the interview process."

The colonel asked if he understood. When the Cadet answered "Yes," the colonel told him again that he would summon him the following week and dismissed him. The Cadet rose and practically ran out of the room, down the hall to a restroom, to relieve himself some more, before going back to the barracks. As a result of that meeting, for the next two and a half years, he would study under the tutelage of the most intelligent, urbane, eccentric, and exceptional man, he had met or ever would meet.

Of course, as agreed, he went on to prepare for the scholarship interview. He studied the colonel's curriculum that covered mathematics, computer science, politics, philosophy, economics, persuasion and propaganda, art, architecture, music, and literature. In addition, he took two years of Russian, plus history and art, so when he was done, he had been educated, unlike any other soldier, since Aristotle taught Alexander the Great.

SOMEBODY TO LOVE

As a result of his promise to the colonel, the Cadet broke all ties with the girls that he knew in high school at home and made no effort to pursue a relationship of any type, with any of the girls that he met in Virginia. The second semester of his third class, or sophomore year, he had to take Introduction to Music and Music Appreciation. As part of his studies, he had to listen to classical music, to be able to identify it. By the end of that semester, he had become fond of going up to that room and listening to music, especially on the weekends, to study by himself. Later, when he was a senior or first classman, after the colonel's death, but at his direction, he enrolled in the History of

Art, Introduction to Fine Arts, and Fine Arts Lab. He even spent some time drawing. By now, he had become quite lonely, and since he didn't have a female friend, on occasion, he would draw a figure on the blackboard, in the music room.

He decided that he would draw the woman that he was looking for, so drew Eve as Adam must have seen her the first time. And because God has a sense of humor, from behind. So, he drew a woman in seven strokes, with thin shoulders, a slender waist, and long flowing hair, like Botticelli's Venus. Over the years that he spent, after being commissioned, sleeping under the stars, in deserts, forests, and swamps, he had often thought of that picture and wondered if he would ever find her. Then, he smiled to himself, as he remembered that after driving nineteen hours to bring the engagement ring that he purchased in Louisville, to his now fiancée, he had woken up the next morning and saw her on the edge of the bed saying her prayers. What he saw took his breath away, because it was her, exactly as he had drawn her years before, a slip of a girl, with long auburn hair, parted on the right, and cascading down to the middle of her back. The woman he had prayed for at midnight. And he knew, beyond a doubt That they were meant to be together, forever, if he could just find a way to get back to the United States to get married.

MY LIFE

After drifting back into a light sleep, he woke with a start when he heard his NCO yelling. Rising immediately, he ran down the hall and checked his watch as he sat down at the duty desk to pick up the phone. It was almost midnight, as expected, so he transcribed the message, waiting for confirmation from his NCO. Grabbing the keys, he hurried down to the basement again. Once he had decoded it, he realized that it was another training message, so he took a deep breath and climbed slowly back up the stairs. Having low given the NCO the good news, he logged in the traffic on the duty log, and stepped out into the night, to see how things were progressing. The noise was still deafening, but not quite as loud. He took that as a good sign and ambled back into the office, where he picked up his book and began reading.

He asked Aguinaldo if he needed a break and the sergeant told him that he would like to go out and stretch his legs and smoke a couple of cigarettes. The Lieutenant quickly agreed. Aguinaldo slipped into the darkness toward the back of the building, trailed by a plume of smoke. When the sergeant came back in, he and the Lieutenant passed the time talking about their past assignments and their future plans. Sergeant Aguinaldo hoped to make Sergeant First Class by the end of his tour in Mannheim, so he could become a tank platoon sergeant when he rotated to the United States. After that, he would be pretty close to twenty years, so he might retire to Guam, unless it looked like he would make First Sergeant at some point, in which case he planned to stay in longer.

The Lieutenant told him that he had come into the Army in hope of becoming a JAG lawyer, like his neighbor from when he was in junior high and high school back in Kansas. Now, since he was transferred to Europe and involuntarily extended, it looked like he would have to put those plans on hold until he was done with this tour and decided whether or not he would stay in. Aguinaldo had brought along a book as well, so he turned to his reading, and the Lieutenant turned to his.

After finishing a few more chapters of *The Great Train Robbery*, he decided he would stay up until 0300. His experience in other units, and in civilian life for that matter, had been that if something bad was going to happen, it would usually happen between midnight and three a.m. He walked to the back corner of the duty office and poured himself a cup of coffee. It was lukewarm and stale, so he asked Aguinaldo if he would make another pot, to get them through the night. The NCO happily agreed. He, too, was of the opinion that the coffee was practically undrinkable after almost sixteen hours. When he left to make the new pot of coffee, the Lieutenant took the NCO's seat, by the window and looked out across the battalion area. He could see small numbers of soldiers moving between the barracks, many of them coming back from a night of hard partying in Mannheim. He could hear the sound of raised voices as they went to their respective units, engaged in greetings, and sometimes arguments, that they carried inside with them.

When the NCO came back with a large urn full of water and fresh coffee, he plugged it in, and soon, the percolator gushed hot water over the grounds and the smell of fresh coffee filled the room. This alone made the Lieutenant feel better, as he impatiently waited for the urn to stop bubbling, so he could help himself to a fresh cup. Since it was going to take some time, he walked the long hallway of battalion a few times to loosen his back and wake up. Coming back and settling in with a fresh cup, as he tasted the coffee, he thought to himself, *here's an NCO who can make coffee.* He always appreciated that, ever since the first time that he ever pulled staff duty at the Fourth Brigade in Kentucky, when he had misread the instructions on how to make coffee that were left with the staff duty officer. And instead of having the NCO make it and trying to make it himself, in the end, he had put in so much coffee that the next morning, the Adjutant, who was the first person to try it, and the colonel both agreed that it practically knocked them out. Consequently, an order was issued across the entire brigade that never again would a Lieutenant make coffee in that brigade. It was fine by the Lieutenant, and since then, he had always deferred to the noncoms who are more experienced in making the one drink that kept the Army running.

Having leafed through every magazine and newspaper in the headquarters by now, he finished another chapter of his book. The Lieutenant was starting to get sleepy but woke up immediately when he heard the telephone at his desk ring. The sergeant picked up the line at his desk and motioned to the Lieutenant to pick up the receiver on his desk to talk to the Headquarters Company CQ. The CQ told him that it was coming up on three in the morning and a large number of the battalion's soldiers, and soldiers who were visiting for the holiday, had come back to the barracks.

The Lieutenant could hear the sound of soldiers arguing, he suspected, about money and drugs and the combination. He told the CQ, "Call the First Sergeant to come in. If you can't reach the First Sergeant, call the Company Commander. If the First Sergeant asks you to call the Company Commander, do that as well. In the meantime, I'll come down and do my best to quiet the troops." The CQ acknowledged his call, and as he hung up the phone, it rang again. The NCO picked it up and yelled that it was a flash traffic message. The Lieutenant thought to himself, three in the morning would be the time that this would happen in conjunction with everything else. He took a deep breath, wrote down the message, checked it with the NCO, and flew down the stairs to break it. Once he had the code book on the desk, he took a deep breath and decoded the message. Then he took another deep breath and decoded it again. Happily, he realized that it was only a training message, so he locked everything up, ran upstairs, told the NCO to log it in, and told him that he would be down in Headquarters Company if he needed the Lieutenant.

The Lieutenant quickly put on his field jacket, strapped on his pistol, and snapped on his helmet. Then he grabbed his ring of keys, slipped them into the left bottom pocket of his field jacket, and double timed down to the Headquarters Company courtyard. The stereos were on full blast and he could hear soldiers yelling inside. He held up there briefly, to assess the situation, before he slipped in the front door to the CQ desk. The CQ acknowledged him and was happy to see him. Most of the troops had gone upstairs, where the hard partying was taking place. The Lieutenant stepped into the hallway a short distance, slipped his keys out, jangled them, and held his pistol in his right hand.

The troops who were milling around in the nearby stairwell began to walk upstairs. Those in the far stairwell stopped yelling at each other long enough to see him. Fortunately, as he was standing there, the First Sergeant came in. Seeing the senior NCO, the troops quieted and began to shuffle upstairs. They moved faster, as the Battalion Maintenance Master Sergeant, and S-3 Sergeant, whom

Sergeant Ascencion had called as backup, walked in. The Lieutenant greeted both of them. After a couple of seconds, the First Sergeant said, "Sir, you can go back to the battalion now." Taking that good advice, the Lieutenant walked through the battalion area, back to headquarters, where he sat down at his duty chair to update the log.

By now, the Lieutenant had resigned himself to the fact that he would just stay awake the rest of the night. He settled in with more coffee and his book at the duty desk. A little after 0430, he asked the NCO to call all of the CQs and get a status report on how they were doing. The NCO reported back, after making all of the calls, that the troops had quieted down, and the Lieutenant, listening through the doorway, could tell that the stereos were either turned off, or at least turned down significantly, to allow the troops who wanted to try and get some sleep. The rest of the night passed without incident.

He fought to stay awake. He did this by taking deep breaths and watching the second hand on the clock in the duty office revolve ever so slowly. He realized by 0500 that he needed to get up. He logged out and told the NCO he would make a check of the battalion area. Grabbing his pistol, keys, and helmet, he went slowly out the door and around behind the battalion buildings. There was no sign of life. It was even quieter as he made his way to the front gate, where he checked in with the motor pool sergeant on guard, and joked a little with a couple of the sentries before walking around behind Battalion Maintenance. By now, he was getting his second wind, looking forward to eating breakfast, and returning to his quarters to sleep. Since it was the weekend, breakfast was at 0700. At 0630, he logged out to make his arms rooms checks. During the next thirty minutes, he visited with all of the CQs, and signed the arms room check sheets, so he could get back by 0700 and relieve the NCO, who he knew must be starving.

When he logged in his return, Sergeant Aguinaldo took off immediately to the mess hall, leaving the Lieutenant alone in the building. It was Sunday morning and eerily quiet. As soon as Aguinaldo came back, the Lieutenant went down to the mess hall and ate a quick breakfast. Then he packed up his things, made corrections to any errors that he found in the log, and waited for his relief. As it turned out, the Bravo XO, Marley, was his relief. He showed up on time and the Lieutenant briefed him quickly on what had transpired that night. Since it had been relatively uneventful, it didn't take long, and the Lieutenant happily signed out as the new duty officer signed in. Aguinaldo had been relieved almost an hour earlier by a staff sergeant from Charlie Company.

With the transfer complete, the Lieutenant grabbed all of his things and practically ran to his room. He took a long, hot shower, to get the smell of cigarettes and bad coffee off him, before collapsing in his bed, to crash hard. He didn't wake up until the middle of the afternoon. After getting dressed, he headed down to the communal kitchen. There was no one there yet, so he decided to go back to his room for an early dinner, and to catch some more sleep before the new week started. Then he remembered that he needed to do something about Major Reardon. So, he hastened out the door and walked over to the PX. After looking around for several minutes, he decided on some bright white masking tape. He purchased a roll and brought it back to his room. Then he laid out the dark blue jacket of his Adidas workout gear, and taped a large M on the back, so it would be readily visible when he was standing in ranks, and even Major Reardon could pick it out.

Realizing that he was hungry, he rummaged around the refrigerator and freezer and discovered, ironically, that he still had a Swanson Turkey dinner. Figuring that he might as well finish off the week with another Thanksgiving meal, he heated it up. After eating it, he went back to bed, to sleep until the next morning. His biological clock was a little off kilter, so he woke about 0430, and decided to have a leisurely cup of instant coffee.

Chapter 21
Da Doo Ron Ron

He still had to turn in Sergeant Robb's efficiency report to the First Sergeant in Bravo Company. So, after he dressed, he jogged over to B Company and went down to the orderly room to give the report to Sergeant Guerra. He left for headquarters company well before 0600 in order to be standing in the ranks before roll call and PT. When he got there, he saw that Major Reardon was early, too. The Lieutenant posted himself in front of the major and said, "Good morning, sir." The Major acknowledged him. Then they stood in ranks together with the other staff officers, while the First Sergeant had the platoon sergeants call roll and report. Then, after Sergeant Asuncion turned the company over to Captain Boris, he led them on a fairly brisk two-mile run. The Lieutenant fell into the back of the formation with Major Reardon. In the first instance, because he did not like running in ranks, and second, to make sure that Major Reardon knew that he finished the run.

When they returned, with their breaths steaming in the cold air, Captain Boris brought the company to a halt, put the men at ease, and ordered the platoon leaders to meet him in his office. He then turned the company over to the First Sergeant, who released the men to their platoon leaders and sergeants, and returned to his desk in the orderly room. The Lieutenant followed close behind him, before making his way to Captain Boris' office. It was a space, with little room for more than half a dozen people to stand in front of the desk. When all of the platoon leaders were assembled, he announced that the entire company would spend the day cleaning the barracks, the vehicles, and the motor pool in anticipation of a visit from Colonel Means, the outgoing brigade commander, and Colonel Tate, the incoming brigade commander, whenever they appeared. In addition, the S-4 section and the Support Platoon would set up the S-4 office space for a reception for the officers to take place in the middle of the afternoon the next day.

As soon as the meeting was over, the Lieutenant jogged back to his room for a shower, a breakfast bar, and another cup of instant coffee, before returning to the company to stand in the rear at morning formation. When he got there, he spoke briefly with Sergeant McNabb and told him that they needed to visit as soon as the First Sergeant turned the platoon over to him. When that was done, he spoke briefly with McNabb, who then ordered the members of the platoon to go to the motor pool and wash their vehicles, clean them out, and do whatever maintenance needed to be done. He would catch up with him later. The Lieutenant then requested McNabb to come with him to his office, where he explained the layout of the room for the reception the next afternoon. He then asked Sergeant McNabb if he had a training program for the drivers, and what training was scheduled. The sergeant was noncommittal, and in fact did not have a training program in place. Apparently, he was operating the platoon on a day-to-day basis based upon the last-minute needs of the battalion.

The Lieutenant thought to himself that he would need to take a look at the files that he had available to him and address the training issues in greater detail with McNabb, once the more pressing matters were handled. He then ordered Sergeant Neil back to the motor pool to take charge of the troops there and get things in order, while he checked with the S-4 to see what else needed to be done. As McNabb left, the Lieutenant swore that he heard the S-4 in his office growling. After five minutes, he confirmed to himself that indeed there was a growling noise that had started out low but was growing louder in the office beside him. He knocked on the door. When asked to come in, he entered and took a seat next to Graham's desk. The Lieutenant told him, "I couldn't help but notice what sounds like growling when I was sitting at my desk, so I came in to see if there was anything I can do."

Graham replied. "Being a Mormon, I can't curse and I don't scream, but I engage in what is called growling therapy. Whenever I am overwhelmed by the incessant demands of Brigade, Battalion, and Major Reardon, in particular, I growl, out of frustration. I have found it to be an effective means of relieving stress, so you can expect to hear it from time to time."

The Lieutenant didn't say anything, but kept a poker face until he finally asked, "Is there anything I can do to help you prepare for the brigade commander's visit tomorrow?" Graham told him that since he had been made mess officer, he needed the Lieutenant, as his assistant, to coordinate with the Mess Sergeant, to make sure that the list of food items, such as sandwiches and cookies, coffee, iced tea, and lemonade were laid out in sufficient quantities and prepared in time for the reception. He also needed the Lieutenant to inspect the mess hall, both today and tomorrow, to make sure that it was ready and the cooks squared away.

Once Graham had finished, the Lieutenant rose, saluted, and took off to accomplish his various tasks. The Mess Sergeant, as usual, looked like he had just got out of bed after a bender. His eyes were bloodshot and his speech was slow. However, he had a good grasp on the S-4's needs. Also, the cooks appeared to be in good order. Leaving the mess hall, he walked quickly down to the motor pool. Once there, he realized that Sergeant McNabb was gone. The next ranking man was an E-5 in charge of the five-ton trucks. The senior GOER driver was a specialist five, but he was not a noncommissioned officer. The Lieutenant asked the buck sergeant, "Where's Sergeant McNabb?"

"I don't know, sir. He took a five-ton to go for parts for battalion maintenance." To the Lieutenant, this seemed odd, but apparently that was how the platoon functioned. He would have to investigate during the following week.

Feeling more frustrated by the minute, he trooped the line, inspecting his vehicles and drivers. The vehicles were in good shape, but the drivers would need to be cleaned up for the next day. He went to the E-5 and told him to have McNabb call him when he returned so that they could inspect the platoon and their rooms. He then told the sergeant that he would be at his office in the S-4 shop.

He wasn't really hungry, so he ate a couple of pieces of red licorice and continued reviewing the files that he had on his drivers. He also started making a training chart and roughing out a training schedule. He continued until the middle of the afternoon, when McNabb called him from the HHC Maintenance Bay, where he had his desk. The Lieutenant asked that he meet him in his office. McNabb appeared about ten minutes later. When asked where he had been, McNabb told him, "Battalion Maintenance needed some parts for wheeled vehicles. I have contacts in the long-haul Supply and Transport battalions in downtown Mannheim, so I went there to retrieve them." He went on to say, "I have to do this from time to time, to keep the vehicles up and running because the supply system doesn't work." Acknowledging that, the Lieutenant told him to retrieve the platoon from the motor pool and meet him at the CQ desk so that they could go up and inspect the barracks.

McNabb met him at the CQ desk and turned to the stairs. The Lieutenant followed the drivers and his platoon sergeant up to their rooms. This marked the first time that he had been on any of the upper floors at Headquarters Company. Like the hall below, the hallways above were painted black. The rooms were also painted black, for the most part. The walls were generally decorated with black light posters of various types. The rooms are also filled with what appeared to be old German furniture - sofas and chairs that have been picked up by the troops when German civilians threw them out on "junking" days. This was a tradition in Germany, that usually happened at the end of a month. Locals who had furniture or other items that they did not either want or need would put them out on the street corner to be picked up by anyone who needed them. This was a good way for GIs who had no money to come up with basic furniture items and also was a great assistance to the growing Turkish population that performed most of the hard labor in the country in the Mannheim region.

From the smell in the rooms, even though all of the paraphernalia was stashed and old needles thrown out, it was apparent that the barracks were essentially a drug den. In this haven, the troops, when off duty hours either shot up heroin, smoked hash, drank heavily, or used other types of drugs. Traveling from room to room, he spoke to the soldiers, introducing himself as their new platoon leader and asking them where they were from. They also talked about their prior duty assignments and basic training. He noted that the rooms were swept and clean, even though they were depressing. He reminded himself that he was not a narc. He also came to the conclusion that he needed to find more training time for them, so they had less idle time.

By the time the inspection was over, it was time for evening formation, so the drivers, McNabb and the Lieutenant filed out into the courtyard. Once the First Sergeant had announced his orders for the day, he reminded everyone that they needed to be in formation for PT first thing in the morning and that after that they would have to be prepared to be presentable with their rooms and vehicles completely clean for the expected visit of the brigade commander in the afternoon. When the company was dismissed, Lieutenant caught McNabb and told him that he would see him in the morning.

RED, RED WINE

Stepping through the door to his BOQ building, he took a minute to recalibrate his head. After gunnery, a holiday, and two nights of duty, he had to remind himself that it was in fact Monday evening. When he climbed the stairs, he could hear Bigfoot and Buzz talking to each other in between sips of red wine, in an attempt to get the taste of turkey out of their mouths. But they come to the conclusion that we needed an Italian food night, so Buzz was going to get some pizza. The Lieutenant told him that pizza sounded fantastic and fished five dollars out of his wallet as a contribution toward the meal. He then excused himself to clean up for dinner.

He felt the need for a quick shower, so he rinsed off, slipped on his jeans, and hurried back to the kitchen. Tom Storch had stopped in and was listening to Bigfoot, who had a firm handle on all the gossip. The Lieutenant joined them to get in on the conversation because Bigfoot had the best stories. When Buzz walked in with three pizza boxes, the four of them made plates and sat down to eat. The Lieutenant was not much of a red wine drinker, so he only took half a glass. He went on to tell them about his two nights of duty and the upcoming visit from the new brigade commander. He didn't know anything about the new brigade commander, just as he had very little information on the outgoing brigade commander since he had not met him, other than briefly, during gunnery.

His office was across the training area, with the infantry and cavalry soldiers, who could only be found in Benjamin Franklin Village on occasion. Apparently, they normally would go straight into downtown Mannheim to seek their entertainment there. The commander on the other hand spent most of his time at Baumholder and division, so he rarely made it to Sullivan Barracks or the Village except on the rare occasions when he could get back to sleep in his quarters.

When everyone was through eating, the Lieutenant volunteered to wash the dishes and put them away, because he had not been there for several days. Tom left as soon as he was done, since Elizabeth was working late that night. Bigfoot followed close behind. He said he had some company coming for the evening. Buzz sipped his wine, a little distracted, so the Lieutenant asked him what was on his mind. Buzz told him, "Earlier Bigfoot and I were talking about old times and our ghosts. Holidays, red wine, and this dreary weather always seem to dredge them up. It takes a while to get them under control, but I'm okay now, just waiting to see Elizabeth." The Lieutenant nodded, Buzz downed his glass and left.

When he got back to his room, he took a deep breath. The Lieutenant wrote a two-page letter to his girl to bring her up to date and tell her again how much he missed her. Then, he listened to the radio for a little bit, set his alarm for 0500, and turned in for the night. Fortunately, the other residents of the "Q" were of the same mind so it was blissfully quiet. The constant change of location and schedule had pretty much worn him out. He slept until the alarm jangled.

DANCE OF THE SABRES

The next day went as scheduled. After PT and a shower, he and McNabb inspected the troops, their vehicles, and their rooms. With the exception of a few minor corrections, everything was fine. The Lieutenant then went to the mess hall to follow up with the Mess Sergeant on the order for food items for the reception. Then he went back to the S-4's office to report the status to Graham. The Battalion Supply Sergeant had contacted McNabb, so a detail of drivers from his platoon were busy rearranging desks and setting up tables before bringing in the requisite punch bowl and coffee urns for the reception. When that was done, the Lieutenant and all of his troops reported for the noon formation, where they were brought up to date on the schedule for the afternoon. They would have lunch until 1300 and then be on standby, waiting for the brigade commander.

By three in the afternoon, it was apparent that the brigade

commander was running late. Word came down through the Adjutant to release the soldiers to their normal duties while the officers assembled in the S-4 shop to meet the new brigade commander. Within the next ten minutes, the battalion commander and his staff officers came into the shop. Shortly thereafter, a Jeep pulled up and a tall, imposing colonel stepped out. The Lieutenant recognized him immediately. He was the spitting image of John Wayne, and he had been the executive officer of the Third Armored Cavalry Regiment at Fort Bliss when the Lieutenant was stationed there. Although he did not know it at that time, from the large ring on his right hand, he could tell that he was also a VMI graduate.

Upon entering the room, the battalion commander called the officers to attention. The new colonel put them at ease and introduced himself. He then took his place in the receiving line next to Shea so that he could introduce his officers to the colonel individually. The primary staff and assistant staff officers went to the front of the line, followed by the line Company Commanders, and CSC and HHC commanders. Being the Support Platoon Leader, the Lieutenant stayed near the rear, where he could monitor the situation and intervene if there was any problem associated with the food or drink.

Finally, when the entire line was almost through, other than a couple of new second lieutenant platoon leaders, the Lieutenant stepped up. The colonel stuck his hand out to shake with a broad grin on his face, saying, "How are you doing, Tiger?" before Shea could even introduce him. The Lieutenant said simply, "Just fine, sir," after which the colonel let go of his hand and the Lieutenant moved down the line, to where the cookies and punch were located.

When everyone had been introduced, and the Lieutenant was back in the corner, Shea made a beeline for him and immediately asked, "Do you know him?"

"Yes, sir. We were in the Third ACR together."

The lieutenant colonel looked at him quizzically and added, "You were in the Cav?"

"Yes, sir. For about a year, on TDY, I worked for Charles F.

'Black Horse' Riley, before returning to Fort Knox and getting shipped here. He was Regimental XO at the time and commanded Third Squadron before I got there. He also went to VMI."

The lieutenant colonel blanched visibly, then said, "Thank you," as he went back to the side of the brigade commander. The Lieutenant smiled and thought to himself, *Stock in me and VMI grads is going up.* By now he was feeling pretty good about himself, so he took some iced tea and a cookie to eat while making small talk with the tank platoon leaders. As a general rule, they were a gregarious bunch, but given the surroundings, they were as quiet as mice.

He joked a little while with them before he saw that the S-4 was motioning for him to come over beside him, near the door to his office. Graham turned to him and said, "Well, Bob, I know you just came off of two days of duty, almost back-to-back, but the Adjutant just told me that you're slotted to be the battalion's pay officer Friday."

He looked at Graham and said, "Okay. I've been pay officer several times in the training center. How does it work here?"

"You will pay out of battalion, at the Duty NCO's desk through the window. An armed guard will bring the payroll to you, in cash, and you will need to draw a sidearm before you report. Normally payday will begin at 0900. When you're through, you'll take the endorsed checks and any remaining cash to Finance and turning it into Master Sergeant Camacho. He's the Finance NCO for this installation."

As the S-4 was finishing, the brigade commander, with a smile and a salute, walked out the door to his Jeep so that he could proceed to their sister battalion before the end of the day. Once he was gone and the other officers had left, the Lieutenant and the S-4 sergeants, together with the cooks, put the office back together.

Since it was the end of the day, the Lieutenant walked over to Headquarters Company for evening formation. He asked Sergeant McNabb how things had gone and was told there were no problems, so he told him he would see him in the morning. When the troops were dismissed, he thought about going back to his room, but decided to divert to the officers club for a real meal and a little relaxation.

LOVE IS IN THE AIR

As he walked up to the door of the club, he happily noted from a sign placed at the entrance, that it was "Steak Night." On the far side of the dining room, he also saw Petrovich, his VMI classmate, and erstwhile roommate. They made eye contact and his Brother Rat motioned for him to join him. As he walked across the room, he looked to his right and saw Lieutenant McCallister, having dinner with a young woman. Apparently, they were on a date, so the Lieutenant stopped by, just to say hello. She was introduced to him as McCallister's date for the evening.

As he remembered, she was also the daughter of the brigadier general who commanded all of the MP units in Mannheim. They made a lovely couple, so he wished them a good evening before sitting down with his friend. Once he had ordered, he excused himself and went to the bar for a bourbon and seven. He brought it back and had a nice chat with his friend regarding their classmates, who were serving in Germany and elsewhere around the world, and some of the funny things that happened to them when they were in the field. His buddy was in a particularly good mood, as he was hoping to get home for Christmas, and then rotating back to the United States after another year. After dinner, they parted company, and the Lieutenant made his way leisurely back to his room. After a shower, he penned a short letter to his girl, and relaxed before turning in. He thought, *Maybe things are settling down, and I can fall into the rhythm of this place.*

SUSPICIONS

The next morning after PT, he caught Sergeant McNabb and asked him about the schedule for the rest of the week. The sergeant told him that the wheeled vehicles in HHC were coming up for semi-annual maintenance, and that all of the drivers would be engaged in maintaining their vehicles before turning them over to HHC Maintenance for service. If a vehicle needed major repairs, it would go on deadline and move to Battalion Maintenance. This was unusual with wheeled vehicles, unless there was a problem with the transmission, or a U-joint. The Lieutenant told him that he would be in his office, working on a training program, and he would check in to the motor pool, at least a couple of times during the day to see how everyone was doing. McNabb initially told him that that would not be necessary, but the Lieutenant insisted, as he liked to see his soldiers and more, he liked them to see him. He also needed a break away from the desk. McNabb grudgingly agreed.

The balance of the day was spent trying to get maintenance done and training planned. Finally, after evening formation, thinking ahead, he decided to walk to the PX to pick up his normal stash of red licorice, breakfast bars, and a few frozen dinners. He just wanted to eat in his room alone to take a night off. Having done that, the next morning he opened his eyes wide, before the alarm went off, dressed, and jogged over to morning formation.

After PT, he asked McNabb if he could meet him in his office at 1300 to begin sketching out the training schedule for the men. McNabb nodded in assent. The Lieutenant took off for a quick breakfast in his quarters. When he was cleaned up and dressed, he went back to Battalion to be present for morning formation. When the Support Platoon headed down to the motor pool, he went to his office, opened the desk, and picked up a final packet of papers that appeared to be a major project that his predecessor was working on before he left. In it was a copy of the table of organization and equipment, including every truck in the battalion and the specifications for how much tonnage that vehicle could carry. There was also a long list of munitions that would have to be uploaded to take to the border, in the event that the battalion had to go to war. Each munition was cataloged, to include the dimensions of the box or crate in which it was packed.

The challenge was to try to figure out a way to load the maximum amount of munitions into the limited carrying capacity and placed in a way that would facilitate using them, when necessary, on the border. The biggest obstacle appeared to be the large number of anti-tank mines, because they were numerous and bulky. In the end, it looked as though each tank would have to carry several mines strapped to the sponson boxes and turrets. After about an hour, the Lieutenant came to the conclusion that he would have to give this more serious thought and then run it past Graham and Reardon in the near future.

Turning his attention to the training records for his drivers, he decided to skip lunch and prepare for his meeting with Sergeant McNabb. Shortly after noon, the telephone at his desk rang. The E-5 in charge of wheels called to tell them that McNabb had left to make a run for Battalion again down to one of the line hall units in Mannheim, so he would not be available for the meeting. The Lieutenant thanked him and asked him to call and let him know when McNabb was back. At about 1630, the Lieutenant finally received a call informing him that McNabb had just pulled in and was turning the five-ton truck that he borrowed over to the driver for maintenance.

By now it was almost time for evening formation, so the Lieutenant decided that he would confront Sergeant McNabb afterward. A little before 1700, he walked over to Headquarters Company to the rear of the formation and waited for First Sergeant Ascencion to turn the platoons over to their noncommissioned officers. When that was done and the troops were released, he approached McNabb and asked him why he had missed his meeting. McNabb told him that he had been tasked again by Battalion to go pick up some parts at the last minute because they were needed for the semi-annual wheel maintenance. There were so many trucks and jeeps in Headquarters Company that these parts runs would take up a lot of his time for the foreseeable future.

The Lieutenant took a deep breath, to quench the cavalryman's temper that was starting to rise in him, and then simply said, "Very well then. I'll meet with you after first formation tomorrow morning for about an hour. That way when you get the tasking to run for parts mid- morning, we will have already had a chance to visit."

McNabb said, "Yes, sir," but looked a little cross, back at him as he walked away. From that, the Lieutenant could tell there was something going on with McNabb that he needed to get to the bottom of. But, the first order of business was to get his soldiers trained, so he would focus on that.

The next day, after PT and morning formation, he corralled Sergeant McNabb and walked with him back to his office. They spent about an hour discussing the platoon's past performance, including the needs of the various soldiers for training, as well as SQT and PT test scores. At the end of the meeting, the Lieutenant ordered McNabb to have a training schedule for the soldiers prepared by the middle of next week so that they could review it and start executing it, when the Support Platoon vehicles were out of semi-annual maintenance. Sergeant McNabb reluctantly agreed, then stood, saluted, and hurried from the room.

As McNabb left, the Lieutenant looked at his calendar and then removed the notepad from his pocket and looked at it as well. It was only then that he realized that he had made arrangements to get his car on Friday, but since he was now pay officer that was going to be impossible. He picked up the telephone and immediately called the Autovon number. For the man that he had spoken with previously. He picked up on the first ring. And when the Lieutenant explained the situation, was happy to agree to do it the following Tuesday. The Lieutenant thanked him, hung up the phone, and went to see the S4 immediately. Graham told him that Tuesday would not be a problem. The Lieutenant then called the gentleman back and they made arrangements for him to be picked up in front of his BOQ on Tuesday morning at 0900.

Feeling pretty good about himself, he worked the rest of the morning, occasionally visiting with the NCOs in the S-4 shop. Then he went out for lunch at the PX for a change. When he got back, he settled in and worked some more on the loadout plans. When he was done with that, he asked the S-4 if there was an SOP regarding alerts that he should read. Graham told him that, in fact there was, and gave the Lieutenant a copy to take back to his desk. He read that until early afternoon. As dusk fell, everyone else left the shop. He decided he would take his time, before walking over for dinner with Buzz and Bigfoot in the communal kitchen after evening formation.

Chapter 22
Ready or Not

Just as he was closing his desk, Kowalski came running through the door at full speed. He told the Lieutenant that he was scheduled for staff duty that night but had just been called for an emergency meeting of all battalion maintenance officers at Brigade Headquarters. Apparently, the new brigade commander was completely unsatisfied with the readiness status of the vehicles in his brigade, and he and the brigade executive officer would move heaven and earth to get the brigade ready, as quickly as possible. This did not surprise the Lieutenant, as the last time that he saw the brigade commander when he was Executive Officer of the Cav at Bliss. His first words when he met the Lieutenant, who had fifteen convicts in the desert by White Sands, were, "What's going on here? It's all screwed up and you're in charge!"

The Lieutenant looked up and silently thanked his lucky stars that he was not going to the meeting. Kowalski asked if he would sit in for him at Battalion until he got back from the meeting that was supposed to last a little over an hour. Not having anything else to do, Lieutenant took a deep breath, shrugged his shoulders, and nodded as

Kowalski flew out to his Jeep. Then he walked over to Battalion Headquarters to sign in.

When he got there, the building was practically uninhabited. Staff Sergeant Fritz was on duty. Fritz was a squad leader in the Scout Platoon, who chain smoked, and drank coffee like a fish, since the cup never left his hand the whole time he was talking to the Lieutenant and bringing him up to speed. The Brigade Adjutant had called down as well, so the S-1 was gone to a meeting along with Shea and the S-3. The Lieutenant explained that he was just sitting in for the Battalion Motor Officer until he returned from his meeting at brigade and then asked if Fritz had eaten dinner. When he said that he had not, the Lieutenant told him to go ahead and take his dinner break, while the Lieutenant manned his phone and desk by the window. The Lieutenant couldn't help but see the irony in that he was going to be pay officer at that very desk in little more than half a day.

With the NCO gone, he looked out across the battalion. It was quiet, and he only hoped that it would stay that way until Kowalski got back. When Sergeant Fritz returned from dinner, the Lieutenant walked down to the mess hall, where he ate dinner in the corner by himself, as quickly as he could. When he was done, he spoke briefly with the Assistant Mess Sergeant and then walked briskly back up to headquarters in the hope of seeing Kowalski. Unfortunately, there was no one in the building but him and Fritz. Since he was only sitting in for Kowalski, he had not drawn a sidearm and did not have any of his gear with him. Lacking anything better to do, he roamed down to the Adjutant's office, to leaf through several back issues of Army Magazine, Armor Magazine, and the Army Times. It was pitch dark when he walked back into the duty office and sat down. He looked up at the clock and realized that it was going on 2100, and there was no end in sight to the meeting at Brigade.

YOU GOTTA BE SHITTING ME

On that note, having watched the second hand of his watch go around and round, for a couple of minutes, he sat back in the duty officer's chair and closed his eyes, hoping to get a little rest, since the next day he would be busy. In addition to the Motor Pool Guard, the Duty NCO and duty officer, the battalion had been tasked to provide a guard for the PSP, or Prepositioned Stock Point, where the battalion's wartime munitions and, most particularly, the new surface to air Stinger missiles were secured. There was a ten-man rifle squad, and a noncommissioned officer assigned to that task. In addition, one of the line companies had a Ready Reaction Force of ten soldiers and another NCO, with vehicles and drivers on standby with the guard.

Suddenly, the radio in battalion cackled to life. Sergeant Fritz picked up the handset and listened. The Lieutenant heard a frantic voice, "Pagan Base, this is Pagan Forward, over."

Sergeant Fritz responded, "Pagan Forward this is Pagan Base, over."

"Pagan Base, they're coming through the wire; we need help right away."

Fritz asked, "They're coming through the wire?"

"Roger. They're coming through the wire. I did perimeter in fire bases in Vietnam. Many nights we have intruders and they're probing and coming through the wire."

The Lieutenant had taken all of the conversation in. He came over to Fritz and asked, "Who is out there?"

"Sergeant Aparicio from Headquarters Maintenance and ten troops from HHC: medics, cooks, clerks, drivers and maybe some radiomen."

He told Fritz to pass him the handset and he called back, "Pagan Forward this is Pagan Base over."

"This is Pagan Forward."

"This is Pagan Base. I'm coming with the Reaction Force." He then turned to Fritz and asked, "Is there a challenge and password?"

Fritz told him that he didn't know of any. The Lieutenant then keyed the handset, and said, "This is Pagan Base, I'm coming. I will use the challenge and password of the 8th ID."

Sergeant Aparicio questioned, "The what?"

The Lieutenant replied, "The challenge is, 'you've gotta be shitting me.' Use it when I get up to you. You'll know the password when you hear it. I'm rolling."

By the time he got off of the radio, Sergeant Fritz had called the motor pool and activated the Ready Reaction Force. The Lieutenant could hear the sound of a deuce and a half truck mixed with another sound, that turned out to be a six-wheeled Gamma Goat. In a hurry, and not having any other gear, the Lieutenant slipped on his field jacket and baseball cap. He was about to go out the door when Sergeant Fritz yelled at him. He turned and Fritz clicked his .45 caliber pistol on safe and threw it like a boomerang across the room, as he said, "Sir, you may need this." The Lieutenant caught it in his right hand and slipped it in the belt at his waist. Only then did he realize that Fritz was visibly shaking and wild-eyed, experiencing a flashback from his time in the Nam.

He stepped out into the night and almost ran face first into Staff Sergeant Jackson, who was still assigned to B Company. He had ten soldiers with him, a buck sergeant with nine troops, all the rank of spec four or less, and most of them barely nineteen. Before he could get a word out, Jackson asked him, "What's the plan, sir?" Jackson had behind him one of the soldiers who was carrying two ammo cans full of M-16 cap and ball ammunition.

The Lieutenant looked Jackson in the eyes and said, "You've been here three years and you're asking me? Give me a minute and come with me." He walked to the hood of the Gamma Goat and looked at the driver. He was a mechanic from Headquarters Company Maintenance. The first thing that the Lieutenant noted was that there was no radio in the vehicle. He was supposed to have a Jeep, but there was none available, since the Battalion Maintenance

Officer had his own, and a driver as well. Unfortunately, they were both at Brigade.

He asked Jackson if he had a radio in the truck. He did not. The Lieutenant told him, "I've read the Battalion SOP for this. and I don't believe it's going to work." Then he asked Jackson, "Have any of the troops with you had any infantry training?"

"None, other than basic training, sir."

"Have they ever been on patrol?"

"No."

"Are any of them scouts?"

"No."

Based on that, he told Jackson, "Here's my plan. The first thing we're going to do is deviate from the SOP and not issue any of the ammunition in those cans. Every one of your soldiers has a full clip, right."

"Yes."

"Very well. Since we have no radios and the probability of us shooting or capturing a terrorist is essentially zero, and the probability of us shooting each other or being shot by the soldiers out there at the PSP is almost a hundred present. You will follow my Gamma Goat and we'll go as fast as we can. When we get to within two hundred meters of the wire, I'll dismount. You leave a rifleman with the drivers, put half of the men on one side and half of the other side of the road and follow me quietly. I'll go to ground, when I can make out the entrance to the PSP and crawl forward. I need you to stay down behind me, until I can advance to be recognized. Once we're recognized, I'll have you come up and we'll coordinate with Aparicio."

"My understanding is that he has Headquarters Company soldiers out there, so they're untrained as well." When the Lieutenant asked if Jackson understood, he said, "Yes, sir," and saluted the Lieutenant, who returned it. Each of them hopped in their respective vehicles, the troops loaded up, and he yelled at the Gamma Goat driver, to move out for the PSP. As he settled into the seat, the driver handed him a set of range earmuffs, to block the deafening noise from the engine of the Goat, that was right behind them. He had never been in one before, and until that time, had never really thought of how truly awful they were. They consisted of an engine, a half-ton trailer on six wheels that operated independently, with the engine in the middle, and some of the most uncomfortable seating in the world, although, at the time, that didn't matter.

The driver could not possibly hear him over the din of the engine, so he had to use improvised hand and arm signals, to relay to him where he needed to go. Since the driver had been to the PSP before, that was not so much of a problem until they neared the wire, and he had to signal for the driver to halt and move over to the right side of the road. The entire time that he had been driven through the woods at breakneck speed, the Lieutenant's one thought was, "My family have been soldiers in this country for seven hundred years and if I have to give my life tonight, please let me die with my feet on the ground as an infantryman, and not in a damn Gamma Goat." He also remembered the sage advice from his old mentor, Uncle Chuck, "Lieutenant, remember: the most important thing in combat is maintenance - the maintenance of one's cool."

Once they pulled up, the Lieutenant jumped out and pulled his pistol from his waistband so that it was in his right hand. He directed Jackson to leave one rifleman with the drivers and the vehicles. Then he signaled for the sergeant and his troops to follow him, about thirty meters behind. When he reached the last fifty meters before the entrance, he got down on the ground and crawled along the side of the road, in the trees, where he called, "Pagan Forward, Pagan Base." Sergeant Aparicio crawled to the gate, his rifle trained on him, and whispered the challenge, "You got to be shitting me." The Lieutenant responded, "Motherfucker," and despite himself, Aparicio chuckled a little, before he told him, "Advance and be recognized."

The Lieutenant crawled forward until he was visible to Sergeant Aparicio. Once the sergeant recognized him, the Lieutenant motioned for Jackson to come forward. He crawled up beside the Lieutenant and Aparicio and the three of them had a short meeting.

First, he asked Sergeant Aparicio if he had any veterans in the detail. He did not. The Lieutenant then looked at both NCOs and told them, "The probability of us getting in a firefight with each other is so high that I'm going to have to break light discipline."

He then told Jackson to have his troopers bring up the ammo cans. He took both and placed them inside the wire. Then he told the sergeants, "If anyone needs ammo, I'll personally crawl to their position and issue it. Right now, I have twenty kids out here who have never fired a shot in anger. If one round goes off, they'll probably start shooting at each other in the confusion. Now, Sergeant Jackson, I want you to crawl back, break your squad into two teams, with you in charge of one and your senior guard member, that I believe is a buck sergeant, in charge of the other. I want you to select a point man for each team. That soldier will turn on his red filter flashlight when I give the signal. Now, I want you to stay down while I go in the wire and brief the team inside."

The Lieutenant then went inside the wire with Sergeant Aparicio. There were five large bunkers, like small hills, with two soldiers on each bunker. One facing toward the road and the gate to the PSP, and the other facing out into the woods. The Lieutenant left Aparicio with the radio and crawled from hill to hill. As he crawled across each hilltop, he stopped at every individual soldier, and asked the questions that all commanders asked, "How are you doing soldier? Where are you from? Is everything alright? Now, here's the plan: I want you to watch, you will see red lights go all around the perimeter. Keep your weapon on safe. Do you know what a forty-five sounds like?"

"Yes, sir."

"No one shoots anything until you hear my forty-five. Do you understand?"

"Yes, sir."

He continued from bunker to bunker, soldier to soldier, until he got to the end. As he was climbing up the side of the last little hill, he realized that he had heard a clattering noise that he had never heard before. He had been on a lot of patrols, in forests, swamps, and mountains, but he had never heard that sound before. He crawled quickly to where the sound was coming from and found the last soldier, facing the wood line. The sound was the magazine in his rifle, rattling because he was shaking so hard.

So, the Lieutenant crawled up beside him, and asked, "How are you doing?"

"I'm scared, sir."

"What are you afraid of?"

"Sir, I'm a medic. I've never shot anybody."

The Lieutenant then said, "You know what, I've never shot anybody either. Do you know what a forty-five sounds like?"

"Yes, sir."

"You don't have to worry about shooting anybody until you hear the sound of my pistol, which means that I'll be shooting at someone for the first time. Do you understand?"

"Yes, sir. I understand."

"Now, you'll see some red lights going around in front of us. That's the guard detail from the Ready Reaction Force executing a heart-shaped reconnaissance to secure the perimeter. That's why it's imperative that we keep our weapons on safe. Oh, by the way, I didn't ask. Where are you from?"

"Outside Detroit, sir."

"Well, I'm from Kansas. And we're in this together. What I'd like you to do now is take your left hand and put it underneath the clip, so it doesn't rattle."

"Yes, sir."

"That's fine. By the way..."

"Yes, sir?"

"In the words of Scripture, 'Be still and know that I am God.'"

The medic looked over at him and said, "Thank you, sir."

The Lieutenant crawled back to Sergeant Aparicio. Once he reached him, the Lieutenant directed, "Keep an eye on the gate. I am going out to start the teams on their recon."

At the gate, the Lieutenant called for Sergeant Jackson and the other team leader. He briefed them again on what they would be doing, which was a heart-shaped reconnaissance pattern around the perimeter. When the teams met at the top of the heart in the woods, they would go to ground, and space out, at roughly twenty-meter intervals, until the Lieutenant gave them the order to fall back and regroup.

He then directed them to tell their troops that if anyone took fire, they were to immediately get flat on the ground, with cover, and that the troops on top of the bunkers would suppress any fire. The NCOs nodded that they understood, and he gave them the order to go back, have their points turn on their red-filter flashlights, and begin to execute their maneuvers. He crawled quickly back to Sergeant Aparicio to brief him. Then he climbed to the top of the center bunker, lay down flat between two riflemen, and waited for the red lights. Oddly enough, at that moment he remembered the statute of General Jackson that stood outside the VMI barracks and thought, *There stands Jackson, like a stone wall.*

As the teams moved around the perimeter, he watched the red-filter flashlight lenses as they moved in two arcs. When they met, and encountered no resistance, the red lights went out and he knew that his recon teams had gone to ground just outside the wire. With the area secure, he crawled down off the bunker to Sergeant Aparicio and had him radio battalion that the PSP was secure and that he and the detail would stay on site for a little while longer. Once they had left, Aparicio was to inform headquarters that they were on the way. With that, he ordered Aparicio to go to the tops of the PSP bunkers and tell the troops that he was going to go outside to walk the wire and check on his guards.

As he did that, he crept from soldier to soldier and whispered to them, asking their names, where they were from, and how they were doing, from troop to troop. Once he had spoken with all of them and the NCOs, he told them that he was going to scout out farther into the wood line, to make sure the area was secure. If there was any fire, he would get to ground and they could concentrate on returning it. Using the woods for cover, he slipped from tree to tree, to get to the edge of the wood line. Looking out across several hundred meters of open space, he saw a German apartment building, at least ten stories, with a glass facade.

Then he realized two things. First, what they were guarding in the PSP, for the most part, because they were the most dangerous, were Stinger surface to air missiles, and they were less than an hour from Rhein Main Airfield and Frankfurt International Airport. If terrorists had those weapons in the mountains surrounding Frankfurt, they could wreak havoc and hold the country hostage.

Second, he realized was that if his soldiers had begun firing, those M-16 rifles would shred the windows of the German apartment building. That would cause a horrific international incident and possibly end up with the battalion losing its training area.

EVERY KINDA PEOPLE

It was only then that he realized it had started to rain, light and cold. He crawled back to his noncommissioned officers outside the wire and ordered them to move to the rear of their respective teams and lead them back to the entry point to the PSP. He followed behind Sergeant Jackson's team to get to the gate and reenter there. As he was making the turn toward the road, he realized there were figures walking through the forest. They were large men, armed with rifles, that looked like they were M-1s leftover from WWII, or maybe M-14s. They were wearing what would pass for American uniforms, from ten to fifteen years ago.

The Lieutenant immediately gave the order for all of his soldiers to halt and get down. Then he crawled forward to Sergeant Aparicio to ask him what the hell was going on. Aparicio told him that it was probably the Polish Labor Service, who had heard of the alert and were coming over to check out the PSP.

The Lieutenant asked him, "What is the Polish Labor Service?"

"Sir, the Polish Labor Service consists of Polish soldiers who fought against the Germans in WWII but couldn't repatriate to Poland because it was behind the Iron Curtain. So, the German government and American government gave them jobs guarding ammunition points. They live in barracks out in the woods, pretty much by themselves."

The Lieutenant looked up, toward the nearest and largest soldier. After slipping his pistol into the waist of his fatigues, the Lieutenant raised his hands in the air and walked toward the large soldier.

The Lieutenant asked the soldier if he could speak English. He did, so the Lieutenant assured him that the site was secure and there was no reason for alarm. The big Pole shrugged his shoulders, and said, "Yes."

"You can go ahead and return your troops to barracks."

The Pole answered, "I will in a little bit," and turned back to his soldiers.

At that point, the Lieutenant realized two more things. It was raining harder and the adrenaline that had been rushing through his body was starting to subside. He was terribly tired and working hard to think straight. As the Poles receded back into the forest, he took a deep breath, recovered his pistol, and was preparing to return to Sergeant Aparicio to gather up his soldiers and get back to battalion, since by now it was after midnight. Then, as he walked down the road, a German Jeep came screaming down the road, past the truck and the Gamma Goat, to the gate and slid to a halt at the Lieutenant's feet.

As he stood there, with his .45 in his hand, a German policeman jumped out of the Jeep, with his 9-mm auto pistol, off safe, waving it around with a wild expression in his eyes.

The Lieutenant looked at him as calmly as possible, raised his hands, and told him in broken German, "Alles ist klar. Alles ist klar."

Unfortunately, this did not have the desired effect on the policeman. He seemed to be even more frantic. The Lieutenant couldn't think of anything else to do, so he yelled to Sergeant Jackson and

Sergeant Aparicio to see if either of them had a soldier who spoke German. Aparicio had a mechanic who was married to a German woman. He came forward to stand beside the Lieutenant to interpret for him.

Through his newfound interpreter, the Lieutenant told the policeman that everything was alright. They had not located any intruders and that the site was secure. The German policeman then relaxed and told the Lieutenant through the interpreter that when he was saying "klar," the policeman thought he was referring to Christian Klar, who was a well-known Red Army Faction terrorist operating in their area and had just been involved in a bank robbery in Zurich earlier in the month. With that, the Lieutenant breathed a sigh of relief. Then he held out his hand to shake the German policeman's hand and thank him for responding.

With that done, the policeman put his pistol on safe, jumped back in his Jeep, and headed back to Mannheim. Now the rain was coming in sheets. The Lieutenant told Aparicio to give him a ten-minute head start to load up the troops and get them out of the general area before he notified battalion that they were returning, so they could try to beat any possible ambush if someone was listening.

Having done that, he thanked Aparicio before he went back to Jackson with the two cans of rifle ammo and had him load them and all of his soldiers back onto the deuce and a half and turn around. He ran back to his Gamma Goat, where the driver had been trying to rest, while observing the road. With the Goat finally pulled in front of the truck, he crammed the earmuffs on and they all headed back for Battalion. After they pulled up in front of the building, he had everyone dismount. They took roll and also made sure that the seals on the ammo cans were still intact, so that they didn't have to inventory the ammunition. He released Jackson and the guard back to the motor pool, cleared his pistol, and walked into headquarters, soaking wet.

I THINK IT'S GONNA RAIN TODAY

As the door closed behind him, the Lieutenant ejected the clip out of the pistol and reinserted the cartridge that he had ejected back into it, before entering the duty office, and handing them both to Sergeant Fritz, who had settled down. Kowalski was sitting in the corner, waiting for his return. He offered to take over the duty, but the Lieutenant told him, "No. It's after midnight, and I'll be busy for at least the next three hours, writing the after-action report and briefing the old man. So please, go home and you can just owe me a night the next time I come up on the roster." Kowalski quickly agreed and bolted out the door.

The Lieutenant took off his rain-soaked field jacket, hung it up, dripping, near the door. He then went into the battalion restroom, washed his face, and dried off, before walking down the hallway to see the battalion commander. After walking into the colonel's office, he came to attention and saluted three feet from the desk. Shea looked up and asked him to have a seat in the chair to his right. Shea listened intently, as the Lieutenant recounted for him all of the events of the evening, and what he believed may have been the true intent of whoever was on the wire that night: someone was trying to get American soldiers engaged in a fire fight amongst themselves, and if possible, fire on a German apartment building full of civilians. The other possibility, of course, was that it was an attempt to steal Stinger missiles, to bring down some airliners, and pressure the governments of the United States and West Germany for some sort of concessions.

When he was finished with the initial update, Shea asked him for more detail with regard to the actions that he took. First, the Lieutenant explained that due to the fact that he had untrained soldiers, both at the PSP and in the reaction force, he had elected not to issue additional ammunition for fear that he would have twenty untrained troops firing at each other, in the middle of the night, in the rain. He also had chosen to break light discipline so that he could maintain control of the converging teams who swept the perimeter and keep the soldiers who were on guard on top of the bunkers from spooking when they heard noises and firing on their own.

Additionally, the biggest impediment the Lieutenant faced was the lack of a second radio to travel with the reaction force. Finally, in the event of a full-blown reaction like tonight, someone needed to coordinate with the Polish Labor Service and the German police, to preclude one or more of those entities engaging in friendly fire with the others. They were lucky that none of his soldiers fired on the Poles, and that the German police did not roll in and fire on anybody.

The Lieutenant then apologized to the colonel for not having his field gear, sidearm, and steel pot with him, but added that he had come up to Battalion to stand in for Lieutenant Kowalski, who had been called to a brigade maintenance meeting that had apparently lasted longer than anyone expected. The Lieutenant then further explained, "I spoke with Lieutenant Kowalski and I'll stay on for the remainder of the night. When I'm relieved in the morning, I'll be coming back because I'm scheduled for duty as battalion pay officer."

At that, the colonel smiled and said, "When it rains it pours, doesn't it, Lieutenant?"

The Lieutenant returned the smile, "Yes, sir." Shea then told him to write up a separate document as an after-action review, with a complete listing, in detail, of what occurred, together with his recommendations. This document was to be provided to the colonel solely. Then he was to update the duty log, with the times and actions taken, in broader terms.

The Lieutenant thought to himself for a minute and added that he would have it for him within three hours, around 0500. The colonel nodded assent and then told the Lieutenant that he would be going back to his quarters to clean up and should be back around 0400. With that comment, the Lieutenant rose, saluted, and went back to the duty office.

Before he sat down, he thought for a minute, then he updated the duty log, and asked Sergeant Fritz if he could borrow his pistol. He hadn't made any arms rooms checks, so he decided that he would make a quick check of the battalion, and the arms room while he was gathering his thoughts for the report. It was still raining hard. He put on his wet field jacket and headgear, and made the rounds of the arms rooms, checking them with his keys in one hand and his pistol in the other.

By the time he had returned to the battalion, he was thoroughly soaked again and took off the dripping field jacket to hang it up again. Then he sat down to write a continuous report of exactly what had happened that night. At the end, he inserted the section on lessons learned and recommendations. By the time he finished, it was after 0400.

Shea came in a little after that and went to his office. The Lieutenant proofed the report a final time, reading it backwards with a ruler, like he had been taught by one of his previous Company Commanders. When he was satisfied, he walked down the hall and knocked on the commander's office door. When asked to come in, he did so and saluted. Shea again had him take a seat, and the Lieutenant handed him the report as he did so.

After reading the report, the colonel looked up and congratulated him on doing a fine job. He then set the report aside and said, "I need to talk to you."

The Lieutenant looked up, thinking he had done something wrong and queried, "Yes, sir?"

"With the new brigade commander coming on, and a new division commander expected in the next six months or so, the training and gunnery schedules have been changed. Instead of going to gunnery in April of next year, we will be going in March, so I'm going to have to cancel your leave."

The Lieutenant looked at him quizzically, before he added, "But sir, I'm going to get married."

"I know," said the colonel, "and that's why I'm telling you now, just as soon as I found out."

The Lieutenant's head was now literally spinning as he tried to come to grips with what he'd just been told. In fact, he had difficulty processing the colonel's next statement: "You, the other bachelor officers, and the married officers and their wives are invited to my quarters for Christmas dinner, beginning at 1700 tomorrow, Saturday."

The Lieutenant acknowledged this last statement, rose, and saluted. The colonel dismissed him and he made his way slowly back to the duty office to sit down and try desperately to gather his thoughts. He knew that he needed a plan. But he was really too tired to make a plan. Especially since he was going on duty as pay officer that morning.

When he got back to the office, he checked in again with Sergeant Fritz and told him that he needed to borrow his pistol again, as he was going to walk the battalion area, to stretch his legs. He also needed some time to think about how to deal with his current problem.

Fortunately, the rain was lighter but combined with a heavy mist. He walked behind the barracks, past the personnel offices, aid station, and mess hall, before circling around the building to the motor pool and then back to make all the arms rooms checks. By the time he was through, he was exhausted. And wet again. When he got back to Battalion, he sat down. He told Sergeant Fritz that he could have the first break for breakfast, and as soon as he returned, the Lieutenant would get something to eat since he was starving.

At 0600, Fritz left his duty pistol on his desk and the Lieutenant took his seat. When the NCO was walking out the door, he filled up a mug of coffee and sat down. Looking at his watch, he thought if he were relieved by 0700, he could get cleaned up and changed and pack some breakfast bars and licorice for the day before coming back to Battalion, to work as pay officer. The problem was that the bank would not be open until he had started paying, so he couldn't get to a telephone to call his girl. Then he also thought, *I'll see what I can do to come up with an alternate plan before calling her on Monday.*

Chapter 23
Money, Money, Money

Having made those decisions, he tried resting his eyes for a bit until Sergeant Fritz came back through the door. As soon as he came in, the Lieutenant rose slowly, stretched to wake up, and headed for the mess hall. As he was walking, he thought about some advice given to him by a grey-haired lieutenant colonel when he was training in North Carolina. The colonel had told him that when you went on patrol, if you didn't sleep, you needed to eat more in order to stay awake and give your body the energy it needed. Looking back, the Lieutenant came to the conclusion that it was good advice, so he ate the largest breakfast he had eaten in a long time, before walking to Battalion and signing out with the Adjutant, who reminded him that he had duty as pay officer so he needed to be back by 0830 to sign for the money. He would also need to draw a sidearm.

The Lieutenant saluted the Adjutant and made his way as quick as he could back to his room. He cleaned up and got squared away before going to HHC, stopping by the First Sergeant's office to alert the armorer to the fact that he was going to need a duty pistol. He walked down to the arms with the armorer, where he drew a pistol belt with holster, a .45, two extra clips, and a lanyard. Luckily, the rain had stopped and the sun was starting to peak through the clouds, so he just carried in his still-damp field jacket, and set it down behind him, when he entered the duty office to sign in and wait for the finance NCO and guards to come by to drop off the cash for payroll.

At 0845, a Jeep pulled up and a master sergeant got out. Both he and the driver were armed. Another armed guard sat in the back. The master sergeant, Camacho, picked up a large metal box that was locked. Coming into the duty office, he set the box down on the Lieutenant's desk, unlocked the box, and had the Lieutenant count the money. It was almost one quarter of a million dollars, mostly in twenty-dollar bills, along with a few tens, fives, and ones.

When he was done counting the money, he signed for it. Camacho also had with him a stack of paychecks that were held together by rubber bands and a piece of paper with the battalion's numerical designation on it, "3-68 Armor." In addition to the checks, the Lieutenant received a payroll roster bearing the name and company of each soldier that he would be paying. The roster and the checks were organized in alphabetical order. Camacho asked the Lieutenant if he had ever been pay officer before. The Lieutenant told him that he had on several occasions at Fort Knox, so the sergeant had him sign for the checks as well, after he had compared them to the roster. Then Camacho turned and left to get to the next unit.

As the master sergeant walked out the door, the Lieutenant rose to get a fresh cup of coffee. He looked at his watch and at 0900 slid open the window in front of the duty NCO's desk, where the roster, checks, and money were laid out. Each check was printed with the payment amount, the soldier's name, and a line at the bottom to be signed as an endorsement by that soldier. Whenever he had served as pay officer before at Fort Knox, he had been instructed that the soldier was to sign on the line at the bottom of the check, so that the endorsement and the amount could be checked easily, and unlike civilian checks that were signed on the back, so that the finance officers did not have to continuously flip the checks forward and backward to audit them.

Five minutes later, soldiers from the battalion began to line up outside the window. The Lieutenant kept his pistol in the holster at his waist and had one soldier come up to the window with the others remaining at least three paces back. When a soldier came up, the Lieutenant asked for his name and identification card. He then compared the card to the list and found the appropriate check. He directed each soldier to sign the front of the check. Troops that had been in Germany longer continued to sign the back as well. He thought this was strange, since the endorsement line was printed on the front, and the finance code was supposed to be uniform across the Army. In any event, once he had the signed check, he counted out the money to the soldier and had the soldier count the money back to him. Then he gave the cash to the trooper, who departed, so the Lieutenant could serve the next in line.

The Duty NCO, who had come on earlier in the morning, took up his position in the duty officer's chair behind the Lieutenant, where he answered the phone and made the checks required of him during the day. As noon approached, he asked the Lieutenant if he was going to take a break for lunch. The Lieutenant said, "No, I brought some food with me, since I've been up all night. I prefer to go ahead and get this duty done and cash out with the finance office so I can go to my quarters and get some sleep."

"Can I go ahead and take lunch?"

"That's fine," the Lieutenant answered, as he had several other people there at the headquarters and almost all of the checks had been cashed.

By the time, the NCO was back from lunch, the Lieutenant only had a handful of checks left. The Lieutenant poured himself a fresh cup of coffee and sat down to rest his eyes a little while he had the sergeant call the missing soldiers' units, to see if they could come up and get paid, or, if they were on duty, away from garrison, or on leave, could get paid at the Finance Office on another day.

Having made all those calls, the NCO told him, "Sir, we have three or four more troops coming up. After that, there shouldn't be any more." By 1430, the Lieutenant had paid four more soldiers, and based upon the NCO's calls, the others were either away from garrison or on leave. So, he slipped one rubber band over the checks that had been cashed and endorsed, another over those that had not been cashed, and placed both sets and the remaining cash in the metal box. He locked it up. Then, having put the key to the lock in his pocket, he checked to make sure that his pistol was on safe, rose, and grabbed the box. The weather was nice, so he hung his wet field jacket up by the door to continue drying.

He walked down the hall to let the Adjutant know he was going to Finance to turn in the checks and remaining cash. He signed out on the log. When he was dismissed, he hurried down the hall, out the front gate and up the street to the Finance Office. When he arrived there, he asked for a Master Sergeant Camacho. Camacho ushered him into his office and quickly counted the cash and added up the money due for the checks that had not been cashed. Since that balanced, he then added up the money on the checks that had been cashed and endorsed. Since those figures had balanced, he looked at the endorsed checks. Then, to the Lieutenant's horror, Camacho told him that he could not accept any checks that were not signed on the reverse.

The Lieutenant was dumbfounded. He had maybe twenty checks that were signed properly on the front but had not been signed on the back. When he protested to the master sergeant, Camacho simply told him that, "USAREUR was a whole 'nother world. The finance regulations in place in the continental United States simply don't apply in Europe. You have to find each and every soldier who only signed the front of the check and have him endorse the back of the check before Finance will accept them. And, if you can't get that done, any shortage will be considered stolen or lost cash, for which you, Lieutenant, will have to reimburse the government."

The Lieutenant took a deep breath and thought to himself, *It's just like his colonel told him earlier that day, 'When it rains, it pours.'* He snatched up the roster and the checks that he had to get endorsed on the back and double-timed back to his battalion to try and catch those soldiers before they left garrison to spend their money downtown, or took off for the weekend, since it was a Friday.

Fighting back his sleepiness, he went from company to company, checking in with the First Sergeant, to see if the soldier he was looking for was assigned there and have that trooper report to the First Sergeant's office, to sign the back of the check. In some instances, the Lieutenant had to go down to the motor pool, battalion maintenance, or other offices to find the soldiers and get them signed.

It was coming up on 1700, and he had to get back to turn the money into Finance, so he literally ran from hole to hole until he had all of the checks signed. Looking at his watch, he realized to his horror he had about twenty minutes to go, so he double-timed into the Finance Office, darted directly back to Master Sergeant Camacho's office, and laid the roster and the checks on his desk. Camacho smiled, as he audited the checks against the roster and the amount of money that the Lieutenant had signed for. Everything balanced, so he smiled again as he thanked the Lieutenant for handling the duty. As he left the room, Camacho remarked, "I imagine you won't make that mistake again, right, sir?"

The Lieutenant said, "Absolutely!" and shot out the door to get to the arms room to turn in his duty pistol so that he could finally go to his quarters.

When he got back to Headquarters Company, it was time for evening formation. He waited until the First Sergeant released the soldiers to their platoon sergeants before he walked to the side of the formation. He told Ascencion that he needed to turn his pistol in to the arms room so he could finally go off duty. The First Sergeant looked at him quizzically and told him, "I expected you to be able to turn in your .45 earlier." When the Lieutenant explained, he shook his head in amazement at the things that lieutenants do before he called the armorer over and ordered him to lead the Lieutenant down to the arms room so he could return his pistol.

Once that was done, the Lieutenant took a deep breath. He was completely exhausted, but he thought he should probably check in with the S-4 to make sure that he knew what he had been up to and make sure that he could go home for the day. He caught the S-4 just as he was coming out of his office and after explaining what had happened to him, Graham told him to go directly back to his quarters and get some sleep. He also reminded him to be sure he was at Shea's quarters early for the annual Christmas Party, because you did not want to keep his wife waiting. The Lieutenant responded, "Certainly."

He walked out the door to head back to his quarters as quickly as he possibly could. Then he remembered his soggy field jacket in the duty office, and he slogged back to Battalion and picked it up. It was still practically soaking wet, so he carried it under one arm and made it back to his quarters. He felt as wet and whipped as the jacket looked.

HEARTACHE TONIGHT

When he walked up the stairs to his building, he ran into Captain Boris, who was taking his little, long-haired Dachshund out for a walk. Boris looked at him and told him that he had heard what happened and imagined that he needed some sleep. The Lieutenant said, "Absolutely," as he made his way up to the second floor.

He passed the empty communal kitchen and walked down the hall to his room. Having finally slipped off his wet clothes, he piled them up in the corner. When that was done, he began to think again about what he was going to do with regard to his wedding day. Once he had his shower, he thought some more, and decided that he really needed to sleep, and put together some sort of plan the next morning but headed for the kitchen instead of to bed.

Since there was no one in the kitchen when he came in, he

assumed that Buzz had taken the night off and perhaps was out for dinner with Elizabeth. The Lieutenant realized he was simply too tired to eat anyway, so he wolfed down a couple of breakfast bars and drank some of the brackish Mannheim tap water, before slipping into a T-shirt and shorts and diving under his poncho liner to try and get some sleep. His only hope at that point was that there wouldn't be too much noise from either next door or above him. He shouldn't have worried. He passed out cold and slept until the middle of the next morning.

When he came to, he looked at his stainless-steel Caravel watch and realized with amazement that it was after 1000. He thought to himself that he needed to get some things tended to. So, he took another quick shower, shaved, drank a cup of instant coffee, ate a breakfast bar, dressed in civilian clothes, and shot out the door, with his dirty laundry and wet field jacket in a bag. Once he turned those items into the laundry, he considered calling his girl but then decided he would deal with that on Monday. So, he went to the bank and picked up some marks.

WHEN THE GAME GOES AGAINST YOU

He went back to his quarters. sat down in one of the black easy chairs in his living room and thought seriously about his situation. First, he was outgunned. However, being a cavalry lieutenant, he was going to need to use indirect fire. Second, he thought back to the words of one of his favorite authors, George McDonald Fraser: "When the game goes against you, keep calm, and cheat." He remembered having seen a German flower shop behind the Finance Office, so the first thing he did was take some marks and go over and buy some flowers to take to the colonel's wife at the party. He then decided that since she, and at least some of the other wives would be busy getting things ready, he would appear early and try to be helpful.

After that, the best he could do was play the hand he was dealt with and hope for the best. He didn't have a vase, so he borrowed a large dish from the kitchen, and filled it with water. He removed the decorative paper from the flowers and stood them up in it, on his small dinner table. He then decided that he would go ahead and eat lunch at the snack bar.

He ate a cheeseburger and passed on the Coke so that he could go back for a nap. Having set his alarm for three in the afternoon, he passed out cold again until he heard the mechanical clatter next to his right ear. Having shot up out of bed, he took another shower and dressed carefully in his best civilian clothes. Then he rewrapped the flowers and by about 1615, was walking out the door to Shea's quarters. Field grade officers below the grade of colonel in Mannheim shared duplex quarters, in a separate small housing area, back behind the company grade officers' quarters. It was actually a pretty nice day: a little blustery, but fairly warm, so the Lieutenant only needed a light jacket over a sweater, as he walked past Captain Coyle's old quarters and made his way back through the woods where his colonel and the other battalion commanders lived.

He quickly located the address from the number on the outside of the building. He walked up the driveway and then across to the sidewalk to the front door. When he rang the doorbell, the colonel's wife, Jeanie, answered and opened it to him. As soon as he met her, he liked her immediately. She took the flowers and thanked him, as she walked back toward the kitchen to put them in an appropriate vase. As he followed her, he was pleased to see that Dana, Graham's wife, was there early, and a couple of other ladies whom he did not know.

He knew that the S-3 was a bachelor and the S-2 was waiting for his wife to come in from the United States, so he assumed that at least one of the other ladies was the Adjutant's wife.

SMOKE FROM A DISTANT FIRE

When he looked around the quarters, he realized that there was a fairly large dining room next to the kitchen, and a living room, that was filled with card tables and chairs to accommodate the fairly large number of officers and their wives who would be attending that evening. He noticed that the ladies were still working, so the Lieutenant offered to help. Jeanie quickly rejected that offer. She asked him to join her and the ladies at the dining room table where they were filling out cards for the attendees so that the officers could locate where they would be seated, and she would know who they were as well.

While they were doing that, she turned to Lieutenant and said, "I understand that you're getting married soon."

"That was the plan. And in fact, I was supposed to be married on the eighth of March, but yesterday, while I was on duty, the colonel told me that my leave would have to be cancelled and I will have to rearrange my marriage because gunnery had been moved from April to March."

When Jeanie heard this, she looked intently at him, very much like her husband, and a couple of the younger wives practically gasped.

Now, seeing that this might be a good time to call for fire, he added, "I haven't told my fiancée yet, and I'm not looking forward to it, especially since her uncle is performing the ceremony and he is the Monsignor of the Cathedral in Kansas City, Missouri. He had to alter his schedule to appear for the wedding in Abilene, Kansas, and that my fiancée had to get special dispensation to be married during Lent. On top of that, her uncle was badly injured in an accident. He'll have to come out late on the Friday before and leave pretty early in the afternoon of the wedding to get back to Kansas City to perform his duties in his parish." Having walked in the door and seen some of the portraits and crucifixes on the wall, the Lieutenant was betting that Shea and his wife were Catholic. So, the Lieutenant thought, *I'm calling for eight inch*, and added, "I am marrying into the true faith, even though I'm a Methodist myself."

By now, the Lieutenant could tell that Jeanie was smoldering, so he just smiled and told her, "I'll just have to make the best of it."

She looked at her watch at that point and excused herself to take a place looking out glass of the front door toward the drive. Once she saw what she was looking for, she stepped out into the driveway. The Lieutenant excused himself to go into the kitchen for a glass of water. From that position, he could see out into the driveway and couldn't resist watching what happened next. Shea drove up and got out of his car. Then as he approached the edge of the drive, he could hear his wife, speaking firmly but loud enough, "Bill Shea!"

The Lieutenant thought, *Shot over.*

The colonel looked up quizzically and said, "What?"

The Lieutenant thought, *Shot out.*

Jeanie added quickly, "How dare you?"

The Lieutenant thought, *Splash over.*

The colonel looked up and asked, "How dare I what?"

The Lieutenant thought, *Splash out.*

FIRE FOR EFFECT

"How dare you cancel that lieutenant's leave?"

The Lieutenant thought, *Fire for effect.*

"He has to go get married. Have you forgotten what it was like? Don't you remember? You diverted to come to Hawaii and marry me and go on honeymoon for two weeks; and you were on your way to an actual war!"

After that, despite the colonel's best efforts to explain, Jeanie was having none of it and almost marched him back into their quarters. As they walked through the door, Shea saw the Lieutenant standing near the edge of the kitchen entrance. He slipped quietly over to him, "Lieutenant, I've reconsidered your leave request. You can leave March first, to get back to the States and get married, with a week to get back to Germany and a week here to get settled in. But when we go to Baumholder, you'll be ready and on the road march that day. Do you understand?"

The Lieutenant answered immediately, "Yes, sir," and excused himself out of the kitchen to find his place at one of the card tables, as far away from Shea, as he could get. He found his name and happily seated himself with Lieutenant Matthews, his old buddy from Bravo Company, and Captain Boris, from HHC.

Having accomplished his mission, the Lieutenant happily got in line with the other officers and their wives. He helped himself to roast beef, vegetables, and salad, as well as a holiday sheet cake. The food was wonderful. Jeanie Shea, like many an Army wife, had developed over the years a recipe book and an innate sense of timing that allowed her to feed any number of people, from three to a hundred on almost no notice.

As the evening progressed, he had a pretty good time, joking and making small talk with his dinner companions. After dessert, he rose and went over to the colonel's wife to thank her for a lovely evening. The colonel, standing beside her, cut him a look that could have welded steel. The colonel was Black Irish and well known for having an extremely hot temper. The indicator that he was upset was that his right foot would begin to stamp if he was angry. The Lieutenant looked over at the foot to make sure that it wasn't stamping and realized that Shea was under control because his wife was standing next to him. It also meant, however, that the colonel had marked him up for payment later, and he could hardly wait.

On that happy note, the Lieutenant beat a hasty retreat back to his quarters. It was still fairly early in the evening on a Saturday night, so he stopped by the communal kitchen to see if anybody was up. Not finding anyone, he helped himself to a beer from the communal refrigerator, before walking back to his room. Taking a sip, he settled into the closest easy chair and listened to the radio for a while. Although he had staved off the wedding date disaster, it looked like he was going to incur the wrath of the battalion commander, and at some point, it would surely be visited on him, through the battalion XO, since he worked for the S-4. However, he still had a smile on his face, because no matter what, he had not wanted to make that call to tell his girl that they would have to wait longer to get married. As he sat and drank a little more, his spirits rose. He retrieved his copy of *The Great Train Robbery*, and read a couple of chapters before turning in.

LOVE WILL FIND A WAY

After the past harrowing week, he woke up late Sunday morning. It was almost 1000. The first thing he realized was that he could swear he smelled fresh coffee and bacon. After he rinsed off his face and slipped into some clothes, he made his way to the kitchen, where he was more than happy to see Buzz hard at work, making pancakes and bacon. The Lieutenant said, "Good morning," and slipped in behind him to get to the cupboard and retrieve a coffee mug.

After having a sip of fresh coffee, he sat down in the corner and asked Buzz, "What inspired him you to get up and make breakfast, since you usually sleep in on Sunday?" Buzz just smiled, as Elizabeth came in, dressed in a long T-shirt, shorts, and sandals. Apparently, she was hungry, and as a result, Buzz had slipped over to the PX to buy pancake mix, bacon, and syrup. The Lieutenant greeted Elizabeth and asked how she was doing. Since the holidays were approaching, she was all smiles and told him, "I'm literally counting the days before I leave for England and my family."

As she sat down beside him, Buzz announced that breakfast was ready, so they stood and made plates, before retaking their positions. Buzz finished making the last of the pancakes and put them on a platter, before pouring another cup of coffee and sitting down next to Elizabeth. Then the Lieutenant, since he had completely forgotten, asked her when she was going back.

"I'll be departing the afternoon of December twenty-first, since exams will be over and the Army will be going on a half-day schedule. The school holiday also starts then, so I look forward to being home, for at least two, maybe even three weeks." Buzz did not mention, and the Lieutenant did not ask if he was going back home for Christmas. In fact, from the tenor of the conversation, it seemed more likely that the colonel would catch a military hop up to an Air Force base in England and find time to connect with Elizabeth there.

The Lieutenant told his tale of how he had narrowly escaped having his wedding leave cancelled, so that by the grace of God, and the battalion commander's wife, he would in fact be going back to the States on March first to get married on the eighth. They both congratulated him. Then he told Elizabeth that he would have to take her cue and start a short time calendar, looking forward to his departure from Germany. Buzz then asked him, "Do you have any plans for the rest of the day?"

"No, I really just needed to get some rest. After back-to-back days of duty, between having to go on alert, and then hunt down soldiers as part of my pay officer duties, followed by the showdown at the battalion commander's holiday party, I'm pretty wiped out."

He also told him that he had made arrangements with a DA civilian to drive up to Bremerhaven on Tuesday to finally get his car, a Volkswagen Scirocco, of all things. He laughed and told him, "I'm probably the only soldier in the history of the Army who actually imported a Volkswagen to Germany. But based on my numerous bad experiences at Fort Bliss with tanks in the field, I wanted to have a vehicle I can get parts for here. So, I traded in a Toyota Celica and bought a new Volkswagen, over the strong objections of my fiancée, who really liked the silver sports car with a smoked glass moon roof."

Elizabeth's eyes brightened and she said, "That sounds great to me, too."

"I understand, but it had a warped flywheel and I just didn't have the time or money to fix that, get it squared away, move to Germany, and hope for the best while getting engaged all at the same time. In any event, it gladdened my heart when I picked up the Volkswagen. I saw what appeared to be a retired colonel and his wife who had bought this Celica. They had the keys and were jumping in it like two teenagers."

The Lieutenant smiled again and thought, *With the addition of the air conditioner, that was demanded by his best friend from*

Pennsylvania, it truly was an awesome ride for El Paso. And, once they had the flywheel replaced it would probably run forever since there wasn't any rust in the Texas desert.

When he was through with his story, the Lieutenant offered to clean up. Buzz agreed, so he washed the skillet, plates, silver, and the coffee mugs as well. After placing everything back in the cupboard, he put the leftover pancakes and bacon in tin foil, and placed them on a plate in the refrigerator, for some lucky, hungry scavenger to find later. By that time, Buzz and Elizabeth excused themselves and headed down the hallway. The Lieutenant decided that he would take a nap and then get up to write a long letter to his girl, letting her know that they would be getting married on March the eighth and asking her to be very nice to the colonel's wife when she got to Germany, because she had taken such good care of him.

After catching up on his sleep and writing the letter, he looked out the window and saw that it was going to be a cold and damp evening. He thought to himself that he really needed something to lift his spirits. He decided to take the streetcar up to Nick's, for dinner and a Greek brandy. He told himself that after all he'd been through that week, he deserved it.

GOODNIGHT TONIGHT

By the time he was ready to leave for dinner, the sun was down, and heavy fog had drifted into the city, with light drizzle as well. He put on his German civilian clothes that were warmer and slipped on a white ski jacket that he had picked up in the German department store where he bought the clothes, since it was light and warm, and would certainly come in handy with the winter coming on.

When the streetcar pulled up in Viernheim, the mist was even heavier. He could barely see the yellow lamp lights near the windows when he made his way down the street, away from the streetcar station. As he stepped through the door into the restaurant, he could feel the warmth from the grill, the bright lights above the bar, and the cozy candlelight in the windows. This was the one thing he had grown to like about Germany: even when the weather was completely crappy, a good guesthouse provided shelter from the storm and warmed the soul. When Nick saw him, he asked where he wanted to sit. The Lieutenant said, since he was alone still, he would just go ahead and sit at the bar if it was alright with Nick. It was, so he pulled up a high stool and sat at the bar, with a glass of white wine, watching Nick work for a while before he finally ordered. Having eaten there a few times before, he had come to the conclusion that his favorite meal in the world now was something called Bayern Speise with French fries, a cold vegetable salad, and the best beer in the world.

Once he had ordered, he slipped the wine glass to the side. Nick immediately brought him a large glass of Kulmbacher Mönchshof Brau. This beer was made in Kulmbach, a small West German town, near the Czech border, where the Second Armored Cavalry Regiment patrolled, day and night, to keep the peace, like their brothers in the Eleventh Armored Cavalry Regiment did along the East German border to their north. This was the beer that the officers in his battalion drank from the boot when they visited the border. It was mellow, smooth, and went down like milk.

When Nick brought his meal out, he congratulated himself on making at least one good decision in the last week. The Bayern Speise was one of Nick's specialties because it combined the two cultures in which he lived. He took thick slices of pork tenderloin and skewered them on a steel kebab, with onions, tomatoes, and peppers. He then seasoned everything in Greek spices and grilled it to perfection before covering everything with a soft cheese that melted over it and flowed down on the bed of French fries that it was served on. Cold salad, mostly carrots, and cucumbers, that were lightly pickled, accompanied the meal and set it off perfectly. While he was eating, Nick came back to him and told him, "You look particularly happy tonight."

"Yes, I am because I finally nailed down the leave for my

wedding. I'll be going back to the United States to get married March the eighth. After that, I'll bring my wife back here to meet you and enjoy your great food." Nick congratulated him. Then he asked if there would be anything else. "Can I have a glass of warm Metaxa, to fortify me against the cold before I have to go back?"

Nick came back right away with a glass half full of brandy, and what looked like a little brass lamp that he lit, so a blue flame came out of it. He slipped the brandy glass into a holder above the flame and rotated it so that it was evenly warmed. After about a minute, he lifted it up and lightly touched the side of the glass. He then gave it to the Lieutenant, who kept it in his hands to warm up a little more and then took a sip. It was wonderful. Nick looked at him and asked, "Since you're American, why have you developed a taste for Metaxa? It's Greek brandy."

The Lieutenant looked at him and told him, "When I was a boy, I studied the great generals, and I particularly loved Alexander the Great. The label says that this is the brandy of Alexander the Great. He never lost. That's why I drink it. Besides that, it tastes great after Greek food and it will keep me warm all the way back to my apartment." Nick laughed a little and then wrote out the ticket. The Lieutenant covered the ticket with marks rounded up to the nearest mark, as was customary in Germany, and then bid farewell to Nick as he plunged out the door, into the even colder and darker German night.

When he got back to his room, he hung the ski jacket up, away from his other clothing as it was pretty damp. Then he slipped into a T-shirt and shorts and crawled back in his bed on top of the poncho liner to just relax and enjoy the evening. In his head, he was going over the schedule for the next week. First, he would check in after PT with his Support Platoon Sergeant and set up a meeting for Monday. Tuesday, after PT, he would wait outside his BOQ for the American teacher who was to pick him up and take him to Bremerhaven. Apparently, that man drove a gray Ford Taunus that he had picked up while living in Europe. Beyond that, he really didn't have any other items on his calendar that he knew of, other than the fact that he wanted to make time to catch Sergeant Robb somewhere on Monday to try to figure out what the hell had gone on Thursday night.

By now, the wine, beer, and brandy had caught up with him, so he set his alarm for 0500, and drifted off to sleep. The next morning, after he looked out the window to see what the weather looked like, he realized that it was cold and dark, and even more bitter than usual. He looked through his things and fished out his navy-blue watch cap from his VMI days. He slipped into his PT gear, put on the cap, and added a pair of olive drab wool glove inserts to warm his hands.

When he was ready, he jogged over to Headquarters Company, and hurried downstairs to the First Sergeant's office to snag a cup of coffee to warm up before the morning run. When he heard the first sounds of soldiers, walking, and in some cases stumbling, at the front door, he joined them and went to the back of the formation, near the executive officer and the staff officers so that Major Reardon could plainly see the large white "M" taped on his back.

After light calisthenics and a fairly short run, he waited for the First Sergeant to release the platoons. Then he caught up with Sergeant McNabb and asked, "How are you coming on the training program?"

"I really haven't had an opportunity to work on it, since I've been so busy dispatching details for the battalion staff and First Sergeant." Based upon that response, the Lieutenant decided that he would wait until he'd returned from Bremerhaven, so he asked him if he would be able to meet him with the updated training schedule on Wednesday. McNabb agreed, so after getting dressed and snagging a breakfast bar, he walked back to the company for morning formation. When the platoons were released to their sergeants, his drivers went back down to the motor pool and HHC maintenance to continue working on their trucks. He walked over to the S-4's office and reported to Graham.

SIMPLE TWIST OF FATE

Graham asked him to come back into his office and have a seat by his desk. He told the Lieutenant that he had heard about the call out of the Ready Reaction Force on Thursday and asked him a couple of questions about that. The Lieutenant recited back the issues that he had identified. Graham nodded gravely because he was a serious man. Then he smiled a little and told the Lieutenant that he had heard about the battalion commander's wife pulling his wedding date out of the fire for him.

The Lieutenant tried to keep a straight face, but eventually a smile broke out. He took that opportunity to remind Graham that he would be away the next day to pick up his car in Bremerhaven. Graham said that he had already put that on his calendar and was aware of it. He then told the Lieutenant to go ahead and take out his notepad and take down a few notes.

Graham told him that he had been called in for an early morning meeting with the executive officer, and that by the end of the next week, Reardon expected a new loadout plan completed and on his desk. The Lieutenant simply said, "Very well."

Graham then told him, "It's a good thing that you have your car, because next Saturday, the division is having a Christmas ball and reception for all officers and their ladies. You'll have to drive over to the officers club in Heidelberg and attend in blues, formal dress, to meet the division commander and to greet the brigade commander again, as well as the other colonels and generals that are on staff. The Lieutenant sighed a little but decided that it should be a fun evening. Also, from what he heard, the food in Heidelberg was wonderful.

Graham, however, continued, "There's a couple of other things. Thursday, Brigade is going to inventory all of your fuel and ammunition. Friday, there will be a division-wide Pegasus war game. Our brigade officers will attend at Third Brigade Headquarters. The other brigades' officers will appear in their brigade headquarters, in Mainz or Baumholder. It will also include the Fourth Infantry Division's brigade that has been rotated into Germany and is stationed at Wiesbaden. You should recognize some of those guys' names, because that's the brigade you were attached to for Reforger."

The Lieutenant was still processing this information, when Graham told him he had received a call from the Adjutant that morning to relay to the Lieutenant that, "First, you will have battalion staff duty on Christmas Eve. Second, you'll have Brigade Staff Duty on New Year's Eve at Brigade Headquarters in Sandhofen."

The Lieutenant's first thought to himself was, *Damn. Shea works fast*. But he bit his tongue and said, "Very well."

Graham immediately followed up, explaining that single officers, especially those newly assigned, habitually got staff duty on the holidays, so that married officers could spend time with their families. He was simply paying his debt in advance, so he would be able to enjoy holidays with his wife when she came to Germany. At that point, the Lieutenant simply said, "Alright," and after being excused, went back to his desk to begin marking up the calendar with all of his new activities. It looked like he was going to have a hard time getting a training schedule set up, but he sketched out a rough one anyway, before turning his attention to the logistical nightmare that was trying to put together a loadout plan with too few vehicles, and too many crates of munitions.

TAKIN' CARE OF BUSINESS

He sat down at his desk to another cup of coffee. He finished and walked down to the motor pool to check on his troops. When he got there, he found out that McNabb had just left on an errand requested by the First Sergeant, so he made some small talk with the junior NCO in the platoon, before going back to work the rest of the day in the S-4 shop.

He spent some time visiting with Sergeant Dryer, who was the Battalion Supply Sergeant. He was an amazing sergeant first class, African American, and funny. He had his hand firmly on the pulse of the battalion, and there was nothing that he wasn't aware of. The assistant S-4 NCO, Staff Sergeant Brown, was their "Mr. Fix It." If anything needed to be done or something needed to be found, he was the guy. He didn't drink a lot of coffee, but smoked almost incessantly, and played some solitaire to pass the time. Every once in a while, he would get up and walk through the battalion area to visit with different non-commissioned officers. He was particularly close to Sergeant First Class Peeler, who was the S-3 NCO and Battalion Master Gunner. They had known each other for a long time, ever since they came back as kids from Vietnam. As he would learn later, they both suffered severely from PTSD.

Chapter 24
Fire and Rain

In the middle of the afternoon, the sun came out of the clouds. The Lieutenant told Sergeant Dryer that he was going to step out and get some fresh air before coming back to tackle the loadout problem again. After slipping on his field jacket and baseball cap, he walked across the line companies' area to the motor pool and glanced over to Bravo Company. Apparently, Lieutenant Marr was off on some sort of a detail for the Company Commander, so Sergeant Robb was left with his old tank platoon. The Lieutenant walked over towards Sergeant Robb and motioned for him to join him in the tree line, along the sidewalk, next to the ready line.

Sergeant Robb walked up and saluted the Lieutenant, who returned it briskly. Then he asked Sergeant Robb, "Did you hear about the Reaction Force being called out Thursday?"

"Yes, sir."

"Alright, what the hell was really going on? Who was out there?"

Sergeant Robb looked at him seriously, and then said, "If I was a betting man, sir, I would bet it was the neo-Nazis."

At this comment, the Lieutenant's eyes widened, and he said, "Then I have more questions."

Robb nodded gravely and said, "Yes, sir."

The Lieutenant queried, "So, besides the Bader Meinhof Gang and the Red Army Faction in Germany, Action Direct in France, the Red Brigade in Italy, and sometimes the IRA out of Ireland, we have neo-Nazis, too?"

"Yes, sir, we do. Who do you think spray paints 'Yankee go home' in our training area?"

"Alright. So why did they attack the PSP?"

"I think," Sergeant Robb said, "they were trying to execute a false flag operation to make us and the German police believe it was the Red Army Faction that was trying to steal the Stingers. They would do this to get the Polizei to believe that their deal with the Red Army Faction had been broken."

"What deal?"

"Sir, there are two things you need to know. First, here in Sullivan Barracks, we are very close to the intersection of four German states. There's no national police force allowed in Germany, after the war, so the individual states' police forces are not allowed to share information across state lines. That means the terrorists can live in one German state, commit a crime in another German state, slip into a third German state to escape, and then slide over into a fourth to hide. That makes it almost impossible to catch up with them. And, in addition to that, because they are historically headquartered out of Heidelberg, and come into Mannheim regularly, the Mannheim police have a deal with the Red Army faction that their members will not be arrested, surveilled, jailed, or prosecuted, as long as they do not engage in any terrorist activities in the city limits of Mannheim. Mannheim is their safe haven. Why do you think the drug trade comes out of it?"

The Lieutenant took a deep breath, thought to himself a minute, and then said, "Thank you. I understand." Then he asked, "Do you think it was effective? Did the Mannheim police really believe it was German terrorist from the Red Army Faction Communists?"

Sergeant Robb added slowly, "Unlikely, sir. We're in a suburb of

Mannheim. And besides, the only people really put at risk were Americans. They don't care."

The Lieutenant shook his head, and then said, "Thanks again. But I have to ask. With all that's going on – why do you keep coming back here?"

"Dawn likes it here, sir."

"How did you two meet?"

"Halfway through my second tour here, after Saigon fell in 1975, I saw her coming out of the PX and went over to introduce myself. She had been working as a translator in the embassy and was evacuated to take a job here teaching French and English. We just hit it off and have been together ever since, other than a year in the states for me attending ANCOC to make E-7."

"She's stunning."

"Yes, sir, she is."

"What keeps you two together?"

"We respect each other's ghosts."

"I noticed you don't wear a combat patch. Ghosts?"

"Yes, sir. I did a hard tour near the DMZ with the Fourth ID, after I was drafted in '68. Ghosts."

"Any chance that the two of you will get back to Tennessee someday?"

"I don't think so, sir. The posts stateside have large Vietnamese communities and she is shunned by her own people."

"Because she's a child of the dust?"

"Yes, sir. At least the Germans get along with us pretty well. They just generally dislike all Americans equally. As for me, I went home a couple of times but did not feel welcome. The kids who were too young to get drafted detest me for going to Vietnam, and the older ones who lost kin over there resent me for coming back."

"You know, I hope someday America is a country worthy of you two."

"But that's not today."

"No, sergeant, not today."

Switching subjects, the Lieutenant turned the conversation to platoon issues. "I also wanted to ask how things are going in the platoon?"

Sergeant Robb smiled and answered, "Fine, sir. I'm just waiting for First Sergeant Guerra to rotate out to CONUS and I'll put in my bid for First Sergeant."

The Lieutenant looked over at him and then said, "I hope you get it. Did you see your efficiency report?"

"Yes, sir, thank you very much."

"Good."

"You signed it and Captain Borman hadn't been here long enough to endorse it, so I got a max."

The Lieutenant followed up, "Very well," and excused himself, to head back to the S-4 shop.

By the time we got back to his office, the S-4 had left and the NCOs were locking up, so he followed them to HHC for evening formation. When the First Sergeant called the company to attention, he noticed that Sergeant McNabb was not back yet, so when the platoons were released, he walked up beside the E-5 in charge and asked him if McNabb had returned. The sergeant told him, "No, sir, but he's expected a little later. He had to deliver something to Kaiserslautern."

The Lieutenant looked at him for a few seconds, before stating, "Kaiserslautern is a long way away, isn't it?"

"Yes, sir. But the Mess Sergeant had something he needed to be handled, so he did it for him."

2,000 LIGHT YEARS FROM HOME

The Lieutenant saluted, the sergeant returned it, and the Lieutenant turned to walk back to his BOQ. On the way, he realized that he hadn't put anything into the communal kitchen fund or purchased anything lately, so he went to the commissary and picked up a pork roast, which he took back to the BOQ. When he walked through the kitchen door, he could say that Buzz was hard at work, making bratwurst with potato chips and green salad, of all things. He gave Buzz an odd look, but he just shook his head and shrugged his shoulders, saying, "It's something different."

Then he looked at the roast and said, "Maybe tomorrow we can have that." Both smiled and Buzz smiled even broader when the Lieutenant fished twenty dollars out of his pocket and laid it on the counter, as payment toward the communal beer, wine, and food fund. Buzz grinned again and told him, "Dinner will be ready in about a quarter of an hour."

After a quick shower and change into his habitual T-shirt, jeans, and sneakers, he made his way back to the kitchen. There he helped himself to a beer from the refrigerator and sat down to wait for Buzz to finish making dinner. The smell of bratwurst that filled the hallway was like a magnet. Bigfoot joined them, almost immediately followed by Bernie, Tom, and Susan and Chris. The ladies enjoyed their white wine while everyone else had beer, and Buzz held court in the kitchen until everything was wonderfully ready.

When he was through cooking, Buzz motioned for the ladies to come and make plates, with a flourish of his hand. As they were sitting down, Buzz filled his plate, followed by the other gentleman in the room. Buzz looked back over his shoulder, but apparently Elizabeth was running late, and he turned his attention back to the other diners. The conversation flowed back and forth across the table, as they recounted the things that were happening in their units, as well as the plans that they had for the Christmas holiday.

On that note, Buzz interjected that he would be with them in the BOQ on Christmas Day and asked how many of the others there would also be present. Tom was the only one that said he would not, as he was going home. Buzz then suggested that they should have a communal Christmas dinner amongst themselves, for anyone in the building who wanted to join. He planned on making a turkey and a ham, with accompanying vegetables, and suggested that Bigfoot might make his way to a good German bakery on Christmas Eve to bring a cake or assorted pastries for dessert. Bigfoot agreed and the officers chimed in around the table as to what they would be bringing.

Bernie immediately said he would be responsible for the wine. Susan and Chris said that they would get with one of their friends who would come for dinner and the three of them would cook various vegetable dishes to go with the main course. The Lieutenant, feeling awkward, since he did not cook at all, volunteered that he would bring condiments: pickles, olives, both black and green, and German cold salad as well. He added, in his defense, that he had just learned he was tagged for battalion staff duty on Christmas Eve, so he would be pretty tired in any event before Christmas dinner started.

Buzz looked at the Lieutenant in shock and asked, "Didn't you just have staff duty?"

"Yes, on Saturday, before that, on Thanksgiving Eve, and then again last Thursday. In addition, I'm also slated for duty at Brigade on New Year's Eve."

In response to that listing, Buzz's eyes only got wider and wider. Then he asked, "Gee, Lieutenant, who did you piss off?"

"Let's say that I managed to get myself on the wrong side of my battalion commander. But I don't care because I have my wedding date and I get to go to Bremerhaven tomorrow to pick up my car. For the rest of it, I'm happy to do it, and I'm really looking forward to having Christmas dinner with all of you."

Buzz then set about explaining the logistics, "Everyone bringing dishes will need to get them into the communal kitchen by 1130 on Christmas Day. Everyone participating needs to their small dining table out into the hallway, so we can put them together and make one large banquet table with chairs on both sides." Buzz told them that he had done it before when he was in Germany alone, and that he had presided over an excellent Christmas dinner last year as well. When he was done, all the officers were beaming, and the ladies promised that they would get out the word to if any other officers in their BOQ or the two across the street that might want to join them.

When everyone was through eating, Bernie and Tom volunteered to clean up, so the Lieutenant wished them all a good night as he walked out the door, back to his room. When he was there, he listened to the radio for a little bit but then decided to turn in early so he would be well-rested for the drive to Bremerhaven and back.

THE ROAD

The next morning, after PT, he decided to have breakfast in the mess hall because he didn't know whether or not he would be able to get any sort of meal going to or coming back from Bremerhaven with his new civilian friend. Before walking to the mess hall, he reminded the HHC Commander and First Sergeant that he would be traveling to Bremerhaven that day.

Once he finished breakfast, and a second cup of coffee, he found Sergeant McNabb in the mess hall visiting with the SFC Sparks, the Mess Sergeant. He walked up to them and told McNabb that he would be out that day, traveling to Bremerhaven and back. And that since he had been tasked by the executive officer to reconfigure the load out plan, he needed to meet with him with regard to that and to also discuss a training program on the following day.

Having finished his coordination, he jogged back to his room to get cleaned up and put on a uniform. Once he was done, he put a couple of breakfast bars and a small package of licorice in his field jacket pockets and went out to stand on the steps of the BOQ to wait for his ride. Winter had arrived with a vengeance. There was a cold, biting wind and a heavy mist in the air when a gray Taunus pulled up and over to the curb in front of him. He waved to the driver and hastened to the passenger door, to dive in, and soak up the warmth in the interior. The Lieutenant introduced himself and Bill, the driver, returned his introduction. Then Bill quickly pulled out, drove around the traffic circle, and exited the housing area to the Autobahn. Once he entered the highway, he accelerated to over one hundred and fifty kilometers an hour, which seemed very fast, until the Mercedes sedans and BMWs passed them like they were standing still.

The road markers flashed past, as they hurtled north toward the great seaport, where, with any luck, the Lieutenant's Volkswagen was waiting. As they drove, they exchanged the usual small talk: where they came from, their assignments, and plans for the future. Bill had been a mathematics instructor in the Army school system for over twenty-five years in Mannheim. He was married to a German national and they had a couple of kids with dual citizenship.

Bill was very excited about getting back to the United States, but he was concerned that life in America would be a complete culture shock for his wife. Although they had visited on many occasions, and the children spoke fluent English, he knew that his wife would be quite homesick and need to return to Germany two, if not three times a year. However, on balance, with the hostage crisis, and the increased frequency of acts of terrorism in Europe, he had decided that it was time to seek a safer home in America.

The farther north they travelled, the colder and darker it became. The weather was visibly worse. They could see snow on the mountains to the west, and they hit flurries as they approached the outskirts of Bremerhaven. When they finally reached the port, the Lieutenant thanked his lucky stars that he was traveling with a civilian who had twenty-five years of experience dealing with the German bureaucracy. For one thing, as they pulled up to the entrance to the port facility, where the Taunus would be handed over to the authorities for shipping, Bill asked him if he had happened to bring along a first aid kit when he shipped his car at Bayone. The Lieutenant told him that he had not, so, Bill quickly added, "In that case, I'll sell you my government approved first aid kit that's in the trunk for fifteen marks and throw in the red plastic gas can, filled with five liters gas for another ten marks." The Lieutenant looked at him quizzically, and Bill answered, "Without a first aid kit that's complete, and at least five liters off gas in your trunk, we won't be able to leave Bremerhaven."

The Lieutenant smiled as he pulled twenty-five marks out of his wallet and thought to himself, *There's always something, and it's a whole 'nother world.* When Bill pulled the trunk release, the

Lieutenant retrieved both the gas can and first aid kit, while his new friend filled out the paperwork for shipping of his car to New Jersey, The Lieutenant walked up to the counter to make arrangements to pick up his car that was parked to the side. Once the paperwork was done, and he convinced the German clerk that he had a proper first aid kit and the required amount of gasoline, one of the dock workers drove his car up and handed him the keys.

The Lieutenant placed the first aid kit and gas can in his trunk and motioned for his traveling companion to join him. Bill's eyes widened, since he wasn't really expecting a small German sports car to be his ride back. But once he had settled in, the Lieutenant slipped the little car into gear and smiled to himself as he pulled out for the highway. It had been a long time since he had been able to drive his own car. He hadn't realized how much he missed it until that moment.

It was now after lunch time, so he pulled over at the first rest area on the Autobahn that Bill pointed out. By now, it was even more bitterly cold than it had been earlier in the day. The wind was fierce, and snow was falling furiously. They made their way into the warmth of the small restaurant and ordered hot currywurst and French fries. The Lieutenant opted for a large German cola, although he disliked the bitter taste, as he wanted to make sure he had plenty of caffeine in him for the drive, and he just didn't feel like any more coffee. Once they had finished eating, they climbed back into the Scirocco. The Lieutenant got on the Autobahn and rapidly accelerated to almost ninety miles an hour. Being from the United States, he made the assumption that over eighty miles an hour was pretty fast. He soon realized, as BMWs and Audis passed him at well over one hundred and twenty, that Germany was indeed a whole 'nother world.

After an hour, Bill started to doze off, after telling him to be sure to wake him before they reached the exit for Mannheim, as he would have to give him directions to where he lived. The Lieutenant then drove in silence, quietly cussing himself for not bringing along some of his cassette tapes to play while he zipped along in the right lane.

After fiddling around with the radio for a while, he was able to come across a classical music station and he played that to break the monotony and continued as quickly as he could to get to Mannheim.

When he saw the first for the Mannheim exit, he elbowed his companion, who came wide awake. Bill gave him directions to drive on to Viernheim and then head west, into the foothills of the Odenwald, where he and his family rented a large apartment in what looked like some sort of German chalet. The Lieutenant thanked him for the lift, and although Bill asked if he would like to come inside for a drink, the Lieutenant told him no, and that he needed to get back to post to prepare for the rest of the week's duties.

They shook hands and the Lieutenant shifted the Volkswagen into first gear and slid through the snow down through the hills back to Viernheim where the roads were clearer. In less than fifteen, he made his way back to Benjamin Franklin Village. His first order of business was to find a place to park his car in front of the BOQ. Luckily, there were a couple of spaces near Sullivan Barracks, so he eased the little sports car in toward the curb, locked up, and almost ran into the BOQ to get out of the wind.

SHARING THE NIGHT TOGETHER

He rushed through the front door and quickly climbed the stairs to the second floor, where he turned the corner, and caught the smell of something coming from the communal kitchen. He poked his head in and saw Buzz was busy, stirring what looked like stew, with a loaf of French bread sliced beside him. The Lieutenant said, "Hello," and Buzz asked him, "How was the trip to Bremerhaven?"

"Cold, wet and miserable, but at least I have my car, so it's all good." Then he slipped his wallet out of his field jacket and handed Buzz twenty marks, toward the meal and alcohol funds, since he was out of dollars.

Buzz smiled, thanked him, and told him, "You have time to get changed and come back before dinner is ready."

"Great. By the way, what are we having?"

"I took off early to beat the crappy weather and put your pork roast on low heat for the afternoon. Then I cut it up to make goulash or 'hunter' soup, with beef au jus, mushrooms, potatoes, butter, brandy, German spices, lots of paprika to fight of the cold, and the last few ounces of red wine left from the bottle Bigfoot and I didn't finish – to remember the ones who aren't with us."

"Sounds perfect. I'll be right back."

When he got back to his room, he took a quick hot shower to warm up, dressed in jeans and a sweatshirt left from his cadet days, and returned to the kitchen for a bottle of beer. As he slid into his seat, up against the window, Elizabeth came in, and Buzz explained that since Bigfoot had told him that he was entertaining, Susan and Chris were visiting a friend in another "Q", Tom had duty, and Bernie was on an exercise, he decided to make a simple, quick meal that would warm them up since the weather was so awful. As usual, Elizabeth's smile lit up the room. She was so excited, with the holidays approaching, and had spent the last weekend doing some Christmas shopping for her family. She told both men all about it. The Lieutenant really enjoyed watching her. She reminded him of his girl. And it was a welcome change to be able to talk to a woman when you had been in and out of the field with five hundred surly, and sometimes violent men in a tank battalion, for the better part of four months.

With his usual panache, Buzz announced that dinner was served. The three of them pulled bowls and plates out of the cupboard, then ladled themselves large helpings of the hearty, spicy soup. Lieutenant thought to himself that there was no better meal on a cold and dismal day than a hot stew and fresh bread with lots of butter. When they were done, he volunteered to clean up, and Buzz and Elizabeth went back to their room to relax for the evening.

After putting the leftovers in the refrigerator, he had washed the dishes and the counters, before heading back down the quiet hall to his even quieter room. There he sat down to write another letter to his fiancée, letting her know that he had finally picked up his car from northern Germany. He could hardly wait to get her in country, so they could explore Europe together. Out the window, he could see sleet coming down in waves. This certainly didn't give him a good feeling about morning PT, so he decided to listen to the radio and read a while, before setting his alarm clock a little later. He wasn't planning on standing around in the back of formation in this kind of weather, getting ready to run.

COLDWATER MORNING

The next morning, as he was getting dressed, putting on an extra layer of T-shirt, he looked out the window and breathed a sigh of relief. Although there was about two inches of snow on the ground, at least the sleet had stopped, and the wind had almost died down. He made himself a hot cup of instant coffee and then braved the weather to get over to HHC in time to see the XO. Once he made sure that Major Reardon knew he was present, he fell in with the other staff officers, behind the formation. Since it was slippery and nasty, Captain Boris only decided to run a slow mile and then forego calisthenics in the snow. The Lieutenant was cheered by this and caught Sergeant McNabb when the company was released to see if he was ready for a meeting later that day. McNabb told him that he could be there at 1000. The Lieutenant confirmed that he would see him in his office at that time. He also asked how the maintenance was going and McNabb informed him that about half of the trucks were through their semi-annual maintenance. It took longer than he had anticipated, because of the large number of wheeled vehicles in Headquarters Company that were given priority over Support Platoon, in particular, all the staff officers' Jeeps, and the mess trucks.

The Lieutenant then said, "Very well," to McNabb who saluted and took off. He went back to his quarters and changed, having decided that he would come back to the mess hall for a long warm breakfast and several cups of coffee before going to his office, that he knew would be cold. As he walked in the mess hall, he saw Lieutenant Marr and Sergeant Robb sitting together in the corner, so he grabbed a cup of coffee, filled a tray, and walked over to see if he could join them. They both welcomed him as he sat back in a chair.

Visiting with them, he learned that the platoon was coming along well, but the Company Commander couldn't seem to focus on actually being a Company Commander. In fact, he was more addicted to automotive maintenance and hot rods, and he would hang out in the maintenance bays with the wheel mechanics working on his Pontiac Firebird on Saturdays. The Lieutenant thought to himself that this was a bit odd but then went on to tell him that he had finally recovered his car from Bremerhaven. Sergeant Robb heard this, turned to the Lieutenant, and quietly told him, "Be very careful if you are ever pulled over by the German police. You may notice, sir, that the German police always travel in pairs, two to a car. If you're pulled over for a roadside stop for a traffic infraction, stay perfectly still, and don't make any sudden moves at all. If you look in the rearview mirror, you'll see that while one officer has come out from the passenger seat, with an auto pistol at his side, the other will have his Uzi submachine gun up over the steering wheel, trained on the back of your head, off safe and on full auto. If anything spooks either of them, the policeman in the car will simply take out everything in your vehicle."

Well, the Lieutenant thought to himself and then said aloud, "We're just Americans, right?"

"Yes, sir. That's about right." Lieutenant Marr nodded as well. They were both happy to have a well experienced hand with them who would do all that he could to keep his officers alive, both in and out of garrison.

The Lieutenant ate quickly. After he finished, he thanked Sergeant Robb and wished them both well, before he took his tray to the window, and dropped off his coffee cup. When that was done, he walked to the line company courtyard where the S-4 shop was located. He went into the offices to say good morning to Graham. He stopped short at the door for a couple of seconds because he could hear the increasing sounds of growling coming out of the office. When he knocked on the door and was told he could enter, he came in and saluted, before taking a seat by the S-4, and asked what was going on. Graham told the Lieutenant that he had just received several new taskings to support meetings of the brigade staff with the battalion staff, as well as a Pegasus exercise the next day, that they all would be attending. On top of that, they would also have to support the annual Battalion Christmas party for the soldiers and their wives and children. In addition to that, Graham was being ruthlessly hounded over numerous delays in delivery of supplies that had been ordered by the companies over which he had no control. And he also needed to begin planning and scouting positions for an upcoming winter exercise.

FORTUNATE SON

"From what I've been told," Graham continued, "my understanding is that the brigade and battalion staff officers, Company Commanders, and platoon leaders will be going into the Odenwald in the middle of January to run a practice area exercise, known as a tactical exercise without troops or TEWT. Based upon that, the Ops shops will put together a combat scenario for the brigade to push all of the line battalions, and at least one troop of the cavalry squadron into the woods, for a week to ten days in February."

Graham seemed particularly concerned, so the Lieutenant asked what was happening. Graham explained, "The battalion did not go to the field last winter, to conserve fuel and parts, but mostly because of the disaster that was their Blue Fox winter exercise, in December of '77 and January of '78."

Graham continued, "In this battalion alone, seven were killed." That got the Lieutenant's attention, because he had never heard casualty figures like that coming out of a peacetime unit. Graham told him, "Yes, that was before Colonel Shea got here and Colonel Greer was the battalion commander. The winter weather was atrocious. There was black ice everywhere, so the vehicles slipped and slid off the road continuously."

The Lieutenant thought back to December of '77 and the early part of '78 at Fort Knox, and recollected that that winter was brutal there, too, and remembered walking to rifle ranges with trainees through snow that was two to three feet deep. The temperatures were so low that he had a space heater in his quarters that he placed next to his bed to keep it warm enough to sleep, which would still not keep ice from forming on the insides of the windows.

The Lieutenant couldn't overcome his curiosity and asked, "How did they die?"

"Well," said Graham, "first thing, I was in Charlie Company at the time. We were road marching through a German village, and our lead tank slipped sideways on black ice that covered the cobble stone road, into the wall of a German restaurant. It pinned a civilian against the wall, where he was caught by the sprocket and the tracks ground him up. Because of the accident, our Sergeant Major drove his Jeep up to take charge of the situation. While he was standing outside directing traffic, another tank slipped on the ice and careened into him and his Jeep, crushing him and grinding him up as well. Over the next week, a deuce and a half with CSC's Supply Sergeant and his assistant slid on black ice covered with snow, head-on, into a tree. Both men were killed by the impact. As you know, there are no seat belts or shoulder harnesses in our vehicles, and a helmet won't keep you from breaking your neck."

The Lieutenant nodded in silence, and Graham went on, "There were lots of vehicles stuck or broken down and the 88s were working around the clock recovering them. It was so brutally cold that when one of the 88 crews finally had a chance to rest, the crewmen locked up the turret to keep the cold out, and all three died of carbon monoxide poisoning. Since I was the senior tank platoon leader, at one point, the XO of Charlie had me take Jeep to drive to Battalion to pick up new orders and acetate overlays for the next day's maneuver.

God! I will never forget seeing Colonel Greer down on one knee, crying beside the body bags that were stacked up beside headquarters. As I approached him, he looked up and declared, 'Lieutenant, I never had anything this bad, even in Vietnam.' He didn't move, so I walked past him into Battalion to pick up the orders and then quickly got back in my Jeep to go back to the company. Then, when we were finally back, and the tanks safely road marched out of the mountains back here to the motor pool, one of the Support Platoon truck drivers was walking through the motor pool with his A-bag slung over his shoulder. The motor pool was terribly muddy, and the weather was awful sleet and freezing rain. As a GOER passed him, he slipped. The rear tire of the GOER caught his duffel bag, dragged him up into the wheel well and crushed him to death. Right here. So, as you can tell, I'm not enthusiastic about a winter exercise and I'm telling you now, you have to be extraordinarily careful because once the roads in this country are covered in snow and black ice, these tanks are nothing but sixty-ton toboggans."

Graham was through. The Lieutenant could sense that the conversation was over, so he stood and left the room. Sitting down at his desk, he continued to work on the loadout calculations. At 1030, he looked up and saw Sergeant McNabb coming through the door. McNabb joined him at his desk, and the Lieutenant told him to have a seat beside him. McNabb had a folder with him, with the beginnings of the training program. The Lieutenant reviewed what he had done so far, showed him his notes on training, and then asked if he had any questions or suggestions. He did not, so the Lieutenant asked him, "Has the Support Platoon ever had to load out the bunkers in your time here?"

"No, sir, that has never been done."

"Have you ever worked on a loadout plan?"

"No."

The Lieutenant then asked what he had done with the prior Support Platoon Leader.

McNabb told him, "The Lieutenant spent most of his time in his office, unless the S-4 or XO had some errand that they needed him to run."

The Lieutenant thought for a moment and then said, "You will see a lot more of me than that. I plan to visit the platoon at least daily when I'm in garrison and available. Also, when the weather is better, we will need to do some training before we have to go to the field again for the upcoming winter exercise."

When he heard the words "winter exercise," McNabb's eyes widened and he asked, "Did you say winter exercise, sir?"

"Yes, sergeant. We'll have a small tactical exercise for the staff and commanders to support with food and fuel. And then a larger exercise for the battalion in February."

McNabb looked at him and said, "Sir, in February, with ice and snow, the roads will be terrible."

"Yes, that's what I've heard. We will have to do the best we can. By the way, do we have chains for any of our trucks?"

McNabb looked at him and said, "A few, for the five-tons, but nothing for the GOERs."

"How do you get them across frozen roads?"

McNabb answered, "We simply deflate the tires a little more, and move very slowly."

The Lieutenant continued on, "I have to go to a Pegasus war game exercise at Brigade tomorrow, so I will need my Jeep and driver at 0700. As I understand it, Specialist Harris will be rotating back to the States at Christmas, do you have a new Jeep driver identified to take his place?"

"Not yet, I will ask around and see if I can get a volunteer from one of the line companies. That's usually what happens, since anyone who can drive a standard transmission can handle your quarter ton even if he's not qualified to drive a five-ton or a GOER."

He nodded his head to show that he understood. McNabb looked at his watch and then said, "I really need to be going. I have to make a run for the Mess Sergeant and the battalion Motor Sergeant."

The Lieutenant looked at him and asked, "Can't you send someone else?"

McNabb quickly answered, "No, sir. This particular mission requires a noncommissioned officer with experience. And anyway, my drivers are all busy working on their trucks."

The Lieutenant told McNabb to give him a call when he was back from his errands, and to plan on meeting him Friday afternoon to review the progress of the training plan and discuss the loadout procedures, since he was going to have to review those with Major Reardon the following week. With the conversation at an end, McNabb stood, saluted, and promptly left the building, for the motor pool.

When he was gone, Graham asked the Lieutenant to come back into his office. He had two things to discuss. First, he reminded the Lieutenant that the brigade Christmas Party was going to be at the Heidelberg officers club on Saturday night, and the Lieutenant was expected to be there in dress blues. The Lieutenant nodded in assent and thought to himself he was so very glad to have picked up his car to make the trip, as opposed to going by streetcar in the cold.

Then Graham told him that he needed to study, in detail, the battalion SOP as it pertained to S-4 operations. He pulled that notebook out of his safe and handed it to the Lieutenant so that he could read it and commit it to memory, in preparation for the Pegasus exercise the next day. Graham added that he needed to be ready to make calculations of how much ammunition, fuel, and food he would have to get transported to the border. And, once the game kicked in high gear, he would have to devise a way to deliver all of those things to the line companies.

The Lieutenant saluted sharply and went back to his desk with the notebook. There he began to read it and match it up against the enclosed maps, so he could begin to understand the problems that he would face. When he was done with the notebook, he returned it to Graham and informed him that he would be taking lunch.

After a quick lunch in the snack bar, he came back to the office to pick up where he left off on the loadout calculations. They required him to get manuals for all of the wheeled vehicles from Master Sergeant Fykes, and revisit many of the calculations that have been previously done. He also sat down with Fykes to try to get a feel for the number of additional items they would need to haul. By late afternoon, not having heard from McNabb, he called down to the motor pool. When the NCO answered, he was informed that McNabb was not back from his errands.

The Lieutenant assumed that McNabb had tasked him and reminded the sergeant that he needed his Jeep and driver at 0700 in the morning. When he was done with that, he asked Graham if there was anything else he needed to do. Graham told him no, and that he should be dressed and ready at 0600, so he could leave at 0700 for the exercise. That meant he could skip PT. The Lieutenant had to admit that he was not heartbroken by that and looked forward to a change of pace at Brigade.

With the day at an end, he went back to his quarters and set the alarm for 0500. He needed to give himself plenty of time to get dressed and squared away, so his uniform would be perfect when he arrived at Brigade. He also decided to turn in early after heating up a frozen dinner in his toaster oven, while he wrote another letter to his girl. When he was done with that, he read a little bit and turned in early. Fortunately, the other officers who lived in the BOQ were busy in their various units or had come back after a long day's work in preparation for the upcoming holiday. The Lieutenant, thankful for peace and quiet, fell asleep until the alarm went off, startling him the next morning.

Chapter 25
The Loser in the End

After he woke up, showered, and shaved, he shined his best set of combat boots, and put on his best fatigues and field jacket. He also took his rubber duck with him, since he anticipated that the weather would turn ugly at some point during the day. Just for safety, he put a package of licorice and a couple of breakfast bars in his field jacket pockets and took off for the company. Since none of the officers were engaged in PT, he went down to the orderly room for a cup of fresh coffee. He checked in with the First Sergeant to make sure that his Jeep driver was up, since McNabb had not yet come in.

Once the company formation was called, he went outside to the back to remind McNabb that he would be at Brigade all day, and that he wanted to catch up with him the next afternoon to go over the training schedule. McNabb and the company then took off for their daily two-mile run, led by the battalion Motor Sergeant. The Lieutenant took the opportunity to eat a breakfast bar, and when his driver, Harris, pulled up, he offered him one as well, because the mess hall wasn't open until 0700. The driver told him no, but it was all right. He would get something to eat over at the mess hall that served the units where Brigade Headquarters was located. The Lieutenant then jumped in his Jeep and closed the plywood door beside him. The heater, lifted from a five-ton truck, was cranked up on high, so it was already warm inside the cabin, even though there were no windows.

Harris looked at the Lieutenant, who smiled and thought of one of his heroes from the days of his youth, when he watched *The Lone Ranger*. He always liked Tonto, who, in his view, was the patron saint of Support Platoon Leaders since he was always having to go into town to get things and getting into trouble. With that in mind, he looked at the driver, smacked the outside of his Jeep door twice and said, "Get 'em up, Scout." The driver laughed, slipped the Jeep into gear, and drove to the entrance to the training area to take the shortcut to Brigade.

After they arrived at Brigade, he released Harris to find something to eat. He walked into Brigade and was directed to a large conference room where we saw Graham in the corner. He went over and sat down beside him, as members of the brigade staff put up large map sheets that showed their area of the border and the locations of various units. The Brigade S-2, with help from an intelligence officer from Division, would play the opposing force. There were various and sundry red units that covered the battlefield from the border north and east. The battalion's intelligence officers provided the reports from the covering force cavalry squadrons and moved their opposing unit. The operations officers of the battalions, task forces, and companies kept up with the movement of their units as well.

The fascinating thing about the exercise was that as the units moved across the board and the battle unfolded, even though it was a war game, the reactions of the officers in the room became more and more intense. The commanders of the line units requested artillery fire, air support, and increased logistical support. These requests, passed through the S-4s and battalion maintenance officers down to their subordinates, and tasked the support platoon leaders to try to move food, fuel, and ammunition as quickly as possible to various parts of the battlefield, as those items were allocated by the commanders. It was a daunting task, and, as in real situations, tempers flared as the operations officers and the logistics officers tried desperately to handle the onslaught of red unit symbols. Throughout the exercise, the officers kept logs of their actions for the after-action review.

They all worked through lunch, until early afternoon, when fresh coffee, tea, and cookies were brought in. After another two hours or so, the exercise came to an end. Shea came over to Graham and let him know that all of the staff officers who participated would have to attend an after action review the next afternoon. This would take place after command and staff call in the morning. Having heard this comment, the Lieutenant realized that he was not going to be able to catch up with McNabb until the following week.

As the officers filed out to their respective Jeeps, he was happy to see that Harris was up and ready to go. The two of them bundled up against the cold and made it back to the battalion in record time. Harris pulled up outside the S-4 office, so the Lieutenant could jump out. When he reached his desk, he called the motor pool and was informed by the sergeant that McNabb was not yet back from another one of his many errands. The Lieutenant told him that wasn't a problem. He would catch up with McNabb at formation. He took a minute to catch his breath and Graham called him back. After saluting the Lieutenant sat down and Graham asked him what he learned during the exercise. The Lieutenant replied that he really had not appreciated the speed of the battle and extreme logistical demands before.

Graham nodded and then added, "And the life expectancy of a platoon leader after the covering force pulls back is six to twelve hours, Company Commanders twelve to twenty-four, and battalion commanders twenty-four to thirty-six."

The Lieutenant reflected for a couple of seconds before he said, "Wow."

"Yes. Everybody loses."

When evening formation came around, the Lieutenant took his place behind the Support Platoon, while Ascencion read out instructions to the platoon sergeants. The formation was released. Then he caught McNabb and told him that, regrettably, he had a meeting tomorrow afternoon and he would not be able to go over the training program with him until the following week. Consequently, he needed to see him Monday morning, to complete the training program, and begin finalizing the loadout plan, so he could meet with Major Reardon the following Thursday. McNabb nodded, saluted, and took off like a rabbit.

BAD CASE OF LOVING YOU

He hurried through the cold to his "Q" and after pounding up the stairs, stopped by the communal kitchen to see if anyone was there. Happily, Buzz was enjoying a glass of wine with Elizabeth before dinner. Buzz looked up and said, "I have been so busy with dental work since the troops are out of the field, I haven't had a chance to think about dinner. But we have enough bread and leftover meats to make sandwiches, and thanks to a couple of cans of chicken soup, that are heating on the stove, we won't starve."

The Lieutenant grinned at both of them and said, "I'll be back."

After changing, he came back and sat down across the table from Buzz and Elizabeth. Buzz informed the Lieutenant that Bigfoot had staff duty, that Susan and Chris had a unit party, and Bernie was currently living off post again, so they would probably be dining alone, with the exception of Tom Storch, who had stopped by earlier and promised to come back. The Lieutenant stood back up and helped himself to a beer from the refrigerator before returning to his seat. As he did so, he asked Buzz, "What are the holidays like in Benjamin Franklin Village?"

Buzz told him, "It's usually laid back, since the units will go on a half-day schedule, from the middle of December until New Year's

Day. The other thing you can expect is a Christmas party every Saturday, at some level."

The Lieutenant chimed in, "Yes, my brigade is having its party this Saturday. I also understand that Division is having its party the following Saturday, for battalion commanders, brigade commanders and their primary staffs, and the division staff. Fortunately, I don't have to go, since I'm a mere lieutenant and a Support Platoon Leader at that."

Buzz chuckled and added, "I appreciated that, since I'm a mere dentist. Which was good in some respects, but with so many troops having half their days off in the holidays, many of them choose this time to catch up on their dental work, which means that the half-day schedule has the opposite effect on me and my other dentists."

The Lieutenant smiled a little and said, "I never really thought of that."

"Nobody ever does. But it's a good trade off, because when all of you are going out on alerts, I get to stay back in my warm office and wish you the best."

As Buzz was chuckling to himself about that last remark, Tom came in, poured himself a glass of German Fanta, and sat down to join them. The Lieutenant watched closely, as Tom sank deeper and deeper into his unabashed worship of Elizabeth. He could see in his eyes that he was infatuated with her, but the Lieutenant could read in hers that it was hopeless. In any event, the soup had started to boil, so Buzz took the leftovers out of the refrigerator, and began making a sandwich for Elizabeth. The Lieutenant hadn't eaten with them in a few days, so he was heartened by the ham, roast beef, and leftover chicken that was available, as well as some various cheeses and lettuce. They always kept a full stock of condiments, as well, so when Buzz and Elizabeth were finished, the Lieutenant made himself a sandwich and ladled out a bowl of soup to stave off the incessant cold draft flowing down off of the back windows. Tom topped of his glass Fanta and followed suit.

Elizabeth, as usual, was bubbly and excited. She could hardly wait to get back to England. Buzz, although a little more subdued, was also happy that he would be able to get back to Fort Hood after the holidays, when his schedule tapered off. Tom was due to fly out in about a week, for over three weeks of leave in the States, so he was not as morose as usual, although you could tell that he really wished he was going to England with Elizabeth. As soon as they were finished eating, Tom offered to clean up, so the Lieutenant went back to his room to rest and prepare for another day.

TOO MUCH, TOO LITTLE, TOO LATE

The next morning, he made sure that the XO knew he was in ranks for PT. Once he had breakfast in his room, and stopped by morning formation, he went to the S-4 office, where Graham informed him that they had the first orders and logistics overlays for the upcoming winter exercise. The S-4 laid the set of maps and overlays out on a large table in his office. The two of them looked at them and began calculating the number of vehicles they would need to support the TEWT, and the later Mini FTX, with a limited number of armored vehicles and the staff. When they were done, Graham told the Lieutenant to go to the mess hall and sit down with the Mess Sergeant, so he could begin putting together a roster of cooks and drivers to support the upcoming exercises.

The Lieutenant said, "Very well," and departed the office to head toward the mess hall. On the way, he decided to detour down to the motor pool to see if McNabb was around. When he got to the motor pool, he learned Sergeant McNabb had to go on another of his many errands. The Lieutenant then went to the mess hall and found that the Mess Seargeant was gone as well. Being unable to complete his mission, he returned to the S-4 offices, where he found that Graham had gone to Battalion, for Command and Staff call, which left the Lieutenant back in his office to complete the loadout plan.

By noon, since Graham had not returned, he slipped over to the mess hall for lunch and to see if he could find the Mess Sergeant.

After learning that he was still not back, he hurriedly finished his meal and returned to the S-4 office, to await Graham's return. Just as he sat down at his desk, Graham came in on the verge of a major growling episode. He slipped back into his office, closed the door, and then emitted several long, loud guttural sounds. When the growls had subsided, the Lieutenant knocked on the door and requested permission to enter. When that was allowed, he entered and at the S-4's request, sat down in the chair by the desk. After taking a deep breath, Graham informed him that the entire staff meeting was primarily an attack on him and occasionally the battalion maintenance officer, with the XO piling on, since he refused to accept responsibility for anything.

On that happy note, he told the Lieutenant, "We'll be going up for the Pegasus After Action Review in an hour. I expect it to be a continuation of the earlier session that will probably take the balance of the afternoon. You need to just keep calm and quiet, and don't interject anything unless specifically asked to comment. In my experience, anything a support platoon leader brings to the table can only cause more problems."

The Lieutenant was happy to report that he had his load-out calculated, and with the exception of anti-tank mines, he could move the balance of the munitions. Graham told him, "That's not surprising, that's usually what happens whenever a support platoon leader has to configure the loads. We'll simply have to present that to the XO, hoping that he's settled down by then." Graham then checked his watch and told the Lieutenant they would leave in thirty minutes.

As they walked up to Battalion, Graham said, "Our meeting with the XO to go over the loadout is still on for next Thursday. With any luck, he'll be in a holiday mood, and we'll get through it pretty quickly."

"Good. I'm also working on a training program for the drivers, but I'm having a hard time finishing up due to the pushback from the Support Platoon Sergeant, who does not seem to be interested in it. Also, do you know anything about the errands that McNabb has to go on almost every day, for the Mess Sergeant, the First Sergeant, and sometimes Battalion Maintenance?"

Graham responded, "I'm not aware that. Runs for various items need to be made occasionally, and in many instances those taskings go through the NCOs."

The Lieutenant then asked, "Can the Support Platoon Leader at least request to be advised in advance?"

"That's never been done by a Support Platoon Leader, but it sounds like a reasonable request, and I'll get back to you if you make it."

When they reached the headquarters, they walked up to Shea's office. The Lieutenant took a seat with his back against the wall, away from the command and staff table. As the battle unfolded, he could see that the Combat Support Company and Alpha Company Commanders were doing a good job of defending themselves. The Charlie Company Commander, Captain Butler, who was new to the battalion, was also a pretty good infighter. The Bravo Company Commander was having a hard time keeping up and, apparently, was in disfavor with the colonel, who asked him an incessant number of increasingly difficult questions.

Fortunately for the Lieutenant, no one was very excited about asking a Support Platoon Leader anything, and the S-4, a veteran staff infighter, was able to fend off any questions that might have made their way to him. By the time the smoke cleared, the sun was setting in the west, and it was after 1700. The officers were released for the weekend, and licking their wounds, went back to their respective offices. When he got back to his desk, the Lieutenant looked at Graham and told him, "After that, it's going to be some kind of fun at a Christmas party in Heidelberg tomorrow."

Graham just smiled and shook his head and said, "Don't worry. Their wives will be there, and the liquor will be flowing, so things will sort themselves out."

The Lieutenant laughed a little, and responded, "I'm not much of a drinker myself. I'll probably have one drink to be social, and switch to club soda and lemons after that." Having said that, he suddenly realized that Graham was a Mormon anyway, and couldn't drink, so he didn't have anything to add.

TRYIN' TO GET THE FEELING AGAIN

When he got back to his quarters, he rounded the corner for the communal kitchen and looked in. Buzz said, "I've been working late again." There was nothing on the stove, although the colonel himself was in the corner nursing a bottle of beer. He continued, "Elizabeth has to tend to some errands and will be in later."

The Lieutenant asked, "I haven't been able to get out much so far, but is there a good Italian restaurant close by where I can get pizzas like the ones you brought in for us before?"

"Yes, that one is really a small takeout trailer off the street opposite the Class VI liquor store. The Italians there make excellent pizza, as you know."

The Lieutenant answered right away, "Don't you worry then, I'll grab some marks out of my room and bring back a couple of pizzas so we can have a low-stress party. Between the beer, pizzas, and cola, we should be able to avoid starvation." Buzz just looked up and chuckled a bit, "Go ahead, Bob. I'll hold down the fort while you're gone."

He thought back a little and realized that it had been weeks since he'd had a pizza, and how much he missed it. He ran down the hall, opened his door, and picked up fifty marks from his top dresser drawer. Then he shot out the door to his car. Sliding into the driver seat, he slipped a James Taylor tape in the cassette deck, backed up, and drove off to the Class VI store, where he kept an eye out for the little pizza place. Once he was in the vicinity, it was easy to spot, and even easier to smell. It was wonderful.

He went up to the counter of what was, in essence, a large shed, like a concession stand at a football game. He ordered two large pizzas, one pepperoni and another ham. He was told it would take about ten minutes if he wanted to sit in his car out of the cold. He paid for them and sat in his Volkswagen, watching the window to the shop. When they were ready, the owner raised his hand and pointed to him. He jumped out, grabbed the pizzas, and stuffed them and their heavenly smell in his car. He hadn't realized until just then how hungry he was, after being bounced around for days, to various locations.

Arriving back at the BOQ, he juggled the boxes in one hand, turned his head against the cold, and hastily made his way up the stairs and into the building. Once he was in the building, he climbed the flight of stairs up to his floor and made the turn into the kitchen. The usual suspects were there. While Susan and Chris had gone out, Bernie was back again to join them for a change, as was Tom. In addition to Tom, Elizabeth had returned and Buzz was in great spirits. As he sat the pizzas down, he could hear the pelting of sleet on the windows and was glad he had got out just ahead of the storm front. With the weather turning crappy, Pavel decided he wasn't going out on the town and joined them as well, for the food and to share the latest gossip.

After opening the boxes on the counter, Buzz went to his room and came back with paper plates. Attracted by the smell, Bigfoot followed close behind him to stand near the door and take his place at the end of the ad hoc serving line. Once everyone was eating, the Lieutenant was amazed at how much happier people were when they shared comfort food that reminded them of home. Tall tales floated around the table, with beer and pizza, and laughter. The Lieutenant thought back to a line from one of the Flashman novels he read when he was much younger, about officers serving in British Army in the early nineteenth century, "It was good to be young and English then."

That was exactly the way he felt, since, although he was always accused of being a German because of his last name, he was in fact three-fourths British. He laughed a little at himself at the thought, and helped himself to another piece of ham pizza, savoring each bite. Then he remembered how much his girl loved pizza, and when she got off work late at night, often they would try to go out and get one.

This brought on a pang of loneliness, so he backed off that thought, finished his pizza, and opened another bottle of German beer.

Bigfoot was in a particularly good mood. Having finished with staff duty, he figured he wouldn't have to do it again for at least a month, well after the holidays. This brought the Lieutenant back to reality, since he knew he was going to be on duty Christmas Eve and New Year's Eve, and would probably be tagged for another night in the intervening two weeks, although Kowalski owed him one. On top of that, since Christmas Eve was a training holiday, he would have to come on duty in the morning and pull a full twenty-four hours.

When everyone was done, Buzz laughingly volunteered to clean up, which he did by throwing away all of the plates and the boxes, rubbing his hands together, and heading for the doorway, where he beckoned Elizabeth to follow him. The two of them took off down the hallway, giggling like a couple of college kids, leaving an off-campus party. While he listened to the two of them practically running to Buzz's quarters, the Lieutenant excused himself and went back to his own front door.

IN ANOTHER LAND

When he arrived in his room, he opened another beer from the small refrigerator by his bed and sat back on the black easy chair in what passed for his living room and listened to the radio a little. Armed Forces Network was providing updates on the news worldwide. There was continued concern about the Soviets massing more troops on the border with Afghanistan. The Carter administration was caught in the throes of the Iran hostage crisis, and was apparently paralyzed, so the world looked like a more dangerous place. Reflecting on that, he remembered the book that he read when he first discovered Frederick Forsyth. That book was the *Odessa File*, although *The Dogs of War* was his favorite. *The Odessa File* was a close second and he thought to himself that he needed to check out a copy, if he could find one, from the post library and reread it. It had more poignancy for him now, since he had experienced what Sergeant Robb believed to be Neo Nazis attacking his perimeter. He also resolved that since he had a car at last, he would get out into the countryside, to sharpen his German language skills by talking to the residents. He wanted to get a better handle on how they gauged Americans and truly felt about them.

The next morning, the first thing he did was to lay out his dress blues, to make sure everything was in order. He shined his shoes and located the required black bow tie. When he was done with that, he dressed in jeans and went to the PX for a quick breakfast. After that, he walked over to the post library. He wandered through the stacks in the library until he located the books authored by Mr. Forsyth. After bending over to the bottom of one of the shelves, he smiled to himself as he pulled out a hardcover copy of *The Odessa File* that was printed in 1972. Looking around and not finding anything else that appealed to him at the moment, he went to the counter and showed his identification card. He signed the book out, and the librarian slipped a piece of card stock with the due date in a pocket inside the cover.

He left the library, with the sun shining brightly. The weather was temporarily almost balmy for Germany in December. But he could tell that the wind was shifting north so he increased his pace to get back to his quarters as quickly as possible. Upon arriving, he searched through his uniforms for his raincoat. He had only rarely worn this uniform, but was glad to set it aside for use, in case the weather was even more awful than usual, when he was in Heidelberg in the evening.

Having done everything he had hoped to, before leaving for the brigade party, he made a cup of hot tea and settled in, planning to read *The Odessa File*. Doing that in Germany, in the gloom of winter, with the near constant presence of fog and drizzle, was a very different experience from reading it in Kansas, that was generally sunny and bright. It was also very different since he now lived fairly close to the Taunus Mountains, that played a big part in the book. After devouring almost four chapters, he decided to take a nap. That way he would be well rested before trying to drive to Heidelberg. He had never been there, and was looking forward to it, since the university was there, along with a huge castle that was world famous. He also knew that the headquarters of the Army in Europe was located in a military compound that was formally occupied by the SS. This fact also made the book practically leap to life.

By the end of the afternoon, as anticipated, the weather had changed and a light drizzle was falling. Based upon that, he quickly decided to get dressed and leave early, so he could arrive before the sun had completely set and the roads became slick or icy. Tossing his raincoat into the passenger seat, he turned the heater up and eased the little German sports car out into traffic before driving around the circle to enter the Autobahn. Mannheim was fairly close to Heidelberg, so he took his time and enjoyed the drive. He especially liked the little German villages. They were all beginning to decorate for Yuletide, and small Christmas markets had opened up in the narrow cobblestone streets.

I DON'T WANT TO SPOIL THE PARTY

He pulled up to the Army Headquarters compound, showed his ID to the gate guards, and asked for directions to the officers club, as he had never been there before. The guards patiently gave him instructions. He drove slowly to the building, pulled in the parking lot, and listened to a Gordon Lightfoot tape for a while, until it was late enough for him to enter the party. The drizzle was heavier, so he put his raincoat on over his blues, donned his armor officer's blue and yellow hat, and hurried to the entrance. There, he was greeted by a doorman who directed him to the ballroom, where the brigade was having its event. When he arrived at the doorway, he checked his raincoat with a member of the staff and entered the room with his hat under his left arm.

Thanks to a large seating chart, he located the place where he was expected to sit, along with Lieutenant Borovich over in a corner, generally, as far away from the head table as one could get. The Lieutenant made his way to the seat and smiled. Borovich had not appeared yet, but he was heartened by the fact that the other officers from his battalion and their ladies were seated around him. As he looked across the sea of blue uniforms, glittering brass, and glittering evening gowns, he could see that his sister tank battalion was on one side and the infantry on the other, with the cavalry squadron close by. The division artillery officers and other support staff were generally in a corner by themselves but still close to the head table.

Looking over to his right, he smiled as he saw Captain Borman and his wife, flanked by Lieutenant Marr and Lieutenant Matthews, seated with Captain Butler from Charlie Company and his wife, with the executive officers from both companies. The Lieutenant found himself a little lonely for Bravo Company because he had quite a bit of fun there. And it was much more rewarding to be on the line than to be in a staff slot, even though he still had a platoon of truck drivers.

Feeling a little abandoned, he looked around the table to the other name cards and realized that while Borovich would be on one side, the new Alpha platoon leader, Lieutenant Heinrich would also be sitting with him on the far side of the room. Since all of them were single, it looked like, for the most part, they were seated at a leftovers table on the edge of the frontier this year. The Lieutenant thought that was just fine, as he wasn't looking to be engaged with anyone, other than exchanging courtesies, pleasantries, and light party chatter.

At that moment, the room was called to attention for the brigade commander, who entered with his wife and the brigade Adjutant. After the colonel told everyone to take their seats, the Adjutant informed the officers that he would be calling them and their wives up by battalion, after the staff, to the reception line, to make sure that the colonel and his wife, Karol, met all of the officers and wives, and could exchange holiday greetings with them. The Lieutenant smiled to himself, because Colonel Thomas still looked like the very embodiment of John Wayne.

Out of the corner of his right eye, he noticed Borovich as he slipped in behind him and sat down at his left. He looked over and asked him where he'd been. Borovich just sheepishly smiled and told him that he had been on a cultural exchange but had his dog tags around his neck. The Lieutenant laughed quietly to himself and looked to his right just in time to see Lieutenant Heinrich approaching. He was tall and slender, with long Gallic features and a thin mustache. The Lieutenant rose when he came over to introduce himself. With that done, being senior, he suggested that they all try to get to the bar to pick up a drink before the reception began in earnest.

The Lieutenant went to the bar with his new compatriots. After a few minutes, he was able to secure a whiskey sour. He liked that drink because it reminded him of the "I" Bar in the basement of the Fort Benning officers club. There, at the age of nineteen, he drank one, illegally, in the company of a SEAL Petty Officer, with whom he was shooting pool, on the weekend before his first parachute jump in 1975. That night ended with his SEAL involved in a bar fight over a topless dancer and their hasty retreat through the club's kitchen to quarters in order to avoid the wrath of the infantry.

He came back to his seat, to visit with Lieutenant Heinrich a little about past assignments and where they went to school. By that time, the Adjutant called his battalion forward to the receiving line. So, he rose with the other officers and ladies of the "Pacesetters" to meet the colonel and his wife.

As usual, the Lieutenant stayed in the rear of the receiving line, so he was the last officer in his battalion to greet Thomas. When the colonel looked up, he gave him a bright smile, noticing the large ten carat gold VMI ring with a sparkling blue sapphire that the Lieutenant wore on his right hand. The colonel had on his class ring as well, and as they shook hands, once again, he said, "Merry Christmas, Tiger! How are you doing?"

"Just fine, sir. Brave Rifles. How about yourself?"

"Very well, thank you. Veterans," replied Thomas and introduced him to his wife. The Lieutenant moved forward and took her hand and shook it. She told him that she was happy to see another VMI graduate. He told her that there were several in his battalion and at least one in the sister tank battalion by the name of Barclay, whom he imagined she would meet shortly.

Regaining his seat, the Lieutenant listened, as Borovich and Heinrich traded information, and brought each other up to date on the woes of being a tank platoon leader, surrounded by drug addicted soldiers and miserable weather. The Lieutenant smiled to himself and just thought, *Wow. You should have been with me at Fort Bliss, where my entire first platoon came out of confinement, and didn't even have uniforms due to their incarceration in Mexico and the Disciplinary Barracks, before they worked for him in the desert in New Mexico.* With that as a baseline, he decided that this battalion wasn't so bad. Particularly, since there was a hard core of really good non-commissioned officers who could be relied upon in times of greatest trouble.

After telling his dinner companions a few of his Reforger stories, the favorite of which was the German villagers banging on the sides of his tanks with wooden pitchforks and torches, the wait staff began serving dinner. The hubbub wound down as the various partygoers turned their attention to their meals. It was pretty standard club holiday fare: chicken, rice, and vegetables, with an apple tart for dessert. After the whiskey sour, he stopped drinking alcohol and switched to club soda and lemon. Turning away the wine that was offered, he eventually finished with a cup of strong coffee to accompany his dessert.

When that was done, he could see the various participants heading for the dance floor, if they had partners. Seeing his classmate, Barclay "the Mad Dog," in the far corner of the other side of the room, he made his way to him to catch up concerning their various adventures since they were assigned to Mannheim. Barclay's date, an MP officer, was visiting with some of the other guests so the

Lieutenant sat down beside him. The last time that he saw Barclay, he was huddled in a BOQ room he shared with another of their Brother Rats, when he and one of his old VMI roommates came over to their place to ring in the new year, fairly drunk and shivering, because it was brutally cold. Barclay was an amazing guy. He was an erstwhile athlete; football, lacrosse, rugby, and veritable legend on the party circuit at VMI, known for his amazing adventures, and quasi crazy temperament. In other words, he was a lot of fun to be around.

The Lieutenant thoroughly enjoyed regaling him with tales from his many crazy episodes at Fort Knox, Fort Bliss, and now Mannheim. Barclay brought him up to date on his assignment as well, since he had arrived about a year and a half before the Lieutenant, and was poised to become executive officer in one of the line companies. He could see that Barclay's girlfriend was headed back so he rose. Barclay asked him where he was living. He told him and Barclay said, "I have the room on that floor next to Pavel. That place was too crazy, even for me, so I moved in with my girlfriend. Good luck."

The Lieutenant broke off from Barclay to make sure that he made the rounds to visit, at least briefly, with his battalion commander, the XO, the Adjutant, and their wives. When he was done with those courtesy calls, he needed to pass on holiday wishes to Lieutenant Graham and Dana, who were thoroughly enjoying the evening. Dana told him, with a slight smile, that she was glad to hear he was going to make his March wedding date. He smiled back and told them that he planned to leave early, because he wanted to do some traveling in the countryside the following day. He then turned and made his way to his table to retrieve his hat, before picking up his coat, and making his way back to his little gold Scirocco.

He was glad that he had stopped drinking early in the evening since the weather had gotten markedly worse. Snow was drifting across the road in several places, but he made it back to his quarters without incident and climbed the stairs to his apartment. It appeared that many of the officers in his building were in the holiday spirit, too.

Stereos were blaring in the halls. He could hardly hear himself think before he dove through the door and closed it behind him. Eventually, after 2200, the noise died down so he could sleep.

DAY TRIPPER

The next morning, the Lieutenant slowly opened his eyes and was surprised to see actual sunlight coming through the windows of his bedroom. It was after 1000. The sun was shining on what appeared to be a balmy day for that time of year in Mannheim. He could make out the smells of strong coffee and what he swore was bacon permeating his room. He dressed quickly and went down to the communal kitchen to find Buzz, Elizabeth, Susan and Chris having pancakes for breakfast. He walked in and poured himself a cup of coffee before asking if he could join them. Buzz, of course, said, "Yes," so he made himself a plate, poured syrup on everything and sat down. They asked him how his evening was, and he gave them a brief synopsis of what a Christmas party in a line brigade was like. They were all from support units, so they were fascinated by the rigid formality of the event involving the officers from the cavalry squadron and combat battalions of the brigade. The colonel was a cavalryman, so there was also a bit of boots and spurs regalia at the event as well, particularly since the commander of the cavalry squadron was a good friend of the brigade commander who had served with him in the Pentagon, in personnel.

When he was through painting the picture of the party for them, he told them, "I want to drive out into the countryside to look around. Do you have any suggestions as to where I should go, on say a forty-mile round trip, to really appreciate the German countryside?" Buzz and Elizabeth told him that they liked Worms because it was a little more cosmopolitan. Susan and Chris suggested Heidelberg, because of the student population, and the many bars there. He then told them that since he had been in Heidelberg the day before, and had driven through several small Christmas markets, he was interested in some of the villages. "In that case," Buzz told him, "you should cross the Autobahn to Weinheim, and take Highway 38 up into the Odenwald." The Lieutenant thanked him for his advice and offered to clean up the breakfast dishes. In the absence of any objections, he washed the griddle and plates and put the leftovers in the refrigerator for the inevitable afternoon scavengers.

Back in his room, he consulted the ESSO map book to locate the highways that Buzz suggested. Then, after cleaning up, he dressed in his German student clothes, snatched his ski jacket, just in case, and ran down the stairs to hop in his car. He was genuinely excited to get out into the countryside and see something other than endless rows of drab barracks and government housing. It was almost noon by the time he crossed the Autobahn and picked up Highway 38. He drove down into Weinheim but found it was really too large a city to interest him much, so he went on up a smaller road to Birkenau. Once there, he found a place to park his Volkswagen around the corner from a German bakery and pastry shop. Apparently, it was the custom, after church and lunch, for Germans to go to such a shop and enjoy coffee or tea with amazing cakes and pastries.

Entering the shop, he felt a little underdressed, but was welcomed immediately, because he looked like a German university student. After being shown to a table, he was served a glass of water and given a menu. Based upon his limited German, he selected what appeared to be a multi-layer cake with chocolate frosting. While trying to order, he explained that he was an American and struggling with the language but would try. The waiter was patient and eventually came back with a cup of dark roast coffee and the most amazing cake he had ever eaten in his life. He savored it and enjoyed watching the locals as they chattered and he imagined gossiping as they took their time enjoying a relaxed atmosphere and sweets. By the middle of the afternoon, he finished and settled up with the waiter, before working his way back to his car and driving further up into the mountains to the little village of Mörlenbach.

When he arrived in Mörlenbach, he was struck by two things

initially. One was a wild game restaurant that had what appeared to be a stuffed boar, deer, duck, and pheasant hanging from the sign outside the establishment. Once he had parked, he walked by the windows and read the menu that was posted on the sidewalk. It was fascinating. He promised himself that someday he would come back and have a meal there. But he was also struck by the center of the village, where many booths were set up to sell Christmas ornaments, mulled wine, and pastries. He passed those and noted that in most of the shop windows, he could see various small objects made out of what appeared to be ivory.

ON THE BORDER

Once he had explored most of the village, at last, he located what he was looking for: a good German guesthouse, with music and lights. He walked back to his car and drove up to the intersection where the guest house was located. Then he drove his car around the corner so that the green US license plates were not readily visible. Then he walked back to the front door and entered the guesthouse to enjoy its warmth as he left the cold outside. The customers in the restaurant looked up briefly and then simply ignored him as he took a seat at the bar.

Sitting down, he ordered a German pilsner, in German. Although he didn't realize it, he spoke mostly with a Hessian accent since his battle position was in that state, and most of his time during Reforger and on the border was spent there as well. When the waiter came back with the beer and a small menu, he ordered a wurst salad with French fries. He was served promptly. While he was waiting in silence, another German, more middle-aged, came in and sat at the bar beside him. When the German had looked at him to try and figure out where he was from, the German made the assumption that the Lieutenant was a student who had traveled over from Heidelberg and asked how he liked it there. When the Lieutenant told him that no, he was an American, the German looked at him incredulously, and asked, "You are a child of the blood?"

The Lieutenant showed him his identification card, and answered in German, "Yes, I am a child of the blood." Hearing this, the German smiled broadly, so the Lieutenant added, "I am a good German."

His new friend caught the waiter's attention and the Lieutenant noted that he asked for a hunter schnitzel and salad. The waiter returned to the kitchen. The Lieutenant asked in broken German if the schnitzel was any good. He was assured that it was so the Lieutenant thought to himself that one day he would be back to try that as well. When the meals were delivered and both of them were eating, the Lieutenant asked, "How do the German people you know feel about American?"

The German thought for a couple of seconds and then told him in fair English, "Americans are to be endured, but they aren't liked. In fact, I have done some work from time to time, for the US Army in Europe. And actually, I have some friends on the four star's staff." He went on to add that, "I tried to get my mother to meet some of them, and one general in particular, who wanted to visit her and take her a gift because he thought so highly of me. She simply wouldn't hear of it; she told me that although she didn't hate Americans, she just didn't want one in her house." The Lieutenant nodded and both of them turned to their meals in earnest.

The Lieutenant, having promised himself not to have more than one beer, asked for a cup of hot tea to steel him against the weather that had turned brutally cold again, even more so since he was in the mountains. He finished the tea, said goodbye to his new German friend, and walked out the door into light snowfall. After giving his car a little time to warm up, so the heater could break the cold, he drove back to Mannheim, feeling pretty good about his first real adventure into the countryside. Driving through the darkness, he could feel the dark primal nature of the forest of the Odenwald. It reminded him of *The Odessa File*.

Back in his apartment, he made himself a cup of Swiss Miss hot cocoa, a habit he picked up when he was a cadet back in Virginia. As usual, it warmed him up and made him think of happier times, when he and his roommates would lay back in their cots to swap tall tales and listen to the latest music on their stereo. Then, he retrieved the copy of *The Odessa File* he had checked out from the post library, and read a few more chapters before setting his alarm and turning in.

Chapter 26
Will You Ever Win

The next morning, he woke to a cold and drizzly Monday. After dressing in his PT gear, he jogged over to Headquarters Company to first make sure that Major Reardon saw that he was there, before joining at the rear of the formation for a short run and a few calisthenics after it. In all, PT was only about twenty minutes due to the abysmal weather. Following a quick shower, he downed a breakfast bar and a cup of instant coffee in his quarters, before he returned to the S-4 office. Over the next couple of days, he completed the loadout plan, and played cat and mouse with McNabb, who always seemed to have something better to do rather than sit down to go over the training schedule. By Tuesday afternoon, completely frustrated, he finally caught him in the motor pool, and walked with him back to his office. There, McNabb finally produced a training program that was a little more than half complete. From what he saw, the Lieutenant knew that McNabb was bright and that he also knew what needed to be done. So, he thanked him for what he had completed so far and told him that he needed to see the final plan by Friday.

He told went on to tell McNabb that he would be busy going

over the loadout plan with the S-4 the next day and meeting Major Reardon on Thursday morning. After that, he figured that he would have several revisions to do and wanted to have the training program finished by Friday afternoon, so that they could begin it in earnest after the holidays. That should give the drivers an opportunity, even with the half-day schedule, to complete maintenance on their vehicles, and support the quarterly services for the tanks of Alpha and Charlie that were to be finished by year end.

On Wednesday morning, the Lieutenant came in to meet with the S-4 at 0800. As he walked into the office, he noted that Graham was not there. Sergeant Dryer told him that he had been called up to Battalion, on emergency basis, to meet with the XO. Other than that, he didn't have any idea what was going on. A little after mid-morning, Graham finally came back and practically ran to his office and slammed the door, before growling uncontrollably for a couple of minutes.

When the growling subsided, the Lieutenant knocked on Graham's door. When given permission, he reported, and at the S-4's behest, took his normal seat by the desk. Once he had composed himself, Graham told him that Reardon was in a particularly foul mood for some reason, and that it had much to do with the food service in the mess hall. Graham then went on to tell him that, given the constraints placed on him by the battalion commander, and the number of foods that could not be served, and to keep the mess hall within budget, he and the Mess Sergeant had to come up with a variety of meals, many of which the soldiers simply did not like. Consequently, he was stuck and frustrated. But he was also interested in seeing the Lieutenant's plan, so the Lieutenant excused himself to get his binder and brought it back into the room, placing it on the S4's desk.

The S-4 had been through the loadout plan exercise at least twice before with the previous support platoon leaders and a previous executive officer. Give that experience, he was able to quickly review and critique the Lieutenant's plan. With the exception of a couple of items, he did not have anything to add.

One item in particular came as no surprise: he and the Lieutenant simply did not have the space or tonnage available to haul anti-tank mines. The only real solution was to issue them to the tank crews and have them strapped to the tanks to be conveyed to the front lines and emplaced in front of their positions. The Lieutenant nodded his head in agreement. He had already come to the same conclusion, and after talking with Master Sergeant Fykes, was assured it was the only realistic option.

With the review complete, Graham asked him if he had ever caught up with the Mess Sergeant yet to go over the taskings for the remaining of holiday season, the upcoming TEWT, and the mini-FTX. The Lieutenant told him no, he had not, so Graham told him to catch the Mess Sergeant that afternoon and cover those items, as they might need to be addressed with Reardon the following day.

Feeling pretty good about himself, the Lieutenant decided to grab a hamburger at the snack bar, before coming back to chase down the Mess Sergeant. A little after 1300, he came back to the office and called the mess hall; he was excited when the Assistant Mess Sergeant told him that in fact the Mess Sergeant was there. The Lieutenant told him to have him stand by so he could come meet with him. Then he practically ran to the mess hall, and into the Mess Sergeant's small office, beside the serving line. When he arrived, he asked the Mess Sergent to join him in one of the booths to go over the upcoming taskings.

The Mess Sergeant was a very large man, with huge arms and hands, and the most bloodshot pair of eyes the Lieutenant had ever seen. He spoke well, but a little dully. Nevertheless, he addressed the Lieutenant's concerns and assured him that everything would be ready for the upcoming battalion Christmas party and Field Day for the soldiers and their families. He also identified the cook assigned to Alpha Company and his trailer with one assistant cook to support the tactical exercise in January. In addition, although it had not been finalized yet, he assumed that those two and the same field trailer would be used to support the mini-FTX as well.

After finishing with the Mess Sergeant, the Lieutenant went back to his office to make a few minor adjustments to the loadout plan, particularly some of the diagrams he had to free hand, with a mechanical pencil and a ruler. When that was done, he called down to see if McNabb was in the motor pool. Apparently, he was on another errand for the battalion, so the Lieutenant walked down to check his soldiers in the motor pool. Finding that they were busy and well taken care of by the Assistant Platoon Sergeant, the Lieutenant returned to his office to lock up.

WE'VE GOT TONIGHT

Returning to his quarters, he wrote a quick letter to his girl. After thinking about it, he decided to walk over to the officers club for a real meal. It was "Steak Night," so he enjoyed an Irish sirloin, and a small glass of red wine at a single table, before coming back to the "Q" to turn in. Before long, he drifted off to sleep, thinking about his upcoming wedding. He also thought that things were just going too well. The next thing he heard was a horrific bang and pounding on his door. He looked at the clock and saw that it was a little after two in the morning. When he went to the door, he asked, "Who's there?" He was told it was a soldier with a rifle from Headquarters Company who had been sent to get him because they were going on full alert.

The Lieutenant hollered out to the driver that he would be there as soon as he could dress and grab his bags. Once that was done, he shaved and jumped in the shower to wake up. Then he hurriedly dressed and slipped on his field jacket. before strapping on his A bag and slinging his B bag over his left shoulder. Locking the door behind him, he loped down the hall and down the steps, before running, in that slow and deliberate pace known as the airborne shuffle, to the gate guards. There he showed his ID and trundled to Headquarters Company. When he got there, he was practically overjoyed to see that Harris had his Jeep already parked near the edge of the courtyard. He slung his bags into the back seat, next to the driver's bags. He also noted that Harris had already drawn his rifle and ammunition. The Lieutenant told him he was going to the arms room to draw his weapon and also coordinate with McNabb to make sure that the drivers had their rifles and fifty caliber machine guns humped to the motor pool to be installed in the ring mounts on the trucks.

The Lieutenant reached the arms room, where he had a brief encounter with the armorer, who reinforced upon him the fact that he was not issued a pistol, but a rifle, and several mags of 5.56 cap and ball ammo. When he had finished signing, he inserted one clip in the rifle and shoved the others in his field jacket pockets. Then, running upstairs, he passed by the S-2, who looped the lanyard with a CEOI over his neck. Exiting the front door, he ran to his Jeep and had Harris drive him to the S-4 office, which was open, so that he could grab his helmet and load bearing equipment that was hanging up on the coat rack near his desk. He slipped the mags into his ammo pouches, strapped on his steel helmet, and ran back to the Jeep where he placed his baseball cap inside the top of his B bag. Then he set his radio to the logistics and command frequencies, and slapping the side of the Jeep, looked at Harris and said, "Get 'em up, Scout." They made their way as quickly as possible to the area of the motor pool where his platoon was located.

When he arrived at the motor pool, he jumped out of the Jeep and walked up to McNabb, who was busy directing the drivers, as they worked in pairs, installing the M-2 machine guns. Once he had his attention, the Lieutenant spoke with him briefly, to coordinate getting out of the motor pool as soon as the Lieutenant received an order from the S-4. The Lieutenant asked, "What's the normal order of march?"

"Sir, your Jeep will be followed by the cargo 5-tons and fuel 5-tons, and then the cargo GOERs and fuel GOERs. The fuel GOERs will be last so that one of them can stay behind to top off any armored vehicles that need fuel. One of the five-ton fuel trucks that dispense

MOGAS will be first out, after the cargo 5-tons to take up the position, just off the road near the gate to the training area, where it can refuel any Jeeps that are low on gas."

The Lieutenant quickly followed up, "How are we supposed to retrieve the refueling vehicles to join the platoon since I only have one radio?"

McNabb smiled and said, "Sir, that's your job. With that Jeep and radio, you have to ride herd all these trucks, rounding them up, leading them, moving them, and bringing them back."

"How am I supposed to coordinate with you?"

"Sir, you'll have to get ahold of the S-4 NCO, who will then get with me. I'll keep one truck driver near him to run as a courier and let me know what's happening."

"Do the drivers of the fuel trucks know what they were tasked for?"

"Yes, sir, they do. It's the normal alert procedure."

The Lieutenant, feeling a little better, peeled the Velcro flap on the OD band back to reveal the face of his diver's watch to see that it wasn't even four in the morning. It still was pitch dark, cold, and drizzling.

With the machine guns mounted, McNabb told the Lieutenant that he would be taking up his position on the passenger side of the lead cargo 5-ton to await the signal to move out. The Lieutenant, feeling chilled and damp, went back and hopped in his Jeep where Harris had the heater on maximum output. Then the Lieutenant remembered that he had a trailer, so he told the driver to go ahead and hook it up, and put their bags in it, so they would have it available when they got to the training area. While the driver was hooking up the trailer, the Lieutenant decided that if he was going to the actual woods for any length of time, he would ditch the trailer because it was just too hard to maneuver and run around with it. However, for the present, it was a good place to stash any items that needed to be hauled and didn't already have a place in his jeep.

Harris drove him back to a position near the entrance to the motor pool, where he could finally observe the happenings. He realized that the officers of the battalion were literally running from vehicle to vehicle in full panic mode. When he saw the S-4, he ran up to him and asked what the hell was going on. Graham told him that the alert had not been called by USAREUR, but by the brigade commander, to test the alert and loadout procedures for all of his subordinate units. Consequently, members of the brigade staff and the commander were patrolling the area, in preparation for a full, and what would undoubtedly be blistering, critique at the end of the exercise. No sooner had the S-4 stopped talking, than the Lieutenant heard a frantic transmission from the XO on his radio, whereupon he jumped in his Jeep and Graham crushed the gas pedal, leaving the Lieutenant in the warm cloud of his exhaust.

As he watched the S-4 disappear into the darkness, the Lieutenant looked over his shoulder to the S-4 577 track, that was parked and running near him. Walking to the ramp on the rear of the vehicle, he climbed up it to see Sergeant Dryer and Sergeant Brown. When he appeared in the red light of the dome lamps in the vehicle, Dryer first gave him a broad, bright smile, and then fished in a large bag beside him. He brought out a large Army-issue thermos and handed it to the Lieutenant. He then told him, "Sir, you most assuredly will need this, because you will run on coffee night and day. My suggestion is that you get it full of hot black coffee as soon as possible and in the future, before you go to the field, lay your hands on some Café Vienna or Café Mocha instant coffee and mix several teaspoons with the coffee so that you have mocha. Because it will have a lot more caffeine and mocha still tastes good, even when it's cold."

The Lieutenant smiled, thanked him, and filed that good information in his memory. He also patted his pockets to make sure he had a full load of breakfast bars and licorice with him just in case. As he made his way back to his Jeep, he thought he could hear the S-4 on the admin channel calling for him. When he got back, he grabbed the handset and answered. He was ordered to move his platoon out through the battalion's rear gate, and into its positions in the assembly area as designated in the SOP. The Lieutenant acknowledged the order, placed the handset back on the radio, and ran up and down the line of his trucks, screaming at each driver as to what he wanted him to do, before returning to his Jeep. He hopped in and Harris pulled in front of the first five-ton truck to lead the platoon out onto the street and toward the training area.

Once he passed through the gate to the training area, he looked over at the five-ton fuel truck parked to the side. He then motioned for Harris to pick up the pace, and looked in the rearview mirror, to see the long snake of tanks following behind him. After five minutes, they came closer to the wood line, so he signaled for his driver to slow down, and the trucks followed him more slowly around the training area to that portion of the wood line where the battalion trucks were to form up and await further instructions. It was still dark and misty, so he dismounted the Jeep and used his flashlight to help guide the trucks into position, so they were spaced out as evenly as possible. When all his trucks were parked, he went to McNabb and told him that he would be driving back to garrison, to lead the two fuel trucks left behind in the assembly area once all of the battalion's other vehicles had departed.

The tank companies had already taken up their positions, as had the various platoons of the Combat Support Company. The staff vehicles had also departed, to set up their field headquarters, so the Lieutenant pulled over by the side of the gate and watched as the trucks and tracked vehicles from Headquarters Company, plus its Maintenance Platoon, and Battalion Maintenance slowly ground through the rear gate into the training area. When it looked as though all of the vehicles had departed, he smacked the side of the Jeep door and told Harris to go back to the motor pool. When they got there, it was deserted, except for the fuel GOER, that he signaled to follow him. He then went back to the gate and signaled for the five-ton fuel truck to follow him as well. With both of those vehicles in tow, he went back to the assembly area, dismounted his Jeep, and parked them.

After they were in position, he breathed a sigh of relief, but then thought about breakfast, and how the mess hall was going to feed the battalion. He knew it was packing up everything to move, while some, if not most, of the food was already cooked. With that in mind, he went over to see Sergeant McNabb to find out where the S-4 should be located. McNabb pointed out an area in the wood line about a hundred yards away. So, the Lieutenant jumped in his Jeep and Harris drove him in that direction. He dismounted when he got to the S-4 577 and went around back to see Sergeant Dryer. The S-4 was busy, in an emergency meeting with the battalion XO and someone from brigade staff, leaving Dryer as the senior soldier in the headquarters for the S4. He asked Dryer what they were supposed to do about mess service. He told him, "Sir, just stand by and wait. The Mess Sergeant was slow bringing up the trailers, to hook them to trucks and link up with their companies. But he kept some cooks behind with him on the off chance that the alert would be called off early so the soldiers can get back and have breakfast, even if it is a light one."

IT'S ONLY ROCK 'N' ROLL

After sitting in the cold for over an hour, the radios in his Jeep came to life, as the first rays of sun appeared on the horizon. The Lieutenant listened as the command net crackled and the line companies were ordered to stand down and return to Garrison. The battalion S-3 then gave individual orders to the companies and they moved out, with Alpha in the lead, followed by Bravo and then Charlie. The Lieutenant, meanwhile, went back to Dryer and asked, "Does the Mess Sergeant have any way of knowing that they were coming back, so he can feed them?"

"No."

The Lieutenant ran back to his Jeep, slapped the side of the door, and told Harris to get him back to the mess hall as quickly as possible. When they arrived, he briefed Sparks, who immediately went into full breakfast mode with the cooks he had available. He was short-handed, since at least some of the cooks had departed with the trailers and trucks to join their line companies; they would just have to fill in when they returned.

Finished there, he and Harris rushed back to his platoon in the assembly area. As he pulled up beside his lead cargo truck, he could hear the various platoons of Combat Support Company rolling across the area, heading for the gate. Then he received a radio call from the S-4. He went to his track and Graham told him that he needed to take his cargo trucks to the PSP bunkers to meet him and the XO. Sergeant McNabb could lead the fuel trucks back to garrison because they would be needed there to replenish the fuel on all the vehicles that had gone out on the alert. The Lieutenant ran back to McNabb and gave him the order. McNabb went from fuel truck to fuel truck, relaying the order, and the Lieutenant went from cargo truck to cargo truck, relaying his. He then began leading the cargo trucks to the PSP as the fuel trucks, under McNabb, went back to the motor pool.

When all three officers had arrived at the PSP and the cargo trucks were lined up, Major Reardon, after a cursory inspection, turned to the Lieutenant and said, "I know we were supposed to have a loadout review today, but I've already been informed by the staffs of the other battalions and the brigade staff, that we'll be engaged in a series of after action reviews and rewrites of the SOPs for the next few days. So, Lieutenant, I need you to completely revise your plan. Due to the upcoming holiday schedule, and the number of meetings that will be taking place, I will not be able to visit with you until after the first of the year."

Inwardly, the Lieutenant felt relieved as the XO took off, Graham chimed in that he would see him later in garrison and released him to take his trucks back to the motor pool so his soldiers could eat. The Lieutenant ran to his Jeep, signaled the trucks to follow him, and they all drove back to the motor pool as fast as they could.

Once he was back in garrison, he could see that McNabb had the fuel vehicles lined up to refuel the tanks and other armored vehicles as they came in from the field. The five-ton fueler for Jeeps was also parked just inside the motor pool so drivers could pull over and refuel as well. The Lieutenant went over to McNabb and asked, "Did you have a chance to eat breakfast?"

"No, sir."

"Alright. I'll spell you so you can get something to eat, and when you get back, I'll go. Also, I need you to rotate the troops who are refueling so they have a chance to eat."

"Sir, I've already done that. Half the drivers are at breakfast."

The Lieutenant replied, "Very well," and felt better about his relationship with McNabb at that point, although he still did not seem to be at all enthusiastic about dealing with training.

As he watched McNabb walking to the mess hall, he turned to Harris and told him to go grab chow but leave the Jeep running, so the Lieutenant could continue to monitor the radio, in case the XO or S-4 was looking for him. He settled back in his seat, in the warmth blasting from the heater, relaxing. In a little less than half an hour, he saw McNabb and those drivers who had eaten come out of the mess hall. They were still carrying their rifles, so they walked over to Headquarters Company to turn them in. When McNabb returned to the motor pool, the Lieutenant ordered him to man the radios so he could go to breakfast with the rest of the platoon and turn in their weapons. When his drivers walked into the mess hall, the Lieutenant decided to go down to the arms room to turn in his rifle and magazines, ahead of the rush. The armorer signed to receive them. When that was done, the Lieutenant slowly walked back down the hallway, which was almost pitch black and then up into the cavernous dark of the first floor of the company. He stepped out into the gray sunrise and took a deep breath before heading to the mess hall.

. . .

As he came through the door, he saw that Lieutenant Matthews and Lieutenant Borovich were back in the officers' mess section, so he took a tray, and after it was filled, went to join them. Since he hadn't had much to drink, this time, in addition to a large mug of coffee, he filled up a tumbler of apple juice and drank almost all of it at once. He was feeling hungry, tired, and trying desperately to rejuvenate. Both of the second lieutenants looked at him and grinned. Matthews said, "You look worn out. Where the hell have you been?"

"As Support Platoon Leader, I've been almost everywhere. I've been out to the PSP and back. I've rounded up and herded trucks, shepherding them there and back to the motor pool for refueling vehicles, as fast as we can. And I met with Reardon for a little prayer meeting as well. Looks like I'll be busy for the next couple of weeks trying to reconfigure the loadout process, to come up with a way to move all of the stuff in the bunkers."

Both lieutenants then chuckled, because, as the tank platoon leaders explained, once their tanks were up and out the gate to the assembly area, it was pretty much a holiday for them, until they got the order to return. Now it looked like the battalion would spend the balance of the day and most of the next, before the weekend, putting the battalion back together. The Lieutenant quickly added, "Yes, and from what the XO told me, the staff can look forward to an entire week of brigade meetings to try and improve our alert procedures, to get to battalions out the door faster." Having said that, the Lieutenant looked at his watch and realized that he needed to get back to his platoon and then see the S-4. So, he rose, said, "Adios," dropped off his tray, and headed for the door.

When he arrived at the motor pool, he asked McNabb if there had been any radio transmissions while he was away. There had not been any, so he asked him what they needed to do for the rest of the day. McNabb answered, "Since all the vehicles are dirty and muddy from being in the assembly area, we need to get them all washed and then finish as many as possible that are going through maintenance, before the half-day schedule starts Monday. Also, I know that I owe you a finished training schedule, so I'd like to present that tomorrow morning. That way it's out of the way before the half-day schedule kicks in, and we have a hard time, with reduced manning, keeping up with any taskings."

For the first time, the Lieutenant smiled, and told him, "I look forward to seeing it at 1000 in the morning." The Lieutenant then hopped in the Jeep and had Harris take him to the S-4 shop, where he dismounted and went in, carrying his bags. The S-4 was back, so the Lieutenant asked him if he could release Harris to do maintenance. Graham quickly agreed, and the Lieutenant went to the door, where he told Harris that he was released to do maintenance in the motor pool. Having said that, he returned and went back to Graham's office door where he knocked, and after being asked to enter, came in to take his normal seat by the desk.

As he sat down in the chair, he finally realized that he was still wearing his steel helmet and his load bearing equipment. He eased out of the LBE and set the helmet down in his lap. Graham was obviously upset, but just too tired to even raise a feeble growl, so he went right to it. Essentially, based on the brigade commander's observations and those of his staff, everything that could go wrong during the alert procedure had gone wrong. And, instead of a coordinated military exercise, it looked more like a Chinese fire drill, without discipline. Consequently, all primary staff officers in the battalions and the brigade staff officers, together with the executive officers from all the units, would be meticulously going over the SOPs and standing orders to improve upon them. Fortunately for the Lieutenant, the S-4 told him that as Support Platoon Leader, he was not expected to attend any of these meetings or be directly affected by them, with the exception of some changes in the loadout procedures, and alert procedures with regard to the Support Platoon and the need to get his vehicles up and out sooner.

In addition to that, however, in anticipation of the brigade meetings, the battalion commander and his staff would have preliminary meetings. Also, the brigade staff officers would rotate through the various battalions for meetings on site. Consequently, as Support Platoon Leader, and as mess officer, he would be in charge of coordinating with the Adjutant to see that all of the meetings were supplied with coffee, tea, Army lemonade, and light snacks.

Graham went on to add that since the primary staff officers were tied up, the platoon leaders and senior NCOs would be pulling duty during the holidays. As an outgrowth of that, he had been told to inform the Lieutenant that he would have battalion duty the following Thursday. Lieutenant couldn't help himself and blurted out, "That's great. Kowalski owes me a duty day anyway, so I'll check with him."

Graham looked at him in amazement and said, "Don't bother, the battalion maintenance officers and the brigade maintenance officer will be working around the clock and in meetings from now, until the end of the year, trying to get all the vehicles off deadline during the half-day schedule." With that, the Lieutenant resigned himself to the fact that he was going to pull another night of duty, in anticipation of his Christmas Eve duty. As Colonel Shea would say, "When it rains, it pours."

Graham looked down at his watch. Realizing that it was past midmorning, he told the Lieutenant to go up and see the Adjutant right away, to try to get a jump on this meeting schedule and what was needed to support the various conferences. Since the first one would probably be tomorrow, he needed to get on it right away and make sure that Sergeant Sparks was brought in the loop. The Lieutenant jumped up and placed his LBE on a coat hanger and placed it on the coat rack at his desk, with his steel pot on the top before realizing that he left his newly won thermos in one of the compartments in the Jeep. Since he still had his field jacket on, he put on his ball cap that he pulled out of his B-bag and made haste to the motor pool. There he saw Harris washing the outside of the Jeep. He walked up and told him that he had forgotten something; he grabbed the thermos from its place in a small rear compartment, and thanked Harris for doing such a great job, before beating feet up to see the S-1.

THE TIMES THEY ARE A-CHANGIN'

When he got to Battalion, he walked past the duty NCO, to the Adjutant's office, where he found Captain Steves in a sullen mood, similar to that he had experienced with Graham. When he walked in, Steves acknowledged his salute and then asked him to take a seat beside his desk. When he had done that, Steves told him he needed to take notes, so he pulled the notepad out of his left breast pocket along with the pen he kept there and jotted down the schedule of meetings at the Adjutant gave them to him, as well as the spot times for the coffee and other food items that would be necessary for each.

When he was through, Steves added that of course, it was all subject to change, so he would need to man his desk by the phone during the half-day schedule, in order to make any adjustments. The Lieutenant then reminded him that since he had duty Thursday night next week, he would probably be off most of Friday to try to sleep. Steves smiled with a wry grin and then said, "Well, Lieutenant, you can try. My suggestion is that you nap next to the phone in your office and be ready, because all next week, we'll be meetings, after which, we will all be trying to enjoy the holidays with our families." The Lieutenant just shook his head, in acknowledgement, then rose, saluted, and took off for the mess hall.

He hunted down the Mess Sergeant and asked him to join him in the booth that was in the corner of the enlisted mess. Sparks sat down across from him. The Lieutenant briefed him on the plan and the necessity for furnishing refreshments for the various meetings. As he went over them with Sparks, he made a second copy on a page taken from his notepad and gave it to him. Sparks looked at him, a little bewildered at first, because he wasn't really planning to support anything other than basic meals during the half-day schedule. When he voiced that concern, the Lieutenant told him, "I don't see any way around this. I know the brigade commander from a previous life in Texas. Colonel Thomas is a genuine, certified, no bullshit individual."

Sparks then acquiesced and asked him what his first tasking was. Again, the Lieutenant looked at the schedule and gave him the news that he needed to be ready with everything in place in the battalion conference room by 0730 the next morning to include coffee, tea, cookies, doughnuts and whatever else was available, so the staff could go to work in earnest.

When he walked out of the mess hall, he was amazed to see that the entire battalion was a complete beehive of activity. Soldiers and vehicles were moving everywhere simultaneously. So much was going on, he decided to walk down to the motor pool and check in with McNabb and the platoon.

When he got there, he could see that his truck drivers were busy washing down their vehicles. A couple of the five-ton truck drivers were changing tires that had blown out after driving through the assembly area. He walked over to the drivers and asked them how they came to have the flats. The drivers told him that the tires were thin and worn anyway, so if they ran over a stump or a sharp piece of underbrush that had been cut back or run over by tank, they would pick up a flat. The bad tires had been rotated into the dual tire sections. That way the flats didn't keep the vehicles from moving. But they still had to be repaired when they returned. Since they didn't have new tires, they would have to patch the flats and place them back on the trucks.

Looking around, he searched for a noncommissioned officer. He finally found an E-5, the five- ton section sergeant in the motor pool, at McNabb's desk. He walked up and asked, "Where is Sergeant McNabb?"

"Sir, he had another tasking from the mess hall to pick up items and equipment, some of which might be needed for the meetings that were planned at Battalion."

The Lieutenant shook his head, silently and thoughtfully, and then asked, "Are there any issues with the platoon?"

"No, sir."

"Very well, I'll be in my office the balance of the day." When he returned to his desk, he decided that he would do his best to stay alert at least until 1700 and work on the loadout procedures until then.

By lunch, having finished two more cups of black coffee, he felt awake enough to eat a piece of licorice and keep working. The S-4 was up at Battalion, in a meeting with the XO and the other staff officers in anticipation of the next morning's meeting with the battalion commander and the following meeting with the brigade staff. Having worked on the loadout procedures until its head was spinning, he put together another series of reports for himself. On a sheet of paper, he listed the bumper numbers and type of each of his vehicles to include the Jeep. He then added columns for the most recent maintenance, upcoming maintenance, and availability of his trucks. Then he called down to headquarters company maintenance and went over the listing with the company motor sergeant, who gave him the status of all of his vehicles. With the exception of the two five-tons that were down for flat tires, and one cargo GOER that was in semi-annual maintenance, all of his trucks were ready to roll. So, he noted that on his report and ended the call.

At the end of the day, he made the First Sergeant's evening formation and was standing behind the platoon when McNabb slipped in front of it. When the company was released, he asked, "Where have you been?"

"I've been running errands for the Motor Sergeant and picked up a couple of things for Battalion Maintenance, sir." The Lieutenant then reiterated to him that he really needed to call the S-4 shop and let him know when he was going to be out of pocket.

McNabb acknowledged this, and then told him, "I'm looking forward to our training schedule meeting tomorrow." He then saluted and the Lieutenant returned it before heading back to the BOQ.

When he got back to his quarters, Bigfoot let him know that Elizabeth was leaving the next afternoon to go back to England, so she and Buzz were out for dinner and everyone else was on their own. The Lieutenant thanked him for the update and walked down the hall to his room, where he surveyed the stock of frozen dinners he had available for just such occasions. Finally, having made a selection, he pulled it out, and inserted it in the toaster oven to warm up, while he took a shower and relaxed. After eating and drinking a beer from his fridge, he decided to write his girl a letter letting her know that he would have duty again the following Thursday, but that on the next day he'd be able to get time off to make it over to the bank and call her on the Friday before Christmas. He hoped to catch her in the morning, because he knew she would be going home for Christmas that weekend.

Chapter 27
Right Back Where We Started From

The next morning after PT, he took his new thermos in the kitchen to wash it out. When he was through, he took it to the office and set it down on his desk before running up to Battalion to check with the Adjutant to make sure everything was ready for the first morning meeting. Sparks was there, and all was in order in the battalion conference room, so the Lieutenant thanked him before returning to the S-4 office.

Back in the office, he filled his thermos and used it for the first time to see if it kept beverages warm. It worked well. He enjoyed the coffee and working on his reports through the early morning, until McNabb came in at 1000. When he arrived, McNabb was squared away, as always. In fact, he looked like a recruiting poster. When the Lieutenant told him to take his seat beside him, he did and laid out his training program for all of the personnel they had at the time. This included training for all the drivers except one, his Jeep driver, who would be leaving over the holiday.

The Lieutenant then asked, "Have you had any luck so far securing a new driver?"

"Not at the present, but I have heard a couple of rumors out of two of the line companies about soldiers who want to transfer so I'll follow up." Having reviewed it in detail, the Lieutenant concluded that the training program was a good one and asked McNabb when he thought they would be able to implement it.

McNabb looked at him thoughtfully, and said, "That may be by the middle of January, if the weather is good. Conditions are usually pretty brutal in Mannheim because of the intermittent ice storms, so I can't advise doing driver training in the maneuver area."

The Lieutenant thought about this for a minute, and then told him, "We'll revisit this after New Year's. If the weather permits, perhaps we can get some training in before the winter exercise coming up in January, followed by the larger winter exercise in February." McNabb agreed and then told the Lieutenant that he needed to absent himself from the battalion for a couple of short runs. When the Lieutenant asked him, "What for?"

"Sir, I need to take a few things over to the mess sergeants in the infantry and cav to support the upcoming meetings in their units. I also need to bring some parts back from Brigade to Battalion Maintenance for the mechanics."

The Lieutenant shook his head and said, "Very well," after which McNabb stood, saluted, and shot out the door. He spent the balance of the day hovering near the phone, with the exception of a short trip down to the motor pool to check on his soldiers right before lunch. Returning to his desk, he had another cup of coffee from his thermos and ate a breakfast bar to get him through the afternoon. The days were getting markedly shorter. It was practically dark during evening formation. After the platoons were released by the First Sergeant, the Lieutenant told McNabb to have a good weekend, and then asked, "How does the company usually handle the half-day schedule?"

"Normally, the soldiers will work in the morning on their vehicles, and then the majority of them will be released in the afternoon, with one or two drivers rotating on a roster, so there is someone in the motor pool available in case there is an emergency need for transport. The mess hall generally operates on a similar schedule, with one set of cooks working in the morning through lunch and the other from lunch until the evening meal."

The Lieutenant thanked him for bringing him up to speed. As McNabb walked away, in the back of his mind, he kept thinking that McNabb must be up to something. And, although he had suspicions, he just couldn't put his finger on it. In any event, with the exception of supporting meetings, the next week looked to be fairly open. Although he would be in every day, he was looking forward to the weekend, and the following Saturday, when the battalion would have its Christmas party. When he was back in the States, he had always enjoyed family days and field days, when the soldiers, wives, and children, could come in and see the vehicles, mingle with each other, and enjoy the holiday. He also thought it would be wonderful next Christmas to have his wife with him and get a chance to show her around a tank battalion for the first time in her life.

LONELY LOOKING SKY

Back in the BOQ, he found Buzz busy in the kitchen, heating up a can of beef and barley soup with French bread for dinner. The Lieutenant could tell that he was in a morose mood since Elizabeth was gone. As he walked in the door, he asked him how he was doing and Buzz mumbled, "Fine. Go ahead and sit down for a while."

As the range heated up the soup, the Lieutenant grabbed a beer and sat down at his normal spot in the corner, like any good combat arms officer, so that his back was secured against the wall on both flanks and he could keep his eye on the door. It was instinctive. Eventually, as time passed, it became apparent that Bigfoot was not coming to join them, so the two of them served themselves and ate a quick meal, practically in silence.

Finally, to liven things up a little, he asked Buzz, "Are you going to try to make it to England after Christmas?" Buzz brightened and appreciably, and answered, "Yes. I'll be flying over for a three-day stay, in a modest hotel in a small village near Elizabeth's hometown.

While I'm there, I'll see her infrequently. I won't be meeting the family." The Lieutenant nodded his head and said, "I understand," before rising and asking if he could do the dishes.

Buzz quickly agreed, and when he was through cleaning up, bid the colonel a good evening. As he was going out the door, Buzz reminded him of the upcoming BOQ-wide Christmas dinner, that he intended it to be a magnificent feast, in which he would star as a chef with supporting roles played by the ladies at the end of the hall and their friends. He went on to add that although he had invited the young second lieutenant, who roomed above the Lieutenant, and her friends to dinner, he didn't know if they were going to make it.

The Lieutenant thought about that for a minute and then stepped out the door to his own room, to rest and plan what he was going to do the next day. Thinking about it for a while, he decided that he wanted to take one of the streetcars into downtown Mannheim. He wanted to visit the Christmas markets, to see if there was anything that appealed to him and enjoy the general high spirits of the German people during the holidays. Then he looked out the window. The night sky was cold and gray, and as he listened to the light sleet pounding against the windows, he thought he would just have to see what the morning looked like before he made any commitments.

MARRAKESH EXPRESS

As was so often the case there, despite the fact that the weather during the evening was terrible, after the sun rose the next morning, it proved to be a bright and almost balmy day. He was also excited because they could hear Buzz in the kitchen. He hurried down and found him frying slices of ham and making French toast with the left-over bread from the previous night. After enjoying a quick breakfast with him and Bigfoot, the Lieutenant told them that he planned to visit downtown that day to see what the Christmas markets were like.

Bigfoot smiled a little, and said, "Avoid Penis Strasse. If you want a real experience, go down to the Turkish quarter, have some Turkish coffee, and enjoy some roasted chestnuts. The women there make the best roasted chestnuts I've ever eaten." Since the Lieutenant had never ever had chestnuts in his life before, he marked that up as an experience he planned to have that day. Excusing himself, he went back to his room and dressed in his German student clothes before going to the streetcar station and purchasing a ticket for the right train.

He thoroughly enjoyed the streetcar ride down into the city. All of the German people were excited and laughing, and if there were any children, they were cuddled, talked to, and spoiled incessantly. It was great. The streetcar finally came to a stop in the center of the city, where there was a combination of old buildings, in two-hundred-year-old architectural style combined with brand new buildings that had taken the place of bombed out structures. After he descended from the streetcar, he decided to just walk throughout the city, to enjoy the sunshine, and absorb the culture as much as he could. The Germans that he met were very friendly, as they assumed that he was one of them, even though the best he could do was exchange pleasantries with a harsh Hessian accent.

Once he got to the Christmas market, he walked past the stalls, amazed at all of the wooden ornaments, Christmas trees, and decorations that he saw. Everything was light and beautiful. He thought, *Christmas is their absolute favorite time of year,* before working his way farther into the city, where he encountered more people, particularly women, in Middle Eastern clothing. He turned down a side street that was lined on both sides with Turkish vendors, who had brightly colored rugs, and various items in brass and beaten silver for sale. He found the promised chestnuts that were roasting on charcoal braziers, like hibachis, tended to by women dressed in black from head to toe. He stopped in front of one lady and used broken German and hand signals to buy a small plate of chestnuts. After burning his fingers initially, he let them cool and then opened them up to devour the amazingly soft and tasty flesh of each nut.

When he was through eating, he nodded thanks to the woman and walked down the street a little further, until he saw a Turkish coffee house. He stopped and took a seat in front of the coffee house at a small table. The proprietor came out and asked what he would like. The Lieutenant, again, in a combination of English, German and sign language, indicated that he would like a small cup of Turkish coffee. The owner shook his head in assent and after a few minutes came back with what, to the Lieutenant, looked like a thimble full of coffee, although it was probably two ounces, with a small silver spoon to stir it. He offered some sugar as well.

Although the Lieutenant did not normally take sugar in his coffee, once he took a sip, he immediately changed his mind and put some in. He stirred the coffee, that had a large amount of sediment in the bottom, which appeared to be the grounds from fresh ground Arabic coffee beans. He took another small sip and then drank the coffee slowly. It was extremely strong. When he reached the end, it left several bits of coffee grounds in his teeth.

The Lieutenant smiled to himself because he was glad that he had tried Turkish coffee but was firmly of the opinion that he probably would not take it again. In the future, he would stick to dark roast German coffee. As he continued to walk through the older portion of the city, he watched the faces of the men and women who appeared in the doorways and stalls. The men outnumbered women by about ten to one, as a result of the German government's policy that did not allow Turkish guest workers to bring their wives and families with them. As a result, there were large numbers of single male Turks, whom the Lieutenant suspected were as lonely and frustrated as he was, and they had no relief in sight. He then thought to himself, over the next few decades, Germany is going to have a problem unless it finds some way to reach a balance with these people who are from a different culture and religious affiliation.

LONELY NIGHT

As the sun started to sink toward the hills to the west, the Lieutenant quickly worked his way back to the streetcar station and purchased a ticket back to Käfertal. When he arrived there, he wasn't ready to spend another night in the BOQ with lonely people. He fished out his car keys and drove up to Nick's. There, the proprietor's wonderful sense of humor, the light and laughter in the bar, as well as the wonderful smell of the Greek food, lightened his mood considerably. After he had eaten, he told Nick that he was going to skip the Greek brandy that night, since he already had two beers, but he would take a good cup of coffee. He then recounted his adventures with Turkish coffee to Nick, who looked at him and told him that it wasn't his taste either. Since Nick was Greek, this didn't surprise the Lieutenant at all.

The Lieutenant thought for a moment and asked Nick, "Why did you and your wife decide to come to Germany to run a restaurant instead of staying in Greece?"

Nick just smiled and said, "Many Greeks have to leave the islands to make their fortune in other countries because there are no jobs and the economy is terrible. If I ever hope to buy a house for me and my family, I needed to work in Germany and make enough money to do that when we are retired. This is true for us, a lot of Greeks, Italians, and even some Chinese, who like working in Mannheim and Heidelberg, making excellent, and to the Germans, exotic food, that you are fortunate enough to be able to enjoy as well."

The Lieutenant smiled and replied, "Well, Nick, I'm glad you did, because I truly enjoy eating here and talking with you. And again, I can't wait for you to meet my wife someday."

Nick brought back his bill. The Lieutenant paid in marks, rose and made his way to the door. When he got there, a couple entered, laughing and smiling, even though there was a light rain falling. For a moment, he wished he and his girl were them. But then he simply dove through the door into the rain and ran around the corner to his car. There he got in and cranked the heater up for a few minutes. Once he was warm, he slipped the Volkswagen in gear, made a tight U-turn, and drove back to his quarters, for what he hoped would be a restful weekend.

Back in his room, he could tell from the quiet that a lot of the officers were already gone for the holidays. He enjoyed himself reading that night, and then again, resting and listening to the radio on Sunday. He smiled to himself as he thought, *If I can just make it through the next two weeks, I'll be into 1980, the year that I'm going to get married.*

JET AIRLINER

After PT and breakfast the following morning, he spent the day at his desk, waiting for the phone to ring, reading the battalion SOP a second time, and exchanging some small talk with the NCOs, who were hilarious. They had seen hard service in different parts of the world and had many stories to tell of the adventures they had been subjected to in their youth. He enjoyed them thoroughly and queried Sergeant Fykes to see if he'd be willing to drive him to Frankfurt International for his flight out to the States in March, and, more importantly, to pick him and his new bride up at that airport when they arrived on the fifteenth of March. Fykes immediately told him, "I'll be more than happy to do that, sir, and I'm looking forward to meeting the woman who is willing to move across the world with you." The Lieutenant quickly rejoined, "You won't be disappointed. I've never met a man who didn't like her. She's not only pretty and funny, but the most engaging person I've ever met."

The rest of the week passed slowly, without event, other than a request to change the cookie menu for a couple of meetings. Thursday, when he went on for a weekday duty, it was a bitter cold night, so the troops fortunately stayed indoors and were fairly quiet. He thought they were probably drinking to ward off the cold or getting high, so he made his checks quickly, just before and after midnight, and spent the rest of the night in the Adjutant's office, resting, until Friday morning.

Once he was relieved by the staff, he slipped back to his quarters to shave and take a quick shower, before returning to the S-4 office. Graham was still engaged in meetings, so he called the Adjutant to make sure the mess hall had taken care of his needs. Everything was fine, so the Lieutenant told Dryer that he would need to go to the telephone in the bank to give his girl a call before she went home for Christmas.

SWEET TALKIN' WOMAN

He ran over to the bank. The line was short, so he slipped in to purchase some marks, before walking quickly upstairs to the phones. He was in luck as there was a phone open, so he sat down, slipped a couple of five-mark pieces into it before dialing her telephone number. It was early morning in Kansas, so she was a little sleepy when the phone rang but immediately perked up when she realized it was him. After reminding each other how much they missed being together, she brought him up to speed on her plans for Christmas. She was going home to be with her parents for the holiday. After that she would be leaving her job and moving back in with her parents until the wedding.

The Lieutenant told her that he had just come off of duty and reminded her that he would have duty again on Christmas Eve and New Year's Eve, so he probably wouldn't be able to give her a call until she was back in her hometown. Looking up, he could see that he was running out of time, so he slipped in another five-mark piece. They spoke a little longer before he told her to go back to sleep and hung up the phone.

Returning to his desk in the S-4 shop, having made his call and relaxing, he had to have two cups of coffee to stay awake. He'd been up most of the night and was practically a zombie, but was determined to stay awake until the evening, when he could get a good night's sleep. At lunch time, he decided to go ahead and have something to eat there, since he was running low on energy and he wanted to see Sparks as well. After he finished his meal, he beckoned the Mess Sergeant to come sit beside him and asked if he needed any support for the upcoming Christmas party. Sparks assured him that everything was under control. They had one every year, and that party, and the field day, were pretty much the highlights of the battalion's year, especially the appearance of the Christmas Tank and Santa Claus, with his driver and elf loader.

The Lieutenant looked at Sparks incredulously at first, but then realized that he wasn't joking, and asked him to tell him more about it. Sparks smiled and laughed a little before he told him, "Usually, the Christmas tank is the battalion commander's tank, 6-6. Captain Boris, because of his position and his size, to a certain degree, will have a Santa costume on and the tank's tube will be wrapped with red and white paper, like a candy cane. He will be the tank commander, have a couple of the smaller tank crewmen, in their olive-green uniforms and elf hats assisting him. They will drive the tank through the battalion area and around the back to the rear entry to the mess hall, where he will come in, sit down to listen to the children's Christmas wishes."

When Sparks finished, the Lieutenant asked, "Is there any sort of a layout done by the mess hall or Support Platoon for the civilians?"

"Not initially. Santa will have candy for the kids, but the cooks will be more than busy making and serving Christmas dinner, to include turkeys, hams, side dishes, and a wide variety of desserts. In addition, sir, some of the staff tracks will be open, and a personnel carrier and a tank from one of the line companies will be displayed. Support Platoon, other than having to refuel those vehicles when they return to the motor pool, shouldn't expect to be tasked for much."

Armed with that newfound information, the Lieutenant went down to the motor pool and was happy to catch McNabb at his desk, so he could make sure that there would be fuel trucks available for the vehicles that had to move at the end of the field day. McNabb told him that he had already received that tasking from the First Sergeant, who was alerted by Sergeant Dryer. The Lieutenant thought for a minute, *It is good to have NCOs looking out for me*, and he thanked McNabb. He told him, "I'll be back at my desk until the end of the day. Then I'll be going to bed, since I've been up for a couple of days."

GOOD COMPANY – A NIGHT AT THE OPERA

When the day finally ended, the Lieutenant made his way back to his quarters. There he was happily surprised to find that Bigfoot had gone out for pizza. Buzz, Bigfoot, and Pavel were eating dinner and asked if he would join them. He quickly made a plate and sat down, just as Bernie came in. Captain Boris, who was enticed upstairs by the smell of Italian sausage and pepperoni, followed close behind.

The Lieutenant asked Boris if he was ready for his Santa Claus appearance. Boris just laughed and told him, "Yes. And I'm looking forward to it, because it's such an unusual duty, and a lot of fun." The Lieutenant nodded, ate one more piece of pizza for the road, and then excused himself to get some much-needed rest. Walking up to his room, he ran into the ladies who lived next door and directed them to the kitchen for pizza; they thanked him, and as they walked down the hallway laughing, he stepped through his door to take a long hot shower before crashing until the next day.

DECK THE HALLS

The next morning was cold, damp, and overcast. The Lieutenant made himself a cup of instant coffee to brace himself before going out into the weather. He put on civilian clothes before he finally decided that he would just go to the PX for a breakfast sandwich and another couple of cups of coffee, before going back to his room to dress in uniform for the party. Mid-morning, he walked over to the company to see how things were going. The sun came out about that time and the day was glorious, for that part of the world. It was a lot of fun to watch the children as they climbed in and out of the APC and the M-577 command track. They could walk around the tanks but were not allowed to go into them. They could, however, climb up to the tank commander's cupola and the loader's hatch and look in. The kids were fascinated by them, and even more excited at the thought of the Christmas tank coming with Santa Claus.

By 1100, the children had been wrangled back to the company courtyards. Once they were in position, the silence of the morning was broken by the thunderous sound of a tank's engine revving up. With the creaking of the tracks as the vehicle slowly moved from the rear of the hardstand toward the exit to the street, the parents and all of the children lined up on the sidewalk opposite the motor pool. The children squealed and waved when they saw that the tank commander was Santa Claus, and that he had a driver in OD fatigues with an elf hat on, and a loader who was serving as ground guide who was also wearing fatigues and a green elf hat, using hand and arm signals to direct the driver out onto the road.

The commander's tank, Headquarters 6-6, with Boris waving to the crowd, slowly lumbered to battalion headquarters, before it took a right and went back behind the barracks to take up a position near the back door to the mess hall. When the tank rocked to a stop, Santa, with a loud "Ho, Ho, Ho," descended gingerly down the front slope of the tank. With the tank parked, the loader climbed up to the bustle rack and handed two large bags of candy down to Santa, who sauntered into the mess hall, where a seat had been prepared for him. He sat down, and his loader elf helped the children come up and get on his lap. The other elf stood guard on the vehicle.

It was really a lot of fun the Lieutenant thought to himself. The children came up to Santa and told him what they were wishing for. Then they beamed when he gave them a couple of pieces of candy. It took about an hour for all of the battalion children to meet Santa. When the last toddler was off Santa's lap, everyone formed up in a line to have their trays filled with Christmas dinner. Sparks and the cooks were more enthusiastic than usual, and the meal was truly a feast. It had something for everyone. There were even pies and cakes of every type for dessert, and once everyone had assembled, Shea and his wife wished them all a Merry Christmas and a happy New Year.

The Lieutenant stood back and waited until everyone else was served, and the battalion commander, Jeanie, and the other officers and their wives had taken their places. Then he quickly grabbed a tray, and asked for a little bit of everything, before sitting down in the corner with Lieutenant Heinrich and Lieutenant Borovich. Then he realized that Lieutenant Marr had come in behind him, so he motioned for him to join the three of them as well. They sat and visited about what they planned to do over the holidays. The Lieutenant told them that for the most part he was pulling duty. Heinrich had family in Luxembourg was planning to visit them. The other two lieutenants simply planned to stay close by and enjoy German hospitality.

The Lieutenant smiled, because he was having a good time, but paused for a minute to reflect as a wave of homesickness came over him. A year earlier, he had been at his parents' house, surrounded by people who knew him and friends. Now he was in a foreign country and enjoying himself, but it was distinctly different. Then he thought again and asked Marr and Borovich if they would like to join him and his compatriots for Christmas dinner in the BOQ. Marr told him that he had other plans. Lieutenant Borovich, however, was all ears, so he told him he would be coming off duty Christmas morning at about 0800, and if he could come by his quarters around 1100, he would have a chance to catch a few hours of sleep before dragging the tables and chairs out into the hallway to set up for the feast.

When they were through eating, the troops of the battalion, with their wives and children assembled around Santa Claus to wish him a Merry Christmas, who, with a loud, "Ho, Ho, Ho," disappeared out the back, to his tank. Then assisted by his elves, he gave all the left-over candy to the children, who had come to the rear door. Then the loader guided the tank as it turned around to head back the way it came with Santa again in the tank commander's cupola, and the driver paying close attention. The children waved and shouted as the tank creaked back to Battalion and then to the street into the motor pool. It was truly an interesting and unusual experience since the tube of the tank was wrapped in red and white crepe paper like a giant peppermint stick, and Santa's beard fluttering in the wind, surrounded by diesel smoke.

With the festivities over, he said goodbye to his fellow lieutenants and headed back to his quarters. On the way, he suddenly remembered that he was on the hook for condiments. So, he walked quickly over to the PX to buy olives and pickles. Having done that, with a large bag containing several jars of each, he went up the stairs to his room and put them all in the refrigerator. Then, since it was Saturday, he decided that he would drive up to the shopping center in Viernheim. There he spent a little time walking around, looking at the Christmas items for sale in the various stores. He wasn't really interested in buying anything, so much as seeing what was there, knowing that next year, with any luck, he would be spending Christmas, for the first time, with his new wife.

RAINY DAY PEOPLE

Light in the afternoon, it was depressing, dreary and damp. He needed to get out so he drove to Nick's for dinner. He was still pretty full from the Christmas party, so he slowly drank a beer and just visited with Nick from time to time. By the early evening, he was finally hungry enough to order his favorite meal, with French fries and another beer. It was fairly quiet in the bar, since it was the weekend before Christmas, and most Germans were spending the time with their families. As a result, he had a chance to talk with Nick quite a while about what Greece was like, and how he liked it in Germany.

Nick liked living in Germany well enough. The economy was

good, but he did miss the sunshine, and beautiful ocean surrounding Greece. He also missed his father and mother, and brothers and sisters who he had not seen in several years. On the upside, he and his wife were exploring the countryside, and making friends, of whom the Lieutenant was one. The Lieutenant looked out the window and saw that a cold rain was falling, so he ordered a warmed Metaxa and drank that slowly to fortify himself before going out into it. By the time he was done, it was approaching 2300, so he walked rapidly out to his car and drove back carefully to the BOQ.

The next morning was Sunday. Fortunately, Buzz was in the mood to cook a monster breakfast. The hallway was full of the rich aroma of dark roast German coffee. As Buzz had relayed to him shortly after they met, his E-6 dental assistant on his first tour in the MACV morgue, always said "I like my coffee like my women: straight, hot, and strong." Buzz confessed that he had also acquired the taste.

The Lieutenant pulled on jeans and his VMI sweatshirt before he walked down to the kitchen to pour himself a large mug of black "go juice," while Buzz worked his magic on the stove. When he was through, Bigfoot showed up, and the smell of bacon even drew Susan and Chris down to enjoy a nice warm breakfast of scrambled eggs, pancakes, fruit salad, and Army gossip. They finished, cleaned up together, and went back to the peace of their respective rooms. He read and wrote his girl a quick note to tell her he missed her and was looking forward to the next year. Then he read a little more, ate a frozen dinner, and went to bed early, since he had to go on duty early the next morning.

Chapter 28
Jingle Bell Rock

The alarm jangled at 0600; it was Christmas Eve and essentially the holiday. He rose, cleaned up, and dressed, so that he could sign in with the Adjutant by 0700. Putting on his field jacket, he grabbed the clock to take with him, and a poncho liner to use as a blanket overnight. Since he would be on for twenty-four hours, he decided to take a new paperback book he had just purchased at the PX, General Sir John Hackett's, *The Third World War*. With those items in hand, he walked up to Battalion and saluted the Adjutant as he approached his desk.

Once he had signed in, the Adjutant told him that he and the colonel planned to be in for at least a couple of hours, after which, the duty officer would be in charge. Also, while it was a half-day, he needed to make two arms rooms checks during the day and two at night since so many of the leaders were on leave. The Lieutenant acknowledged the instructions and went to go draw a sidearm and pick up his LBE. But first, he would have breakfast, after relieving the outgoing duty officer on his way down the hall and then come back to headquarters.

When he reached the door to the duty office, he saw Lieutenant

Borovich slumped over, fast asleep. The duty NCO was wide awake, thumbing through an old issue of the Army Times. The Lieutenant couldn't resist the fun of coming up behind Borovich and whispering in his ear. Borovich jumped, came wide awake, and then chuckled when he realized who was talking to him. He stepped back and asked Borovich how his evening was. He said, "The weather was cold and nasty, so it was a quiet evening, other than an altercation in CSC at about 2200. It was nothing, really, and quieted down quickly, so I had a restful night." Borovich packed up his things and prepared to leave. The Lieutenant wished him Merry Christmas and they both smiled, since they were far from home and anything approaching the holiday spirit.

Once Borovich was gone, the Lieutenant told the Duty NCO that he was going down to HHC to draw a sidearm. After that, he planned to have a quick breakfast, pick up his LBE and helmet in the S-4 office. Then, as soon as he was back, the NCO could go to breakfast, and the oncoming Duty NCO could as well. The NCO shook his head in agreement and the Lieutenant was out the door in his field jacket, to face a light drizzle and a northwest wind.

Once he got to Headquarters Company, he asked the Specialist Five CQ if the First Sergeant was in. He was, so the Lieutenant went downstairs to the orderly room and informed Ascencion that he needed to draw a sidearm for duty. The First Sergeant sent a runner up to get the armorer, and they chatted briefly, while they waited for him to come downstairs. Ascencion was a Pacific Islander, built like a tank, and sharp as a tack. He and Captain Boris were kept busy, trying to catch the drug dealers in the unit and court martial them out of the Army. It was a full-time job. Unfortunately, it was not very effective, because the dealers knew the drug laws better than the officers. With that thought in mind, he asked the First Sergeant how things were going. Ascencion told him, "Reasonably well. But I still had a lot of bad actors on my shit list." The Lieutenant nodded, and as the armorer passed, he excused himself to draw his pistol and three clips that he arranged and loaded, in the usual fashion. Once he signed for his weapon, he went up the stairs quickly and over to the mess hall, to eat a light breakfast and slam two hot cups of coffee, before gathering up his gear in the S-4 office and heading back to Battalion.

When he reached Battalion, he found the outgoing Duty NCO and the oncoming Duty NCO, who was a friend of his, smoking and shooting the ball. When he walked in the door, he told both of them that they could get breakfast if they wanted to, and he would man the desk. As they left, he wished the outgoing Duty NCO a Merry Christmas and settled in behind the duty desk with his book, to read the fictional history of the next World War. When the Duty NCO returned, he asked him which company he was from, since he did not recognize him. He was Staff Sergeant Eaton, newly assigned to the mortars in CSC. And, since he was single, and newly assigned, he was pulling duty on Christmas Eve. The Lieutenant told him, "I sympathize. The same thing happened to me, and, as an added bonus, I'm pulling New Year's Eve duty at Brigade. Anyway, at least Christmas Eve duty tonight will be something different."

The Lieutenant decided he would walk down to the Adjutant's office to see if there was anything he could do, since he was bored beyond belief. He told him, "No. Just keep an eye out. Be particularly alert for flash traffic. Higher headquarters love to send alert messages over the holidays."

The Lieutenant put that thought in the back of his mind and decided that he would go ahead and make a check of the battalion area, since it was almost noon. As he walked through the battalion, the place seemed eerily quiet. There was very little work going on, with the exception of one vehicle in Battalion Maintenance. He walked back behind the billets and into the mess hall through the rear door. Sparks had apparently taken the day off, so he visited with the Assistant Mess Sergeant for a while to see how things were going. Satisfied that all was well, he told the sergeant that he would be back for his evening meal, and looked forward to it, but also asked what he was serving. The sergeant smiled, and said, "We're making cheeseburgers and fried chicken. Those foods make the troops feel like they are at home." The Lieutenant thanked him and walked out the door to the Headquarters Company courtyard. The wind had come up, and it was colder, so he put his head down against it, and walked quickly into HHC to do an arms room check. When he was finished with that, he checked the other companies, and when he was done, returned to Battalion.

He walked into the headquarters and could immediately tell that all of the other officers were gone, leaving him and the NCO alone in the building. He released Eaton for lunch and took up his secondary position at the NCO's desk so he could see out the windows. There he continued to read *The Third World War*, and thought to himself, *How crazy it would be, if the whole thing started on Christmas Eve.* It wouldn't be the first time that an American army was engaged in combat on Christmas Eve. From George Washington on, it seemed to be a tradition. When the NCO returned, he excused himself to the Adjutant's office, where he set his book on the coffee table, laid the poncho liner out as a blanket, having told Eaton that he was planning to rest for a couple of hours, since it was going to be a long night. When he got to the office, he pulled a couple of pieces of licorice out of his field jacket and ate them, before drifting off to sleep.

Bang! Bang! The Lieutenant looked up and realized that Sergeant Eaton was knocking on his door. He simply said, "Flash traffic," and ran down the hall. Once the Lieutenant reached the office, he picked up the phone and transcribed the message. Message written down, he ran out of the doorway and down the stairs to the S-2's office. He unlocked and closed the door behind him, and opened the safe, to pull out the code book. He worked quickly, but calmly to break the message. Once he realized that it was a test message and not a full alert, he calmly placed the code book back in the safe, locked it and the door before he walked up the steps to let the NCO know that they were not on alert. The sergeant breathed a sigh of relief. It was his first tour in Europe. He had spent some time in Korea and a couple of tours with infantry divisions in the States, but Germany was different. The Lieutenant told him not to worry, and with any luck they would get off duty tomorrow for a peaceful Christmas. The Lieutenant then thought to himself, *And a great Christmas dinner*, if I can stay awake for it.

It was 1500. Back in the Adjutant's office, he set his alarm for 1600 and closed his eyes until it jangled beside his head. He shook his head, turned it off, and walked down to the duty office to pour himself a cup of tepid but effective black coffee. He chatted briefly with Eaton who was leafing through old magazines, while looking out the window at the weather. The temperature was dropping quickly and the wind was a little stronger. It was going to be a chilly night. Pulling on his field jacket and scarf, he grabbed the keys and his sidearm to make the second arms room check before 1700. He moved quickly and upon his return, he released the sergeant to get dinner.

When Eaton returned, the Lieutenant strapped on his pistol over his field jacket, grabbed the keys and walked down to the mess hall, where the cooks still had some leftover cheeseburgers and chicken. He paid for his meal at the normal rate without surcharge since he was on duty. He got a tray of food, grabbed a cup of coffee, and sat down in the corner to eat. The troops, for the most part, were in a good mood. Once he finished, it was dark and since it wasn't going to get any warmer, he decided to make his evening arms room checks early. He hurried from company to company into the basements of the barracks with his ring of keys in one hand, shaking it, and the forty-five drawn in the other. He initialed all of the check sheets. When he came out of Charlie Company, he was hit in the face by a strong gust of icy cold wind. It gave him an instant brain freeze, so he dropped his head and almost ran back to Battalion to warm up.

SANTA CLAUS IS COMIN' TO TOWN

When he reached headquarters, he took off his jacket in the duty office and went back to the Adjutant's office to lie on the couch for a brief nap. After thirty minutes, he walked back up to the duty office with his book to read some more and visit with Eaton, who had brought in a small transistor radio. AFN was playing Christmas music that helped them pass the time as they looked out the window and watched the troops as they moved through the unit, to visit each other. Some of them stopped by Battalion to wish him and Eaton a Merry Christmas.

This went on for hours, until about 2100, when an odd little gentleman dressed in green, sort of like an elf, with a large Santa hat, popped through the door and entered the duty office. The Lieutenant calmly looked up and asked him, tersely, "Who the hell are you?"

In reply, the fellow just chuckled. He sounded like he'd been drinking a bit and told him that he was working for the USO and had a large bag of goodies to hand out to the soldiers in the barracks. The Lieutenant asked him if he'd ever been in the billets before. He said, "Yes, LT, I have on other holidays and I know that I need to move quickly and not ask any questions." The Lieutenant told him very well and released him to go to the barracks. But before he did, he gave the Lieutenant and Eaton their own Christmas gifts: notepads fashioned to look like packs of Kool cigarettes. The Lieutenant shook his head in wonder at the never-ending love affair between the Army and nicotine, and laughed softly, as Santa's helper half stumbled and half ran out the door on his mission of merriment.

When the jolly elf had disappeared, Eaton sat their gifts down and the Lieutenant turned back to his book. He found it a fascinating reading and a little chilling, especially the description of the Soviet Spetsnaz soldiers being dropped behind enemy lines to kill duty officers. When he'd finished that portion of the book, he looked at his watch, to see that it was approaching midnight. He told the NCO that he planned to walk through the barracks and hit every floor to wish the troops a Merry Christmas, before descending into the basements and making his second set of checks right after midnight. That way he could spend the rest of Christmas morning at headquarters and maybe get some rest.

GOD REST YE MERRY GENTLEMEN

He bundled up and entered Charlie Company first. He walked up the barracks stairs from floor to floor, talking to the occasional soldier who was standing in a doorway, moving quickly along to wish those who were awake a Merry Christmas. When he got to Alpha Company, things were a little grimmer and there were very few soldiers who greeted him. Bravo Company was different because many of the soldiers there knew him. He had a good time there since it was almost a reunion. Combat Support Company was quiet; the Scouts were drinking hard and telling "no shit" stories but appreciated his coming by.

Then he entered Headquarters Company, still painted jet black, and walked up the first flight of stairs. As he moved from room to room, he could see cigarettes and joints glowing like fireflies in the darkness, with several soldiers passed out, most probably from alcohol or heroin. He continued to walk quickly from door to door. On the third floor, he ran into an entire room full of soldiers who were puffing on pipes full of Turkish hash. The smoke was so thick it looked like Baumholder's fog at dawn. He could barely see three feet, and the smell was noxious and overpowering, but all the troops in that room were in good spirits, and wished him well, so he turned and headed down the stairs into the basement for his first arms room on Christmas Day.

He made that check as quickly as possible before climbing the stairs and exiting to cross the courtyard, almost in a run, because it was colder and starting to snow a little. He dove into Combat Support Company. Once inside, he waved to the CQ and pounded down the stairs to the basement for his check of that arms room. When he was through, he quickly walked up the stairs and out the door to the courtyard, and then around the corner of the barracks to Bravo Company, where the tankers again greeted him. He shot down the stairs to make his check after midnight. When he was done with Bravo, he steeled himself against the cold wind and hustled over to Alpha Company. For some reason, he had never felt at ease in Alpha Company. But on this cold Christmas Day, there was very little movement in the barracks. He waved to the CQ again, wishing him a Merry Christmas as he slipped down the basement steps to make his final check there. When he was done with Alpha Company, he put his head down and walked over to Charlie Company.

The troops in Charlie Company, as usual, were in a good mood. He had always felt at ease there, so he visited with a couple of the younger soldiers hanging around the CQ desk, manned by a buck sergeant tank commander, asking them if Santa had been good to them, that sort of thing, before he descended and made his final check. As he was coming up the stairs, he could hear the sergeant screaming for him. He ran up to the desk and the CQ told him that he had just been called by the Duty NCO: there was flash traffic, so he had to get to Battalion immediately. By that time, the Lieutenant was feeling tired for making his rounds, but the adrenaline hit him and he didn't even feel the cold, as he double timed as fast as he could into the headquarters.

WAR - WHAT IS IT GOOD FOR

As he ran into the headquarters, Eaton motioned for him to take his seat at the duty officer's desk. The NCO had just received the message and was reading it back. The Lieutenant picked up the phone and transcribed and read back the message himself. Then he grabbed the keys to the S-2's office and safe and ran down the stairs. After turning on the light in the S-2's office and retrieving the code book from the safe, he began to meticulously break the code. As he did that, his blood ran cold, because the message addressed every single MACOM in the United States Armed Forces. This included Europe, of course, and the Atlantic and Pacific commands. He then scrolled down to the list of subordinate units. The first words in the code that he broke read, "**Alert – Alert – Alert**. This is not a test. This is not a drill. We are at war. The Soviet Union has invaded Afghanistan and hostilities have commenced."

He read the message several times and then went to the last few sequences of numbers and letters that would tell him whether or not the message was indeed the first message of the Third World War or a test message, sent to catch duty officers half asleep so that they would call their battalions to full alert on Christmas Day. He checked the end twice, and there were two characters that made the message a test message. Feeling relieved, he broke the entire message again, just to make sure, before he put the code back in the safe, and locked the S-2's door behind him.

He was climbing the steps to the hallway next to the duty office when he heard a crash into the door by the Duty NCO's office. He looked up and saw his VMI classmate, Mad Dog Barclay, who apparently was on duty at his sister battalion next door. He had the most wild and crazed look in his eyes as he asked the Lieutenant if he had broken the message, and if it was real. The Lieutenant looked at him and told him to take a deep breath and follow him down to the S-2's office, where he would break the code again with him, just to make sure, because he believed it was a test message. Again, he removed the code book from the safe and sat down to find some inner peace, so he could try to review the entire message, with his classmate breathing heavily over his shoulder. Finally, he convinced himself and Mad Dog that they were in fact not headed for World War III, but it was some staff officer's idea of either a joke or a complete test of the nerves of the officers on duty.

SILENT NIGHT

As Barclay shuffled out the door into the cold, the Lieutenant took a deep breath and let loose a sigh of relief. He then walked over to the duty desk, put a bookmark in his copy of *The Third World War*, slammed it shut, and walked back to the Adjutant's office. There he put it with his poncho liner and other things because he had no intention of ever reading it again. Having done that, he was still feeling pretty juiced, and he couldn't sleep, so he decided to go down the hall and have a cup of coffee with Eaton, who had made a fresh pot. They watched the weather, and swapped stories about other Christmases, most of which had been shared with family, but sometimes not. By now, the wind had died down, and there was a transcendent calm that descended over the battalion. The stereos were turned off, and the soldiers in the barracks were huddled against the cold. Some of them were still smoking or maybe even shooting up.

But, still, it was Christmas and a holiday, a time for peace and relaxation. After visiting for about an hour, the Lieutenant asked Eaton if he wanted to get a little rest so he wouldn't be a complete zombie when he spent Christmas with his wife and kids. The NCO said he would take him up on about an hour and a half nap. He set a small travel alarm clock that he brought with him to let the Lieutenant know when he should come and get him up.

Alone in the duty office, the Lieutenant's thoughts couldn't help but turn to his girl. He wondered how she was doing, and if she was enjoying the holiday with her parents back in Abilene. He hoped so. Christmas was absolutely her favorite time of year. She loved everything about it. In any event, he thought, *once I'm married, I won't spend next Christmas alone or any Christmases thereafter.*

WHITE CHRISTMAS

When the alarm went off, he was jolted out of his reverie and drank another half a cup of coffee before going down to get the Duty NCO. He woke him up and told him that it was going on 0600. He also told him that as soon as the mess hall was open, he should go get breakfast. By 0700, Eaton was starving, so he left to go to the mess hall and see if he could have the cooks make him something in advance of opening for the troops or at least get a better cup of coffee. When he was gone, the Lieutenant again looked out at the battalion that was covered in a light dusting of snow.

By now, the Lieutenant was looking forward to Christmas dinner with Buzz and the gang in the BOQ. He also had no appetite, after sitting with the NCO who smoked constantly. And he was completely and absolutely tired. When Eaton came back from breakfast, the Lieutenant excused himself to the Adjutant's office, and on the way stopped in the battalion restroom and showers to wash his face and try to wake up a little. Then he went down the hallway, where out of sheer boredom, he leafed through some old field manuals that were in the bookcase next to the sofa.

Looking out the window, he could tell that the sky was just beginning to lighten. He glanced down at the steel diver's watch on his left wrist and saw that it was about twenty minutes after seven. Then, a thought occurred to him, and he rose quickly. Almost running to the NCO's office, the Lieutenant told him he was going next door. So, he put on his hat and field jacket and turned the corner to his sister Battalion's duty office. Peering inside, he could see his classmate almost asleep by the duty desk. He walked up beside him and whispered, "Mad Dog."

His classmate, Barclay, immediately came to life with a big grin. He told him, "Merry Christmas," and then asked if he had any plans for Christmas dinner, since the BOQ was having a full Christmas dinner on his floor, with an open invitation to everyone who wanted to come. Mad Dog sleepily looked at him and told him that he had made other plans with his girlfriend to have Christmas dinner with a couple of her friends and their dates so he wouldn't be able to attend. In fact, she was going to pick him up, since his car, a Fiat two-seater that he called "Fix it again, Tony," was in the shop as usual. "I like it though. It goes like a raped ape when it runs." The Lieutenant laughed as he continued, "Be sure and raise a glass for me with Buzz and the gang."

The Lieutenant smiled, "I sure will, BR." Then he quickly walked back outside the building to his battalion's headquarters and sat down behind the duty officer's desk, to relax and wait for his relief.

A little before 0800, he made out the form of Lieutenant Marr walking toward him through the mist of the early morning. When he came in, the Lieutenant told him how happy he was to see him. Then Marr returned the compliment, although he wasn't exactly excited about spending Christmas Day there. However, since he was single and new to the battalion, he knew that's how Army organizations operated. The Lieutenant said, "You bet. In fact, I'm tagged for brigade duty on New Year's Eve." Lieutenant Marr just shook his head at that, and added, "Borovich and I were out late last night so he won't be able to make dinner in your Q." Then he signed in behind the Lieutenant on the duty log, showing that he was going off duty and Marr was coming on.

As soon as he had signed, the Lieutenant excused himself, grabbed his field gear, book, and poncho liner, and practically ran to Headquarters Company, to the familiar CQ desk and black walls. He told the E-5 on duty to wake up the armorer so that he could turn in his sidearm. When the armorer finally arrived, he was bleary-eyed and half asleep, but capable enough to escort the Lieutenant to the arms room where he accounted for his ammunition and clips and turned in his pistol and holster.

WE ARE FAMILY

As soon as that was done, he put the book inside his folded poncho liner to make sure it wouldn't get wet and practically ran up the stairs and out the building to get to his room. When he entered the stairwell of the BOQ, he could smell ham and turkey roasting in the communal kitchen. Turning the corner, he could see that Buzz was flitting back and forth, like a good grandmother awaiting the arrival of her wayward children. Having said hello, he dove into his room, stripped off his clothes, set the alarm clock for 1030, and promptly fell asleep. When the alarm went off, he shaved, took a shower, and dressed in jeans and a dress shirt. Feeling better, he headed towards the kitchen because he could smell Buzz's famous black coffee. After he had a few swallows, he felt like he was back in civilization enough to talk to people and ask about how things were going. Buzz and Susan proudly informed him that everything was going swimmingly, and with any luck, everything would come together just before noon. Susan added, "I'm running upstairs to work with Chris on the sides."

The Lieutenant asked if there was anything he could do. When Buzz told him there was not, he remembered that he had his condiments in his room, so he excused himself and came back with pickles plus green and black olives to set them on the counter by the refrigerator, before putting them in serving dish, or what passed for that in the communal kitchen. Buzz arched his eyebrows a little bit, looking at them, and the Lieutenant explained, "I'm no cook. So, I'm bringing these to help make the meal." Buzz chuckled softly and told him that it was all right, the Lieutenant reminded Buzz of himself, when he was newly commissioned and could barely boil water.

The Lieutenant sat in the corner about half asleep while Buzz continued to move around the kitchen. It looked like he was making some sort of gravy while the turkey and dressing roasted in the oven next to a ham. Around 1100, several other people started to arrive. Susan and Chris had taken over the kitchen directly above him, as they were responsible for side dishes, including mashed and sweet potatoes. They had also invited a friend of theirs, a nice young lady named Robin, who was kind enough to bring the ingredients for green bean casserole, the staple of every American holiday meal.

A little later, Bigfoot, and one of his friends, showed up with German bread, cheeses, and cold meats for appetizers. Then Bernie came out of his room, somewhat bleary eyed, but no longer stoned, and happily appeared at the door, with a couple of pumpkin pies he had purchased from the PX the day before. With all of the food set out, they were soon joined by Meghan, from upstairs and her three friends, who, apparently, did not cook either, but did bring the red and white wine, plus Coke and Fanta for beverages.

The Lieutenant asked Bernie if he'd been over to invite his old girlfriend from across the street. Bernie slyly said, "Not on your life."

He then added, "I think she has found another fellow to hang out with, and from what I was told, they're going over to Heidelberg for Christmas dinner at the officers club there."

The Lieutenant shook his head and thought that it sounded like a pretty good deal, before turning to Buzz and asking, "Where are Pallas and Pavel?" Buzz told him that Pallas and his girlfriend were having Christmas dinner with her German parents. Pavel had decided to go to Heidelberg as well, probably to meet some people and look for a new assignment after he left the battalion and command in about a year.

Finally, Captain Boris came upstairs to join them. He rarely socialized, but on this occasion, it was only appropriate. His little dark brown and black, long-haired German dachshund, Oscar, was yapping in his arms, until he finally set him down to let him out to run up and down the hall. Once it looked like everyone was assembled, Buzz asked the Lieutenant to run upstairs and knock on the door of the lieutenant from 5-68, who was always throwing his furniture out the window. The Lieutenant did not know the man but ran like the wind up the steps to knock on his door. After several attempts, the door finally opened slowly, and he told his fellow armor officer that he was invited to Christmas dinner, if he would like to attend. At first, the man told him that he had no intention of coming down, but when asked again, finally agreed to clean up and join them in about ten minutes.

GOD BLESS US, EVERYONE

By now, the entire second and third floor smelled amazing and Buzz gave the order for everyone who lived on the second floor to pull their small dining tables and chairs out into the hallway to make one large communal table. When that was done, Buzz told everyone to take their seats, with the exception of those who were bringing food to put on the table.

The ladies brought down their side dishes from upstairs. Buzz brought in his gravy, as well as platters of ham and turkey that he had sliced. Bigfoot and his girlfriend put out the platter of cheeses, salami, and bread. Bernie put his pies back in the kitchen to be served later with coffee. Then Buzz, ever the host, came out with plastic wine glasses, one for red and one for white. These were passed down the table, until everyone had a pair. Paper plates and napkins were also passed along, as well as plasticware. The entire thing looked like a church picnic, which of course, it was not.

Once everyone was seated, the 5-68 lieutenant came down from upstairs, and sat with Bigfoot and his girlfriend at the far end. Buzz said a short prayer, thanking God for His bounty, and to grant peace to their friends and family all around the world.

As soon as the prayer was finished, they all passed plates of food and ate. The Lieutenant was happy to see that Buzz had set out the pickles and olives as well. Since he hadn't eaten since the night before, and he'd been up all night, the Lieutenant was starving and fell onto his food with a vengeance. Still, in between bites, he was able to carry on a conversation with the ladies from next door and their friend, Robin. She was a finance officer, apparently, and assigned to Schwetzingen, over by Heidelberg, but had become fast friends with the other ladies. She was overjoyed to be invited to what passed for a family meal. He looked up and down the table and thought to himself, this was one of the most memorable holiday meals he had had, or probably ever would have, given the amazing and odd assembly of different types of people from all over the country, stuck in a foreign land, for their favorite holiday.

Eventually, the three ladies asked him how he felt about the fact that they were homosexuals. As usual, he told them it didn't bother him in the least, and he certainly wasn't the sex police and had no intention of ever being so. They accepted that and felt more comfortable. He did tell them that he was looking forward to the new year because he was going back stateside to get married and bring his new wife to Mannheim. All three of them smiled and began asking him questions about wedding planning, of which he was blissfully ignorant. All he knew was that he was supposed to leave Germany by the first of March, spend about a week there, show up for a wedding on March eighth and then bring her back on March fifteenth. Other than that, he didn't have a lot to do with any of it, and he was glad, since he had enough going on in Germany to keep him busy.

He turned and asked Susan and Chris what their plans were. They told them that they were trying to get assigned to the finance department at West Point. They had hopes that in a liberal state they might be able to get some lasting arrangement like a legal partnership if they couldn't actually get married. The Lieutenant wished them well. Robin added that she really didn't have anyone she was close to at the time but was hopeful that someday she would. The Lieutenant wished her well too and further went on to compliment her because she made the most amazing green bean casserole he'd ever eaten in his life up to that point in time. That made her smile.

They all relaxed and told each other stories about their past Christmases. The Lieutenant added his tale about trying to get back for Christmas during his first year at VMI. That trip ended up taking three days by bus through a blizzard. It also included a riot in the D.C. Greyhound station, with the cops standing on top of the lockers around the room with their pistols drawn, a cold ride to Philly in a cherry blossom tour bus with no heat but a glass top so the falling snow looked like Star Trek, and a near death experience on I-70 outside St. Louis, when the retired driver who was activated when Greyhound went on strike slipped past two jack-knifed semis and narrowly missed a stranded cement mixer on the other side, while playing cards with the passenger seated behind him.

After another half day stuck in Kansas City, on the final leg, he was so tired of riding buses that he asked the driver to pull over and stop at an intersection in his hometown, so he could jump off with his suitcase and walk home through the snow at midnight. Luckily, a family with a son who had been at West Point recognized the cadet uniform and gave him a ride home. Little did he know that his parents and siblings were at the bus station waiting for him, so when he arrived at his house, he had to break in the back door of the garage to get in and thaw out. But it was a great homecoming. He could still smell the bacon, coffee, maple syrup, and buckwheat pancakes that his mother made him when she got home.

O'ER THE HILLS AND FAR AWAY

As the meal wound down, Buzz asked them all to raise a glass for those they knew who had fallen, and for those they knew who were far away. The Lieutenant took a drink of his white wine, thought of his girl, and smiled because he hoped to see her soon. With the toast completed, Buzz told everyone that dessert would be served buffet style in the kitchen, and that anyone who wanted it could come through and pick up coffee and pie. As the holiday participants rose, the Lieutenant looked down the table and saw that Meghan and her three friends were laughing, chattering back and forth, having a great time. Even the armor lieutenant from upstairs was smiling, at least a little, as he got up to get his dessert. The Lieutenant decided to forgo coffee, since he'd lived on it for twenty-four hours and needed to go to sleep. But he did dive into the pie as soon as he got back to the table.

With dessert almost finished, the fact that they were far from home and were not getting back anytime soon struck them. Sadness fell on the participants and they became a little quieter. Finally, when they were done, Buzz solicited some help to clean up. The residents of the second floor took their tables and chairs back into their rooms before going down to the kitchen and helping Buzz put things away. When Buzz saw the Lieutenant, who was so tired that his eyes were practically closed, he told him to get back to his room and get some sleep because he looked like he'd been up all night, or maybe even two or three. He thanked Buzz, and walked quickly back to his room, where he set the alarm for 0500 and passed out cold until the next morning.

Chapter 29
Magical Mystery Tour

The alarm clock clicked and clattered at 0500. His eyes popped open. He realized that he could hear sleet driving against the window and burrowed even further down into his bed because he did not want to go out in it. Nevertheless, he got up, washed his face, dressed, and made his way to Headquarters Company. When he walked through the guard post, it seemed odd, because the entire battalion was a ghost town. Then, thinking back, he hadn't seen anyone moving in the BOQ either. He was cold and getting colder in his PT gear, so he soon pulled the door open, and hurried into Headquarters Company, to the oppressive feel of black walls and the CQ desk that still towered over the hallway. He stopped and asked the CQ if Top was in. The CQ nodded yes, so the Lieutenant walked downstairs to the First Sergeant's office to ask him what was going on.

The First Sergeant just looked at him like he was daft. And then smiling said, "It's Little Christmas, so the battalion has a training holiday today." The Lieutenant simply shook his head and wished that someone had told him. But Support Platoon Leaders, he thought, were always the last to know anything. It's practically part of the job description. However, he was excited at the prospect of a day off and a chance to go back to his warm room and rest up. Before that, he asked Ascencion what Germans did on "Little Christmas." The First Sergeant again looked at him like he was from another planet and calmly said, "That's the day when Germans give gifts to their neighbors. Christmas is for family and Little Christmas is for neighbors." The Lieutenant thought to himself that it was an odd but nice custom. And he was determined to enjoy his day off.

"Top" turned back to his duty rosters and other administrative paperwork, so the Lieutenant turned on his heel, ran up the stairs and out the door to his BOQ room. When he was there, he heated water for instant coffee and took a shower and shaved. It struck him that he really didn't have much to do since he didn't have any relatives in country. But he decided to make a quick breakfast and perhaps read a bit. There was still light snow on the ground, so he thought he might drive around a little, to see what the neighborhood looked like, decked out in white. Then, maybe by early evening, he would drive up to Viernheim, to see if Nick's was open so he could have a decent meal, before the duty week started in earnest.

After he had a cup of instant coffee with mocha in it and a breakfast bar, he realized that he was pretty bored and stood in his doorway listening to see if anyone had come down to the communal kitchen. Buzz had gone to England to see Elizabeth, but he saw Bigfoot heading down to make coffee, so he followed him in. As the strong German coffee was brewing, he asked Bigfoot how his night had been, and he said just fine. About that time, his female guest came wandering in looking for coffee. She curled up in a chair by the window. She wasn't talkative, so once the coffee was made, the Lieutenant poured himself a cup, excused himself, and took it back to his room, so Bigfoot and his visitor could have some time alone.

He savored the coffee and decided he agreed with Buzz before he sat in an easy chair with *The Odessa File*. As he started reading, it felt like he was in it: the dark German forests and cold winter weather were everywhere. He was fascinated by it, because he knew full well that there were many German veterans of World War II at every level of the Army and government, both in the United States and the Federal Republic.

By noon, he was starving, so he put on his German civilian clothes and ski jacket and fired up his Sirocco to drive around and see if there was anything open. Most of the restaurants and businesses were closed, but he did find a currywurst stand by the railroad tracks that was up and running. He stopped and had some with French fries. He was amazed at how the curry sauce warmed him up both inside and out. When he was done with that, he went back to his room and listened to the radio for a while, thinking of his girl, and hoping that she was warm and enjoying Christmas with her parents.

By late afternoon, he decided to see if Nick had opened up his guesthouse for the evening. When he got down to his car, he slipped in a cassette tape of Gordon Lightfoot songs. He listened to his *Song for a Winter's Night* over and over again, as he drove northeast from his quarters towards Viernheim. When he pulled up outside the Hotel Tenne, he saw that the lights were on and Nick was doing a tremendous business. As usual, he parked his car around the corner, and slipped in out of the cold, looking like a German student. The booths were full, so Nick motioned to him to come up to the bar.

Nick smiled at him. He sat down and ordered a glass of Pilsner and fried squid for an appetizer. Nick brought both of them out to him and asked him how his Christmas had been. The Lieutenant told him that it was fine. He had been on duty all night and spent Christmas Eve with the soldiers, before Christmas dinner with his friends in the bachelor quarters. Since no one had told him about Little Christmas, he had gone into work and looked like an utter fool, but he was happy to be here now, to enjoy a meal and company.

Nick just laughed to himself and smiled and asked him what he would like to eat. The Lieutenant, of course, ordered his favorite meal: pork cutlets, on a shish kebab, with French fries and cold German salad. As he sat there, alone, he looked around the room at the people, laughing and having a good time, and it raised his spirits as well, knowing that he would not be alone much longer. And he knew that his girl would love this place.

When Nick came back with his food, he asked the Lieutenant how long it would be before he could go back home and get married. The Lieutenant told him about two months and two weeks now, and that he was counting the days. Nick told him that he understood and guessed that he must be lonely far from home. He told Nick told him that he surely was, but he was blessed to have a friend like Nick who ran a restaurant like this. Nick just looked at him and told him that so many Greeks had to leave the islands to make their fortune that they all understood, like the Swedes understood, what it was like to work, far from home in difficult places, with the promise that someday you would return. The Lieutenant told him that he was surely right, and he couldn't wait to bring his girl in to meet him, because he knew that they would get along. Nick chuckled to himself because he got along with most women, since he could cook, and carry on a colorful conversation.

When he was through with dinner, Nick took away the plate and then came back to ask if he would like a hot brandy. He thought for a minute and then said, "Sure. Why not?" So, Nick brought out the little warmer, a brandy snifter, and a bottle of Metaxa. He poured the Metaxa in the snifter and put it in the heater, having lit the flame, and rotated it slowly.

When the brandy was warmed, the Lieutenant took the snifter from him and gingerly took a small sip, followed by a larger one. The warmth flowed down through him and instantly made him feel better. Nick smiled and said he thought the Lieutenant needed something to raise his spirits because he really looked lonely. The Lieutenant agreed, and then Nick turned, and picked up something from the bar, brought it back, and gave it to him. It was his briar handled corkscrew that he used to open wine bottles. After he handed to him, Nick said "Little Christmas: my gift to you." The Lieutenant smiled and apologized since he didn't have a gift for Nick.

Then he told him he would surely bring him something in the very near future.

When he finished with his brandy, the Lieutenant looked over and thanked Nick again. Then he paid for his meal in marks with a healthy tip and slipped out of the restaurant into the darkness. Surrounded by sleet and freezing drizzle, he made his way to the little car and made the long drive back to his room. When he arrived there, he ran into Captain Pavel, who had been away for a couple of days. They exchanged Christmas greetings and then went to their rooms to try to get some rest, before cold dawn and physical training in the snow.

SLIP SLIDIN' AWAY

When the alarm went off, it was still pitch black, so he had to roll over and check the time, to make sure it was correct. It was, so he got up, washed his face, and dressed in PT gear to head back to Headquarters Company. When he arrived, he walked in the barracks and heard a few soldiers moving around, but otherwise the place was dead. Running down the stairs to the orderly room, he was heartened to see that Ascencion had a fresh pot of coffee going, so he poured himself a cup, and sat down in a chair in the corner, to wait for his day to begin.

When Captain Boris arrived, he went into his office for a while with the First Sergeant. probably to discuss their ongoing operations, thought the Lieutenant. Boris was an intelligence officer, so he was always under the impression that if he used enough informants, he could root out the drug dealers in the company and court martial them out of the Army. Unfortunately, most of the dealers were professionals. And, so far, they had beaten every court martial he had thrown at them. To boot, the informants generally ended up overdosed, convicted, or beaten to the point that they stopped informing.

When Boris came out, the Lieutenant followed him upstairs to the courtyard where the company was forming. The soldiers were shaking, and pretty much spaced out after two days of drinking and drugs. The vast majority of them appeared to still be stoned or wishing that they were. Finally, after Ascencion called the company to attention and formed everyone up, he turned it over to Captain Boris, who led the unit on a fairly short run, through the sand and snow, into the training area and back. It was simply too cold to do much of anything else. When they returned, he turned the company over to the First Sergeant, with orders that details would be done in the morning hours, and the troops would be released at noon since the battalion was on the holiday half-day schedule through New Year's Day.

The Lieutenant looked around his platoon and finally found Sergeant McNabb. He asked him if he could come by his office and go over a training schedule with him after breakfast. But McNabb told him that he had several vital runs to make for the battalion in the morning, and in the afternoon, he had promised to go out with his wife and children to do some shopping. The Lieutenant then asked him if he had any plans for the following day. He told him that he could stop by, first thing, after 0800 on Friday for a short period of time, but that he was committed to making several deliveries on Friday as well, before he had the afternoon off. The Lieutenant was beginning to believe that he was never going to be able to nail Sergeant McNabb down, to the point where he would actually be doing his job as a platoon sergeant. But he wasn't going to get anywhere until the half-day schedule was over, and he and the sergeant could get on the same page.

It was bone chilling cold. Shivering, the Lieutenant decided he would have breakfast in the mess hall in order to get some hot food into himself. He walked through the doors of the mess hall, glad for the warmth, and paid the Assistant Mess Sergeant the going rate for a meal which was sixty-five cents plus a surcharge of another quarter. He proceeded down the chow line until he had a heaping plate of creamed beef, toast, hash browns, scrambled eggs, bacon, and some fruit. When he was ready, he looked around the room to see if there was anyone who might not be offended to have to sit with an officer. As luck would have it, he saw Sergeant Brown from the S-4 section, sitting in the corner drinking a cup of coffee and smoking with Sergeant First Class Peeler, the S-3 NCO. He liked both of them and walked over to sit down beside them. They made some holiday small talk, and then drifted into silence as they ate, smoked, and drank coffee. Sergeant Peeler asked the Lieutenant how he liked being a Support Platoon Leader. The Lieutenant answered that the job had a lot of problems and challenges, but at least he had a Jeep and a driver and a heater in the Jeep, so he wouldn't be freezing his ass off like he was in a tank. Sergeant Peeler and Sergeant Brown both laughed at that, and the Lieutenant rose to go back to his room and dress.

When he got back on post, he went to the S-4 office and sat at his desk, pretty much bored to tears. There was nothing really going on in the battalion, and nothing much to do except to continue trying to improve the load-out procedures and figure out how to get about twenty tons more material into his trucks than they could possibly carry. He fiddled with that for a while before going back to his master training roster for a little bit. When the noon hour came, he checked in with the S-4 and since he had nothing for him to do, went back to his room to make some hot Swiss Miss and read a little bit. As he sat in his room, he thought that a year ago he was cold and bored in his parents' house in western Kansas. But the next day, on the twenty-eighth, his whole world had changed, when he went out for coffee with his girl for the first time. He thought about that and then drifted off to sleep, as he was still trying to recover the hours that he lost on duty.

About 1800, he came awake again and swore that he heard the clattering of pots and pans down in the kitchen. He changed into jeans, a T-shirt, and sneakers, before he walked down to the kitchen. There, almost beyond belief, he found Bernie and a new girlfriend, who apparently were determined to make spaghetti and a salad. The Lieutenant asked if he could be of any assistance. They said yes, he could help make the salad, because they really didn't trust him with the spaghetti. The Lieutenant thought to himself that was a good choice. Then he asked if they could use a bottle of red wine as he had some in his room. Their faces brightened immediately, so he brought a bottle down, opened it with the communal opener, and poured everyone a small glass. As for himself, he was really not a red wine drinker, but to be sociable, he was happy to join in.

It turned out that Bernie's new girlfriend was a young transportation lieutenant, who was a friend of Meghan, the lieutenant whose room was above his. She had arrived in country barely two months earlier and was just starting to get her bearings when she ran into Bernie. Bernie had taken her under his wing and was doing his best to get her accustomed to the country and more particularly the United States Army in that country. When they were done eating, the Lieutenant could sense that they wanted to be alone. So, he washed their dishes, put them away, and went back to his room to try to get some rest.

THE ANNIVERSARY

The next day was the twenty-eighth, the anniversary of his first date with his girl. He thought back to that day and how much it changed his life. He determined that he would do something to celebrate, so when his duty day was over, he decided to drive over to the Class VI store and buy a bottle of Wellers Bourbon, and a gift bag to put it in. Then he drove back to his apartment, put on his German student clothes, and drove quickly up to Viernheim to Nick's. When he walked through the door, Nick's face brightened immediately, his eyes widened when he reached in the gift bag and saw what the Lieutenant brought him. The Lieutenant explained that this bourbon was 107 proof and from Kentucky, where he had trained on tanks. It was horse country and beautiful. But more importantly, it was his girl's father's favorite drink, and he always made him double highballs when he visited. Most probably he did it to try to get the Lieutenant into a weakened state, so he couldn't take advantage of his daughter. In any event, he was a great guy and it was great booze. He also hoped that Nick could enjoy himself and make a few extra marks selling it over the bar.

Once Nick had carefully placed the bottle of Wellers back on the bar, he came over and asked the Lieutenant what he wanted to eat. Since he was still cold, he asked for a beer, and a hot bowl of hunter soup, followed by a rump steak with herb butter and French fries. Nick turned and came back quickly with the beer and the soup. Then he made the rounds of the guesthouse, talking to customers and taking some other orders, while the Lieutenant's food was cooking. In less than ten minutes, dinner was served and the Lieutenant dug into it wholeheartedly, because he was starving. Once he had a couple of bites of steak in him, he settled down and enjoyed the rest of the meal, making small talk with Nick from time to time.

When he was done, Nick asked him if he had any plans for the new year. The Lieutenant told him that sadly he was going to be on brigade duty on New Year's Eve, and into New Year's Day. Then, on New Year's Day, he would have to go to his battalion commander's quarters for New Year's brunch, which was a tradition in the Army to wish the commander a happy New Year. He would have a light meal and either a Bloody Mary or a screwdriver, and champagne. He told Nick that after all of that, he imagined he would be ready to fall face first into bed for the better part of a day.

Nick just laughed and then asked him if he would like some warm Metaxa. "Of course," the Lieutenant said, "and I'll even have some good strong coffee afterwards. I don't have to work tomorrow, so I can stay up later tonight and sleep in." Nick came back with the brandy and went through his usual ritual of warming it up for the Lieutenant, who took a small sip and exhaled slowly, completely satisfied.

When he was done with the brandy, Nick brought out a cup of thick black coffee and a little cream. The Lieutenant savored every bit of it, because it was good coffee, as opposed to the brackish dishwater he could get in any Army mess hall or battalion headquarters.

As he sat alone at the bar, he thought again about his girl and how a year ago, after coffee they had gone to a liquor store and picked up some booze and mixers and ended back in her apartment. There he made himself an amaretto sour, and she drank slow gin and Coke, which apparently was her favorite. After a while, as they were sitting there, his best friend called from Colorado to see if he was there, since he wasn't at home, and he talked to him briefly and then gave the phone back to her. When she was done, she hung up the phone and he told her that it was time for him to go and left, thinking to himself that most probably he would never see this girl again because surely, she would get back together with his old friend.

That didn't happen, and now he was on the cusp of his wedding, counting down the days before he would get on an airplane back to the States. When he was through, he motioned to Nick, who brought him his tab. He put the full amount on the counter and an extra five marks. He then wished Nick a Happy New Year because he doubted that he would be back before New Year's Day. Nick wished him a Happy New Year as well. Then he put on his white ski jacket before walking out into the mist to drive back to his room.

The weekend went by slowly because it was cold and snowy and there wasn't much to do. The Lieutenant decided to go have dinner at the officers club on Saturday night, just to see who was there and mingle a little bit and talk with other officers who were suffering through Christmas far from home. On Sunday, the communal kitchen was quiet, so he made do with what was in his refrigerator, before setting his alarm for five in the morning to get up on Monday, to get ready, draw a sidearm and pick up a Jeep and a driver to go to Brigade for duty on New Year's Eve.

BAD MOON RISING

He woke up at 0430 the next morning, ahead of his alarm, and rose quickly. He made hot water for instant coffee, then dressed in his best set of fatigues for brigade duty. He relaxed a little and at 0600 decided to put on his field jacket and parka to keep warm on the drive. He took along his attaché case with a couple of books and magazines, and some licorice because it was going to be a long day. When he reached Headquarters Company, he went down into the orderly room for a decent cup of coffee and sat and talked to Ascencion about how things were in the company.

As usual, it was rife with drugs and violence. But according to him, he and Captain Boris were starting to get a handle on it. The Lieutenant told him that he was going to go ahead and get some breakfast, so he left his attaché case with Top, behind his desk. He walked over to the mess hall, for his usual creamed beef on toast, scrambled eggs, bacon, and fruit. There weren't many people up, so he sat over in the officer's mess side of the facility and browsed the copy of the Army Times that had been left there. When he was done, he went back to the orderly room and asked the First Sergeant to summon the armorer so that he could draw a sidearm. When he had is forty-five, with three clips, and an extra round in the chamber, he put it on safe, holstered it. He attached it by the lanyard to the pistol belt, before putting it on over his field jacket and parka, and going upstairs to meet the soldier who was detailed to be his driver that day.

When he walked out into the courtyard, he saw that his Jeep was ready, and there was a soldier he had not met before standing beside it. He walked up and the soldier, a Private First Class, saluted him smartly and said, "Pacesetter!" The Lieutenant was a little taken aback, because although that was supposed to be the battalion greeting, in his experience, few, if any, members of that battalion actually said it, as opposed to "You gotta be shitting me."

He saluted back and, gave the required response, saying "Press On!" This made the soldier smile. The Lieutenant asked him, "Are you new to the company?" He said he was and that he had just arrived in country forty-eight hours earlier. Since he was the newest man in the company, he was detailed for New Year's Eve. The Lieutenant told him that that was the normal practice. He then asked him if he knew how to get to Brigade. The driver told him he did not, so the Lieutenant told him, "Don't worry. Once we're in the Jeep, I'll give you directions. We'll get there just fine."

The driver got into the Jeep. The Lieutenant was preparing to slide into the passenger seat, when out of the corner of his eye he saw Sergeant Robb briskly walking toward him from the mess hall, so he stopped. Sergeant Robb came up to him and said "Sir, I'm glad I caught you." Then he said, "I need to talk to you."

He and the Lieutenant then walked back behind the Jeep. Sergeant Robb quietly told him, "Don't go out, sir."

The Lieutenant asked him, "Don't go out where?"

Sergeant Robb continued, "Don't leave Brigade Headquarters for any reason, sir. Don't go out there. If a subordinate unit calls you with a problem at Brigade Headquarters, do not go out of the building. Call the duty officer or NCO in that battalion and have him go to handle whatever the problem is. The troops there will know him and they'll be less likely to be violent. Last night, 3/8 Cav killed one of their own over at Coleman. A trooper who wouldn't pay his drug supplier. So, they shot him up with heroin and killed him, took everything he had, stripped him naked, painted him red and white, froze his body in the snow, with the troop guidon in his hands, and placed him in front of the Squadron Headquarters in the middle of the night."

The Lieutenant looked at him calmly and said, "I suspect that somehow his death was the result of some sort of training accident, if not then, certainly later."

Sergeant Robb looked at him. He didn't say a word at first and then repeated, "Sir, please don't go out. I'm telling you that as bad as things are here right now, they're worse over there. And sir – a few days before you got called out to secure the PSP, Klar and the RAF robbed a bank in Zurich. They attempted to murder a police officer and killed a civilian by accident. You're lucky the Polizei didn't come in shooting that night. With the situation in Iran, things are heating up."

The Lieutenant looked calmly at his faithful platoon sergeant, who had gone out of his way to take care of him, and simply said, "Thank you." Sergeant Robb saluted him and the Lieutenant saluted back before climbing into the Jeep beside his new driver. When he was in, he had the soldier crank the heater up to high, and since it still wasn't cutting it, he told him to drive out of the Kaserne and directed him to the BOQ. Then he told him to stand fast as he was going to run up and get a couple of poncho liners for them so they would have blankets on the way and also to use during the evening when they wanted to sleep. The driver looked at him sort of quizzically, but the Lieutenant told him, "Trust me, you'll be glad we have them."

When they finished picking up the poncho liners, the Lieutenant slipped his right hand out, through the windowless Jeep door and slapped it on the outside twice, with his gloved hand, saying, "Get 'em up, Scout." The driver looked at him quizzically. The Lieutenant told him that he always did that because his hero was Jay Silverheels, Tonto, the Lone Ranger's friend, who was always catching hell. Somehow, he felt an affinity with him, and since he was from the Cav, he liked to call his horses Scout. The driver just chuckled and they took off, through the battalion toward the training area. When they reached the entrance to the maneuver area, he showed his new driver the hand signals he would use, over the gear shift, to tell him which way to turn, when to move forward, when to speed up, when to slow down and when to stop. He explained that once they were driving, the cold wind whistling through the open Jeep would make it really uncomfortable for both of them, unless they had their winter headgear on and scarves pulled up to cover almost every part of their faces. This made speaking really difficult, and it was just easier to give directions by hand. The driver looked at him quizzically again, but as soon as they started moving through the training area, with the frigid wind whipping them, he understood.

After half an hour, they reached Coleman Barracks. He had his driver proceed slowly past the 2/13 Infantry and 3/8 Cav, to where Brigade Headquarters was located. He then had him combat park the Jeep in the spot reserved for the duty officer. They picked up the poncho liners and their belongings and walked into Brigade Headquarters. When he walked through the main door, the Lieutenant recognized the Adjutant's office and proceeded there, after he directed the driver to check in with the Duty NCO and see where he wanted him to stay.

The Brigade Adjutant was a major, who was promotable to lieutenant colonel, and a typical, brash cavalry officer. He had just transferred in from personnel in the Pentagon to work for Colonel Thomas, as he had been in the Third ACR when Thomas was a lieutenant colonel and squadron commander. The Lieutenant had got the background on him from Captain Steves, the Battalion Adjutant, when he was assigned to brigade duty, because this major was particularly irascible. Steves had told him, "Lieutenant, you need to show up early, by at least half an hour, and be immaculately dressed."

The Lieutenant presented himself, saluting, and saying, "Lieutenant Manteuffel reports, sir." The Adjutant looked up slowly, checking his appearance up and down, including his spit-shined jump boots, and then told him to take a seat. Since he had not pulled brigade duty before, the Adjutant went through the SOP in excruciating detail. For the next twenty-four hours, as at Battalion, he needed to be on alert for flash traffic and break the coded messages as quickly as possible. He was also responsible for good order and discipline on the Kaserne, as the brigade commander was the ranking officer. While on duty, he would report directly to Colonel Thomas.

When he was almost through, the Adjutant added, "I'm glad you came in today, because I need to tell you that you have jury duty for a court martial this next week. Do you have a notepad?"

"Yes, sir."

"Then write this down. At 08:45 on Thursday, you will be at Turley Barracks, building 481 in greens. Do you understand?"

"Yes, sir," the Lieutenant replied. The Adjutant asked if he had ever had court martial jury duty before. The Lieutenant told him that he had not, so the Adjutant went on to explain how the military justice system worked, and what his duties as a juror would be. The

Adjutant then escorted the Lieutenant through the building, showing him where the S-3's office was, and more importantly, where the S-2's office was and the safe. He then gave him the keys for the brigade's offices and to the safe and had him open the safe so he could make sure he knew how to do that and locate the codebook within it.

When that was done, he escorted him to the brigade duty officer's office. It was a separate room adjacent to the brigade Duty NCO's office, that had a coffee maker, desk and phone for the Duty NCO and a couch for the driver to sit, and lie down on to sleep, if possible. The brigade duty officer's office also had a desk, chair, telephone for flash traffic, and a sofa that he could sleep on at times during the night, if he got the chance.

With the tour done, the Adjutant asked if he had any questions. The Lieutenant simply said, "No," so he added that he needed to sign in to be responsible for the brigade for the next twenty-four hours. The Adjutant then informed him that he would be the only officer in the building from this time forward since it was New Year's Eve. And, if he needed to eat lunch or dinner, there was a small PX auxiliary snack bar about a block away, or he could take his meals with the infantry battalion. He added, as he headed for the door, that he did not advise going to the Cav, while he was on duty. The Lieutenant nodded his head, showing that he understood, and did not bring up the murder in the squadron. Then he watched as the Adjutant jumped into a Saab sedan and took off for his married quarters in Benjamin Franklin Village.

Once he was gone, the Lieutenant thanked his lucky stars for Sergeant Robb's warning and the fact that when he went in to get his poncho liners, he had snagged some licorice and a few breakfast bars that would tide him over for meals so he wouldn't have to go out of the Headquarters. He took Sergeant Robb's warning seriously and had no intention of getting embroiled in anything on New Year's Eve.

Once he was settled in his office behind the duty officer's desk, he folded up his poncho liner and sat it on the sofa. He arranged the tablet and pencils to write down flash traffic messages, next to the phone, and then picked up one of the books he had brought along with him. He had decided to read his copy of *The Lionheads*, a novel written by Josiah Bunting, one of VMI's most famous graduates. It was a book that was more than a little cynical about the Vietnam War and the Army in general. Given where the Lieutenant was, it suited his mood. He also had with him the hardcover copy of *The Odessa File* that he had not finished yet. Between the two, he figured that he would have plenty of material to read. And if not, the good thing about Brigade Headquarters was that it had a much better assortment of magazines and newspapers than Battalion. He actually found some fairly recent copies of Time and Newsweek magazines, and even the New York Times. The officers at Brigade were apparently more politically astute and interested in current events, and how they might affect their careers. They also had more time for reading since they did not have to deal with the day-to-day problems of command at the battalion and company level.

After reading for a few minutes, he walked over to the to the NCO and introduced himself. The Duty NCO was a Sergeant First Class named Green from 5/68 Armor, the Lieutenant's sister battalion. The thing that was different about 5/68 was that it was equipped with M60A-2 tanks that fired missiles. The 68th Armored Regiment itself was unique in that it traced its lineage back to World War One, and at one time, was the largest tank regiment in the Army, with eight battalions, five in the regular army and three in the reserve. Of the five on active duty in the Army when the Lieutenant reached Germany, two were in Baumholder, two were in Mannheim, and one was the only airborne armor battalion in the entire force. That battalion was stationed with the 18th Airborne Corps and the 82nd Airborne Division back in Fort Bragg, North Carolina.

Because 5-68 had much more sophisticated weapon systems that needed a higher level of maintenance, that battalion tended to draw soldiers from a different talent pool than those assigned to the Lieutenant's battalion, which used straight tanks. These less smart weapons were really the hammer and the anvil of the Army. They could withstand more punishment in the field. Consequently, soldiers that had problems in other units could be transferred back to the 8th Division tank battalions and infantry battalions, and eventually, if they had enough problems, to 3/68, in Mannheim itself, or to 3/8 Cav and 2/13 Infantry that were located across the post from the USARER Confinement Facility.

When he had finished visiting with Sergeant Green, he turned his attention back to his new driver and asked him for his name. The young trooper told him he was Private First Class Hawkins from upstate New York. The Lieutenant couldn't help himself when he asked, "What in the world possessed you to join the Army?" Hawkins looked at him wryly and told him that he wanted to get out of rural New York. The Lieutenant told him that he had been in a small town called Hinesville with one of his college classmates, and he could understand. He then went on to add, "Since you're new to the battalion, all I can tell you is to stay off drugs, steer clear of drug dealers, and thugs. Do your job every day and lock your doors at night. Okay?" The driver looked at him a little quizzically at first, but then realized that he wasn't kidding, and took it to heart. The Lieutenant told the two enlisted men that he would be next door in his office, and to let him know when they were ready to go to lunch, as he would man the Duty NCO's desk while they went to eat.

He sat back on the sofa in his office to read for a while and before he knew it, he had dozed off. He was almost immediately awakened by a knock on the door and stood up quickly to open it. He saw that the NCO and driver were ready to go, so he told them to go ahead and have lunch, and when they got back, they could relieve him. He sat at the NCO's desk and manned the phone. It never rang since it was New Year's Eve. He went back to the office and picked up the book he'd been reading, then brought it to the NCO's desk and continued to read until Green and Hawkins came back and relieved him. Once they took over their positions, he went back to the duty office just in time to hear the phone ring for flash traffic. He picked up the phone and wrote down the message, while the NCO did the same. They then quickly compared notes and he went to the S-2's office to open the safe, withdraw the code book and break the message. As he anticipated, it was fairly lengthy, but in the end just a test, so he went back to the Duty NCO's office and logged receipt of the message, noting that it was a test, before he went back to his office to try and catch a nap, since it was early afternoon and already starting to get colder.

KNOWING ME, KNOWING YOU

By mid-afternoon, he woke up feeling bored, so he went over to sit with the driver and the NCO to visit a bit and have a cup of coffee to wake up. When he was through, he decided to take a tour of the building, to see how the brigade staff lived. It was a beehive of desks and a jumble of notebooks and bookcases. But overall, it was a pleasant enough place. He stopped by and looked in Colonel Thomas' office, that had several photographs from past postings. One was from the Three-Quarter Cav, in Vietnam. Another from a tour in Germany, apparently as a Company Commander, and one from his time as a squadron commander at Bliss. Then a last picture of him with Colonel Cardillo, the Regimental Commander, when he was executive officer of the Third Cavalry Regiment. There were also one or two mementos from his time at the Virginia Military Institute, and the Lieutenant smiled to himself when he saw them.

After walking around, he felt much better. Since it was going on 1730, he released Green and Hawkins for dinner, before sitting down to drink another cup of coffee, eat a breakfast bar, and enjoy a couple of pieces of licorice. He thought to himself that he hadn't lived like this since he'd last been in the field. But, given what he had seen of the Support Platoon Leader job, he imagined he would be living on a steady diet of mocha, licorice and breakfast bars, coupled with pulling over occasionally at a German market for meat, cheese, mustard, and bread to make sandwiches.

When the NCO and driver returned, it was almost dark. The sun went down early in the day in Germany because it was so far north. It was practically like living in Alaska. With the light fading, the cold descended with it, and the wind came up. Coleman Barracks was on the banks of the Rhine, so there was very little foliage to keep the wind from whistling off the water, through the Kaserne. Consequently, it became extremely cold, extremely fast, and the Lieutenant was glad that he wouldn't be venturing out to try to do anything, either before or after midnight.

THE PAGAN

The night dragged on as he watched the large clock on the office wall, waiting for the New Year. At almost 2330, true to form, they received a flash message that, on its face, appeared to be plunging them into World War Three, but was in fact a test. The Lieutenant broke the message and logged it as such. Then he sat down for his own countdown to midnight and 1980, the year that would finally see him back in the States getting married. As the clock's hands approached midnight, he thought how lucky he was to have been at a New Year's Eve party with his girl and kiss her for the first time at the stroke of midnight exactly one year earlier. He also thought back to the strange set of events that had followed him, from a midnight many years earlier.

For, although he had been born and raised to be a Christian, the grandson of a Methodist circuit riding minister, by the time he was a junior in high school, with all the ignorance and arrogance of youth, he had turned his back on God, not only to atheism, but past it, to the dark side. He studied all of the black arts: astrology, numerology, divination, the Satanic ritual, and Tarot. He and his then best friend in high school would run around late at night, and as Halloween approached in his senior year 1972, his friend came to him and said that he had struck up a relationship with a lady who read Tarot cards. And she wondered if they would like to come by at midnight on Halloween and have their futures read.

His friend had picked up a part time job as a disc jockey at the new FM light rock station in their hometown, and he told him that, believe it or not, women would call into the station out of the blue and ask if they could meet him. Sometimes he would take them up on it. One evening a few weeks earlier, he had met this woman that they would be visiting and struck up a friendship with her. That is how she came to invite them over.

He had told his friend that that was fine with him, so after they drove around, and had a couple of cups of coffee at a diner, they pulled up in the driveway at a quarter before midnight at the lady's house. Oddly enough, it was the house where his best friend from freshman year, the son of a doctor, had lived before they moved across town. He knew it well. The lady of the house met them at the door, and as they walked in, he remembered being there many times before. When they were in, she told his friend to go behind the sofa and open the window. She then told him to go to the dining room, open a window, and have a seat in the corner. When he asked why, she told him that it was to allow the spirits to pass through. Since it was Kansas, there was a constant twenty-five to thirty-mile an hour wind blowing. He could hear the leaves rustling across the grass, and the wind moaning through the bare limbs of the trees in the front yard. When the windows were open, she told him to turn out the lights.

While he was walking across the room, she lit a large candle in the center of the coffee table, between the sofa and two chairs. His friend sat down, cross legged, Indian style, on the side facing the sofa, and she sat down opposite him. The Lieutenant sat silently, in a chair in the dining room. He saw that she had his friend hold his palms out to her so that she could read them. Then he shuffled the cards, and she began reading his Tarot future. Since he didn't want to intrude, the Lieutenant just closed his eyes, took a deep breath, and tried to relax. When they were done, she called him to come over and sit down across the table from her.

She was a pleasant enough woman, with dark brown hair and

darker eyes. She told him to stretch out his hands to her, palms up, one on each side of the candle. She looked at his right palm and then focused on the left. She told him that he had a long lifeline. It was intersected in several places. He would live a long time and he would do many things. She then told him to shuffle the Tarot deck. He shuffled the cards and sat the deck down beside his right hand. She then told him to cut one fourth of the deck and set it face down to the left of the remaining cards. When he was done, she picked up the first card from the deck, but before she looked at it, she told him, "This card is your character, your personality. It's who you truly are." Then she flipped it over and said, "This is the Knight of Pentacles. You are diligent, brave, strong, and true. And you will work every day of your life." After she set the card down face up, next to her, she turned back to him and told him, "Cut one third of the cards and set the third that you cut down on top of the others you just cut. This next card is your destiny." She picked up the card, flipped it over and showed it to him. "This is the Four of Cups. I see a girl sitting under a tree reading a book. She is your destiny." She set the card down, turned back to him, and said, "Now cut one half of the deck and put it on top of the others." He did so. She looked into his eyes and said, "This card is your life." She pulled the card up and flipped it over. "This is the Nine of Swords. You were born into the House of Sorrow, and in the House of Sorrow you will dwell all the days of your life. But you are the Knight of Pentacles. Five times your heart will be pierced by sorrow before you die. That is your life." And then she blew out the candle. He sat silently in the darkness for a couple of seconds, before his friend turned up the lights. Then he looked at her, thanked her and they both got up to leave. As they reached the door, she looked at his friend and he told her he would be back in a little while after he dropped his buddy off.

THE PRODIGAL

Oddly enough, after wandering darkness for almost two years, the Lieutenant had come back to God in the spring of his senior year of high school. A young Christian speaker had come to the college in his hometown. It was sort of like a revival, but more modern. The Lieutenant was good friends with the Associate Pastor in his church and his wife, so at their request, he accompanied them. Although he was cynical at first, something moved him, so he decided that very night that he would give his life back to Christ. Years later he would come to appreciate the truth in Dorothy Sayers reflection that, "God wastes nothing, not even sin. The soul that has struggled and come through is enriched by its experiences, and Grace does not merely blot out the evil past but in the most literal sense 'makes it good'."

So it came to pass, years later, he had come back from Fort Knox to get engaged to his girl at Easter time. He went to Mass with her and she went to Methodist service with him. She in a dress and he in his dress blues. After that, they visited each other's parents, and he asked her if she would marry him. She said "Yes," but the Lieutenant told her, "You really don't know what the life is like. So, here's what I will do. I'll leave you my copy of the Officer's Guide that has a chapter in it on what an Army wife can expect. Read it then let me know whether or not you think you can do it. If not, I'll understand, if so, then we'll go ahead and get married." Since he only had a four-day pass, he was back in his car, headed for Kentucky before she could read it.

He had asked her before he left what kind of ring she wanted, so a couple of weeks after he was back to his battalion at Fort Knox, he called her at home on Saturday afternoon. She was out, which was unusual, so later that evening he called again and asked where she had been. She told him that she had taken that funny blanket he always kept with him and sat down in the park where they used to go to read the book. He asked her, "So, you were a girl sitting under a tree reading a book?"

She replied, "I suppose so." Then, he thought for a moment, and asked if she'd made a decision. Unexpectedly, she told him, "I won't wear gloves."

As he would many times in the future, he simply blurted out, "What the hell?"

She continued, "I won't wear gloves. Sorority girls wear gloves and I am no sorority girl."

The Lieutenant laughed a bit and said, "Well, I don't think it's going to be a problem. But in any event, I'm not the glove police. I don't care. So will you do it?"

She told him, "Yes," and the very next Saturday, he drove into Louisville, to the best jeweler that he knew, and proceeded to buy a diamond. He told the jeweler how much money he had and told him that he wanted to purchase the largest diamond that he could for that amount of money, but that it had to be absolutely perfect. The jeweler looked at the figure and shook his head and told him, "I can't get you even a quarter carat for that, but I have one stone that I've never been able to sell. It is .23 carat, and I can sell it to you for that price. But I'll tell you this, even though it's a little smaller than most, it's absolutely perfect."

The Lieutenant smiled at him and said, "That's the one I want. My girl's a little smaller than most, but she's absolutely perfect."

A RIVER ON FIRE

The clock's hands approached twelve. He decided to break Sergeant Robb's rule and step outside the Headquarters to breathe some fresh air and wait for midnight. As soon as the last minute had passed, the entire Rhine River exploded. The Lieutenant was amazed. New Year's is a tremendous holiday in Germany. It goes back to their pagan roots. Guns and fireworks were set off to drive off evil spirits for the New Year. The Germans took this seriously. They put barges out in the middle of the river and set off fireworks not just for twenty or thirty minutes, like Americans would, but continuously until dawn. Consequently, he was treated to a fireworks show of amazing beauty, unlike anything he'd ever seen.

After about an hour, he decided to go back into the headquarters because he was starting to freeze. He sat in the Adjutant's office, looking out the windows, toward the river while the fireworks continued until dawn. As luck would have it, around 0600, when everyone was on the verge of falling asleep, they received another flash traffic message. The Lieutenant had to hustle down to the office to transcribe it and then break it only to find out that it was in fact another test message. Once it was logged, he released the NCO and driver for breakfast and drank another cup of coffee with a breakfast bar, while he waited for the end of his shift.

After checking his watch, the Lieutenant realized he had about an hour before his relief would arrive, so he read a little bit more of *The Odessa File*. Then it occurred to him that this year he would in fact be marrying the girl who was his destiny. Good, he thought. He could hardly wait to see what the future held. He was smiling, as a first lieutenant from the infantry battalion came through the door and told him that he was reporting. He had the new duty officer sign the log before he asked him if he had any questions. The infantry officer told him, "No," and that he had heard earlier from his battalion duty officer that the Adjutant would be by in about an hour to make sure he was squared away. With that, the bone-tired Lieutenant said farewell and Happy New Year to the infantry officer, then snatched Hawkins, and told him to get in the Jeep and drive them back to Mannheim as fast as he could. They bundled up against the cold wind, whistling through the Jeep, leaning low over the heater, away from the windows.

Chapter 30
My Melancholy Blues

When they arrived back outside Headquarters Company, Hawkins climbed out and stood beside the Jeep, saluting. The Lieutenant was surprised and almost shocked by this. As he bundled up the poncho liners and briefcase, he stepped outside the Jeep and returned the salute. When the driver said "Pacesetter," the Lieutenant replied, "Press on." Then, as he walked away, Hawkins took off for the motor pool, and the Lieutenant ran into the hallway to the CQ desk, where he told him to roust the armorer and have him meet him at the arms room, so he could turn in his sidearm. When he got downstairs, the warmth overwhelmed him and he headed for a cup of coffee in the orderly room, before thinking better of it. He knew he needed to get some sleep before he got up at 1030 to put on his blues and drive over to the battalion commander's quarters.

Once the armorer arrived, he cleared his weapon, took the extra bullet that he ejected from the chamber and inserted it in the clip that was light. Then he turned everything in to the armorer and quickly headed up the stairs, after which he jogged back to the BOQ and climbed the steps to his room. As he turned the corner past the communal kitchen, he realized that he recognized a voice, and peering in saw that Buzz and Elizabeth were making coffee. They both looked really tired. He asked them how their visit to England went. She told him that her parents liked Buzz a lot. Buzz shrugged, and said, "I enjoyed seeing the sights of London with her." The Lieutenant then told them that he was glad to see both of them back. He added that he would love to visit but really had to catch at least a couple of hours sleep since he'd been up at Brigade all night. Also, he had to make it over to his commander's house well before noon to wish him a Happy New Year, which was traditional, especially in cavalry units.

As he trudged down the hallway, there were no other signs of life, so he went to his room, stripped down, and practically fell face forward into his bed before he realized that he had to set the alarm for 1030. Having done that, he passed out cold, until the clatter woke him up. Once he came to, he had to clean up and lay out his dress blues. He hadn't worn them since Easter a year ago, when he went to church with his girl. That is, he hadn't worn them socially. At the training center, the executive officer always had to bring in all of his clothes; on the second day of basic training, they had to change into various uniforms, so that trainees could recognize what an officer looked like in khakis, fatigues, the green uniform with blouse, and dress blues. Once he was set, he gave his shoes an extra lick, before walking past the kitchen, to head down the stairs. There was nobody up and around, so he walked outside, hopped in his car, and cranked it up to get the heater going for a few minutes before driving over to the colonel's house, on the street named for Ulysses S. Grant.

ALL ALONG THE WATCHTOWER

As he pulled up to the address, he smiled a bit and congratulated himself on having the foresight to having snatched a pretty good bottle of white wine from his own refrigerator to bring over for the house. When he reached the front door, the colonel's wife, Jeanie, gave him a hug. He wished her Happy New Year and gave her the wine. She thanked him with a broad smile on her face and asked him if he was excited about getting married soon. He told her that he was, and that he couldn't wait to bring his girl back to Germany to meet her. She pointed him in the direction of the dining room where the officers were standing by the window, making small talk. There was non-alcoholic punch available, but also screwdrivers and bloody Marys. The dining table was set for buffet brunch, with ham, scrambled eggs, German bread, and cheese, and, of course, the traditional black-eyed peas for good luck. The Lieutenant walked past the food because he was really too tired to eat and picked up a screwdriver before he joined the other junior officers in the far corner of the dining room. Most of the bachelors looked pretty hungover. They apparently had started with a New Year's party at the officers club in Heidelberg before hitting some of the bars in Mannheim and finally stumbling home. It sounded like fun but given that he was so worn out from duty, he was glad he hadn't joined them. Besides that, he had already paid his dues partying in Louisville and Juarez as a second lieutenant.

Colonel Shea and Reardon were standing to the side visiting when the S-3 came over and joined them. After all of the officers and the wives had arrived, the Adjutant announced that they were there for the traditional brunch with the battalion commander, to wish him a Happy New Year and toast it in properly. He then told the officers and those assembled to proceed to the kitchen to pick up a small glass of champagne. The Lieutenant was beginning to wonder how he was going to make it out of there, running on no sleep, with a double screwdriver and champagne in him, but nevertheless proceeded to pick up a glass. Once everyone had reassembled in the dining room, the Adjutant proposed a toast: "Happy New Year," to the battalion commander. After that was done, they made a toast to the Army and then the United States of America. Oddly enough, the Lieutenant was moved by the scene because it was really touching. They were all on the battlements at the edge of the American empire, in many ways forgotten, but strengthened by each other and their commitment to the country. He had not been in a line unit before on New Year's Day and given his rancorous relationship with the Third Armored Cavalry Regiment when he was there, it was unlikely that he would have been invited anyway. And, fortunately for him, last New Year's Eve he was with his girl, far away from the desert.

With the toast over, the officers and their ladies formed up to make plates for brunch. The Lieutenant came over to the rear of the line as usual. He wasn't particularly hungry, so he picked up some bread and ham, and stood in the corner to eat. When he was done, he went over to see the colonel's wife and told her that he would have to excuse himself because he had been up all night at Brigade and would probably pass out on the couch if he sat down. She looked at him and smiled and then told him he was certainly excused and hoped he had a good day's sleep because the coming year would bring a lot of time in the field, on top of his wedding. He laughed and headed out the door to drive back to his quarters.

AULD LANG SYNE

He dashed to the front door of the BOQ, because it was windy, his blues were lightweight, and he had not taken an overcoat with him, because he couldn't stand to wear it. With his hat in his left hand, he ran up the steps and opened the door. As he climbed the stairs to round the corner, he was practically run over by two soldiers carrying boxes stacked up to their eyes. He moved quickly to his right to allow them to pass, and then he saw that Meghan was following them, with a third trooper trailing her. Looking out the window in the stairwell, he saw that they had placed the boxes on the sidewalk, and one of the soldiers had run across the street to crank up a two-and-a-half-ton truck and bring it around beside them.

When the truck was parked, they put the boxes in the back, leaving one soldier at the wheel. The other two, and Meghan, went back upstairs to pick up additional boxes and carry them out to the vehicle. While they were making that trip, he walked up and around the corner to the communal kitchen to find that Buzz was standing in the kitchen window, watching. Apparently, he and Elizabeth had come in for a late lunch together, and she had gone back to bed. As usual, the Lieutenant, not knowing anything, asked Buzz, "What the hell is going on?"

Buzz looked at him calmly and told him, "Meghan is being transferred from Fifth Corps to Seventh Corps and a transportation unit outside of Stuttgart, about two and a half hours away."

The Lieutenant thought to himself, "Oh," and then said, "Because?"

"Yes," said Buzz, and the Lieutenant then added, "Well, at least her three troopers are helping her move. That's something."

Buzz looked at him, wryly, and said, "Yes, and I imagine that they're looking forward to one last hurrah when they get her moved in before they come back here."

The Lieutenant shook his head, but then added, "At least I won't have to listen to the showers running all night long, especially today when I could definitely use some sleep." He wished Buzz a Happy New Year, since he had forgotten to earlier, and walked down to his room, where he stripped, hit the bed and passed out until a little after 0400 the next morning.

TUBULAR BELLS

He ran past the guards, through the gate to Headquarters Company in his PT gear, and then down to the orderly room, for a morning cup of coffee. When Boris came out of his office, he followed him upstairs, to fall in behind the company when it was called to attention. After some calisthenics, and a brisk run, the First Sergeant let the troops know that they were still on half-day schedule, so once lunch came, they would be released unless they were mission critical. The Lieutenant was relieved to hear this as he could surely use some more rest.

Before he left for breakfast, however, he went over to Sergeant

McNabb and asked him what his day looked like. As usual, McNabb told him that the Assistant Platoon Sergeant would be down in the motor pool assisting the troops with maintenance on their vehicles. But he would be engaged in picking up some items for the mess hall and for Battalion Maintenance. The Lieutenant then told him that he would be on jury duty the next day for a court martial, but that he would be back Friday. He went on to tell him that he needed to visit with him by 0830 on Friday morning to go over some things, as they were going to be involved in an exercise at the PSP the following week. Sergeant McNabb saluted him, turned, and promptly took off. The Lieutenant just shook his head.

He went down to see Captain Boris and asked, "Is there was anything you need me to do?" When he was assured that there wasn't, he told him that he would be in the S-4 office until noon, after which he would be in his room in the BOQ. Then the Lieutenant took off to grab a breakfast sandwich at the PX before getting dressed and coming back to the battalion.

As he stepped through the doorway to the S-4 section, he could hear the sound of growling behind the S-4's door. When the growling was done, the Lieutenant knocked on the door, and when told he could open it, peered in and asked Lieutenant Graham, "What's wrong?"

Graham angrily told him, "The XO's after me again to reconfigure all the loads for the Support Platoon, for ammunition for alerts. He wants you to call some lieutenant named Oden at Brigade for a POL inventory. And he has planned an exercise, at one of the preposition stock points, to practice getting empty ammo boxes onto a GOER to see if there is any way to get more boxes on it." The Lieutenant just shook his head, because Major Reardon appeared to be among the ranks of the truly stupid, since there was just no way you could configure the load of a GOER any better than they had been doing it for the last two years. Still, it would give them something to do, and it would keep the XO busy for at least half a week.

Once he was at his desk, looking out into the line company courtyard for a while, he grew restless and bored. So, he put on his field jacket, cold weather headgear, and scarf for the trip down to the motor pool. When he was there, he checked in with the Assistant Support Platoon Sergeant and then walked the line to visit with his troops, who were busy fiddling with various things on their GOERs and trucks, to keep them well maintained. After he finished with that, he went back to his office to warm up. When noon came around, he excused himself to go back to his quarters to take a hot shower, and rest for the afternoon.

LONELY BOY

The BOQ was silent as a tomb until 1700, when he could hear pots and pans banging, down in the communal kitchen. So, he slipped on his jeans and a T-shirt and sneakers to head down to see what was up. Apparently, after his trip to England, Buzz realized that he was hungry for Yankee food, and he and Elizabeth were busy making a pot roast. The Lieutenant asked if he could join them and then sat down in the corner by the window to watch the two of them work. They got on well together and soon the oven was on high heat, with the pot roast cooking wonderfully. The Lieutenant asked if there was anything that he could bring up for dinner, and they asked if he had any wine. He told him he did and came back with a bottle of red that he put on the counter. Then, having decided he was beyond tired, he slipped downstairs to the soda-machine and bought a can of Coke.

When he got back upstairs, he could tell that the smell of dinner had attracted several of the usual customers. Bigfoot was there, as well as Pavel and Storch, who was back "home" from Christmas leave. When Tom came in, they asked him how his trip had been. The Lieutenant smiled broadly, hoping for a good story from him. He had gone back and seen several of his old friends from college. He had also been in hopes of running across one particular young lady, but she had become engaged while he was away, and so he was, as usual, in something of a funk. After about half an hour, visiting among themselves, Bernie stopped by without his new girlfriend. Apparently, they had already started drifting apart. This, however, had no real effect on Bernie. Then the ladies from next door showed up with some leftover cake and cookies that they brought back from a New Year's Eve party. When everything was ready, they all sat down for their communal meal and chatted amongst themselves. The Lieutenant thought, *It's almost like home, but unlike any home I've ever been.*

After he was done, he excused himself, saying that he would have to get up early and be dressed in greens for jury duty over on the post where the MP Brigade was located. His friends wished him well, and he proceeded back to his room to listen to the radio and read a bit before setting his alarm and falling asleep. The next day was fairly uneventful, although it was the first time he ever sat on a court martial jury. The trial proceeded quickly. The evidence was such that conviction was assured. He was released before 1600.

The following morning was Friday, and after PT, he went to the S-4 office. The S-4 was just finishing growling as he came in the door. Graham told him that he needed to call Lieutenant Odin at Brigade, who was going to set up an inventory of the PSP, and all of his refueling vehicles. About that time, Sergeant McNabb came in, so the Lieutenant relayed that information to him and told him that he needed him to be in the motor pool so that he would be there when he called him once he had coordinated with this Lieutenant Odin. McNabb looked at him, out of the corner of his eye, as though he was ready to say something, but then thought better of it. He then saluted and headed for the motor pool.

The Lieutenant dialed the number he'd been given and coordinated with Lieutenant Odin to inventory the munitions in the bunkers and in the preposition stock point. Lieutenant Odin would also inventory the fuel in his tanker vehicles as well. In anticipation of the inventory, he was told to send Sergeant McNabb to the S-3 office to pick up the necessary forms. After he hung up, Lieutenant called down to the motor pool for Sergeant McNabb. He wasn't there but apparently would be back a little after lunch from running some errands for the Mess Sergeant and the Battalion Maintenance Sergeant. The Lieutenant left a message for Sergeant McNabb to report to him as soon as he was back after lunch.

At noon, he decided to walk over to the PX to have a sandwich for lunch. After he returned to his desk, McNabb came in, saluted, and reported to him. He told him what he needed him to do and that he had to have the platoon ready at 0800 on Monday morning for the inventory. Major Reardon would be accompanying them, and the executive officer was always in something of a mood, so they couldn't afford any foul ups. Sergeant McNabb nodded that he understood, then saluted and promptly took off for the S-3 shop.

DEAR FRIENDS

When the day ended, the Lieutenant went back to his BOQ room. When he passed the communal kitchen, he was happy to see that Buzz, Elizabeth, and Bigfoot were busy making a spaghetti dinner. He told them that he would be right back. After he changed into civilian clothes, he pulled a bottle of wine out of his refrigerator before joining them in the kitchen. As they sat and visited, he told them that earlier in the week he'd made a call to his old VMI roommate, Matt, who was an infantry officer assigned to Baumholder. Matt had invited him out to spend the Saturday night with him and his wife. The Lieutenant was looking forward to seeing them again as he had not seen them since they were married in New York City, in the spring of 1979. Kate was a finance officer, and Matt was assigned to a mechanized battalion.

Since the Lieutenant had to get up early to drive, he excused himself after dinner, to get a good night's sleep before the trip to Baumholder. He thought that was one heck of a place to try to live as a married couple, since the weather was terrible and the accommodations were frightful from what he'd heard. He felt fortunate to be bringing his girl back to Mannheim, although it had challenges of its own.

The next morning, he got up early and had his usual instant coffee and a breakfast bar before packing a small suitcase and jumping in his Sirocco to hit the Autobahn out of Mannheim. Once he got to the outskirts of the city, he opened an Esso map to locate the address where his buddy and wife lived. It was a small apartment that was attached to a small house. The living room and kitchen were downstairs, and a couple of bedrooms were upstairs. It was cold beyond belief. The heater barely kept up, and as Matt pointed out, when it did kick in, had a tendency to belch black smoke onto everything. It was a true test for both of them to spend their first three years of married life in those conditions.

They went out for the evening, and the Lieutenant paid for dinner in a warm and friendly guesthouse with great food. When they got back, they visited for a little more but then decided to go to bed because it was so cold. Since it was even colder the next morning, and after Kate made scrambled eggs and toast, he decided to excuse himself early so they could enjoy the rest of the weekend and he could get back to the relative warmth and comfort of his apartment before his week began in earnest.

STILL THE SAME

The next morning, there was a heavy drizzle and fog. Captain Boris decided to forego PT, so the soldiers could warm up before heading to the motor pool, or in the case of Support Platoon, to the Prepositioned Stock Point to inventory all of the ammunition and practice the load out. The Lieutenant ordered Sergeant McNabb to have the entire platoon in their vehicles, ready to follow him in his Jeep. Sergeant Brown from the S-4 shop had obtained a Jeep from CSC and was going out early to coordinate opening the bunkers, and to help with the inventory. The Lieutenant also ordered Sergeant

McNabb to have the fuel trucks come to the same location so that their fuel tanks could be measured.

The platoon left for the bunkers on time, accompanied by Major Reardon, who was nervous, pensive, and a little clueless all at the same time. When they got there, the Support Platoon drivers began the arduous process of unloading everything out of the first bunker. The men then tried to load the ammunition on to their large GOER vehicles. Soon it became apparent that they simply had too many tons of munitions to move with the relatively small number of vehicles they had at hand. This did not keep Major Reardon from having them unload and reload several times to see if they could make it fit. When it became apparent that this was not going to work, he turned the inventory over to Lieutenant Oden, the Lieutenant, and Sergeant Brown before he headed back to Battalion.

The Lieutenant breathed a sigh of relief under his breath and walked over to Oden to see if he could go ahead and inventory the fuel vehicles so they could accomplish that mission while Sergeant Brown was assisting inventorying the ammunition. Oden came over and proceeded to have the drivers insert a measuring stick in all of the vehicles that showed how many gallons of fuel were in a five-ton or GOER fuel tanker. When they were done, the Lieutenant was happy to note that he did not come up short on any fuel, and in fact, it looked like he had too much in at least one of the vehicles. Lieutenant Oden told him that this was not unusual because fuel would actually swell, on occasion, even if nothing was added to it. The Lieutenant looked at him quizzically, but since he came from Brigade, he took it on faith and they finished the inventory of both the fuel vehicles and the ammunition by noon.

With the inventory completed, the platoon went back to the motor pool, in convoy, and the Lieutenant went back to his office. When he arrived, the S-4 informed him that he would have to write up a full report of the inventory and take it to Brigade the next day pursuant to the XO's order. The Lieutenant sighed a little but then sat down and drafted the report on tablet paper before sitting down and hunting and pecking on a typewriter to complete the report on a disposition form.

The following day, as directed, he took the report to Brigade, where Lieutenant Oden looked at him quizzically, before asking why he had brought a report on a disposition form to him. The Lieutenant told him that he was directed to do so by the Battalion XO. Lieutenant Oden just shook his head and bewilderment because the forms that the S-3 had supplied to his Platoon Sergeant earlier were all that he needed, and those had been completed by the NCOs on site. The Lieutenant just shrugged his shoulders and hopped back in his Jeep.

Sergeant McNabb had detailed one of the truck drivers to be his Jeep driver for the last couple of days. He was enjoying it, but really wanted to get back on his truck, so he asked the Lieutenant if he was going to try to find a new Jeep driver. The Lieutenant told him he would, and when they got back, went down to the First Sergeant's office to see if Ascencion could coordinate with the personnel section to find him someone to be his driver for the long haul.

When he explained his situation to the First Sergeant, Ascencion appeared less than enthusiastic but told the Lieutenant that he had a line on a new driver, and once he had interviewed him, he would send him over to see the Lieutenant around 1500. The Lieutenant thanked him and then walked back to his office. As he came in the door, he could hear the Graham growling again, so he went into the office and asked him what was going on. Graham looked up and told him that he had just received a warning order from Reardon that the Lieutenant would be going along with him for a supply and support reconnaissance at Bad Kissingen the next day.

That meant they would have to be up early to try to be on their way by 0500 in order to get to the borderlands in the early morning for a meeting with the 11th ACR's S-4, and the Brigade S-4, at the behest of the new brigade commander. They could expect to be hip deep in snow this time of year, so it would not be a pleasant outing. In addition, once the meeting was over, they would have to drive all the way back to Mannheim in the brutal cold of the night.

DESPERADO

The Lieutenant thought to himself, *This surely will be a true test of my new driver: his ability to drive in foul weather as well as keeping the heater running.* He looked at Graham and asked him how long the executive officer had been sitting on that order. The S-4 looked up and told him, "Probably the last three days. He just never gets around to doing anything until the last minute because he's too busy running around making work, trying to fix things that aren't broken and refusing to take responsibility for the things that are." The Lieutenant then asked the million-dollar question: whether or not the XO would be going along with them. This was the only time the S-4 smiled because he said, "It doesn't look like it. He and the battalion commander have a command and staff meeting of the battalion executive officers, operations officers, and commanders with the brigade commander, his XO, and the Brigade S-3." Based upon that, and the fact that he really liked Graham, he found himself actually looking forward to the following day, with the exception of the cold and snow.

A little after 1500, a young soldier walked through the door into the S-4 offices. The Lieutenant recognized him because he was a tanker from Bravo Company. He motioned to him to come over to his desk. The young trooper came to the position of attention three feet away and saluted. "Specialist Caceres reports, sir," he practically shouted. The Lieutenant told him to be at ease and take a seat beside him. He asked him where he was coming in from. Caceres told him that he had been in Ramsey's Third Platoon, Bravo Company, but since he was Puerto Rican, he wasn't really welcome there. Caceres did not like the man, or any of the tank commanders either, so he had asked for a transfer. He couldn't transfer within Bravo Company, so he asked the Headquarters Company First Sergeant if he could post him to Headquarters Company to get a job there, either in the three-tank headquarters section, or doing something else. Ascencion immediately asked him if he would be interested in driving a Jeep for a young staff officer. He told him that he would because he had driven a variety of vehicles back in Puerto Rico, and he liked doing automotive maintenance as well.

As the Lieutenant listened to the trooper, he realized that he liked him, and it appeared that he would at least try to do a good job. So, he told him that he needed to have his Jeep outside Headquarters Company, ready to leave at 0500, and he would need to make arrangements with the Mess Sergeant to pick up something to eat. He should see if he could get his hands on a couple of C-rations and also bring some Deutschmarks, as they would try to hit a meat market or guesthouse somewhere along the way to warm up and get some real food. Caceres was not put off by this, but in fact was grinning from ear to ear, as though he had found his dream job. He rose saluted sharply, turned and almost ran out the door to get his new charge ready to go.

Later that afternoon, when Acencion released everyone from evening formation, the Lieutenant cornered Sergeant McNabb and let him know that he was going to be gone for at least the next day. When he was back on Friday, he was tasked by Boris to do an Arms Room inspection in HHC at 0900 and after that, to go see Major Reardon to complete an in-progress report on the inventory of the ammunition in the PSP bunkers. Reardon also wanted an update on progress associated with finding ways to load the materials out, so that in the event of a real alert, they could move the munitions and the tanks to the border. The Lieutenant went on to say that once that was done, he would come down to the motor pool and see if he could visit with him to catch up on how he was coming with the training program for the drivers. He would not be able to spend very much time, however, because he was scheduled to report to pull Brigade Duty on Friday night so he would leave the battalion at 1500, and wouldn't get off duty until the next morning around 0800.

HOT STUFF

Done for the day, he went back to the BOQ, where he climbed the stairs as quickly as he could, to get to higher and warmer ground. As he came around the corner, he could see that Buzz and Bigfoot were busy over the stove, making something that smelled like chili. And it was. This boosted the Lieutenant's morale quite a bit, so he poked his head in the doorway and announced that he would be happy to join them in ten minutes, once he had a chance to rinse off and that he would bring some beer back with him for everybody.

True to his word, ten minutes later, the Lieutenant came back with two six-packs of St. Pauli Girl and sat them down on the counter. He took out a bottle, opened it, and sat down in the corner in his usual place. He asked Buzz if Elizabeth was going to join them and he said, "No, she has to teach an extra class at night school for some children who fell behind in the last semester. I expect her to show up about an hour and a half from now."

The Lieutenant smiled and told Buzz, "Say hello for me when she comes in because I plan to be fast asleep. I have to get up at about 0300 or 0330 to get dressed and packed up to leave at 0500 for the borderlands. And then turn around and come back, since I have Brigade Duty Friday night."

Buzz looked at him in astonishment and said, "Brigade Duty again so soon?"

"Yes, sir. I pulled a weekend Brigade Duty earlier, so now, apparently, I owe them a weekday duty. I'll go over to Brigade at about 1500 Friday and get off at 0800 Saturday."

Buzz looked at him again, quizzically, and said, "So you will be on again, after driving to the East German border and back, in this weather?"

"Yes, sir," laughed the Lieutenant. "It's the way of the Cav. Bad weather and late hours are our hobbies." Buzz chuckled, and so did Bigfoot, as he stirred the chili, that had come to a boil and added more chili powder.

After another half hour or so of small talk, they were joined by Tom Storch who opened a bottle and sat down opposite the Lieutenant. He could tell that Tom was really stopping by to see if Elizabeth was back for the night, and since she wasn't, he went into a little bit of a funk. But a bowl of chili soon lifted his attitude. When everyone was done eating, as usual, the Lieutenant volunteered to wash their dishes and put them away. Buzz kept the chili on the stove in the hopes that Elizabeth would want to eat when she came back.

The Lieutenant put the last dirty dish away, said "Adios," and headed for his room as fast as he could. He set his alarm clock for 0300 and promptly lay down. He was having a hard time sleeping because he was counting down the days until he could get back to the States and see his girl again. Eventually, he relaxed and let himself go to sleep because he knew that alarm clock was going to wake him up before he knew it.

Chapter 31
Glory Road

His hated alarm clock clattered at 0300. The Lieutenant shook his head in disbelief and looked over at it to make sure it wasn't a dream, before silencing it to keep it from disturbing he neighbors. Moving slowly, he rose out of bed, shaved and took a pretty cold shower to try to wake up. Then he laid out all his field gear to get dressed for the cold. Besides fatigues, he had field pants and a field shirt, a field jacket, parka, overshoes as well as a helmet liner and a scarf. He also picked up his map case and a couple of poncho liners so he would have them with him, to put over the windows when they were stopped for any length of time.

As soon as he was dressed and ready, he braced himself to step out of the BOQ into the cold. There was a brisk wind blowing, but at least it wasn't snowing. When he got to Headquarters Company, he went downstairs to the orderly room to pour himself a large cup of hot, black coffee. The First Sergeant was already there. Most of the staff officers, in addition to the S-4 and the Support Platoon Leader, were also going on border reconnaissance so Jeep drivers were everywhere, checking in and grabbing a cup of "go juice," before they went to the motor pool get the quarter-tons fired up.

As he sat in the corner, drinking his coffee, the Lieutenant pulled out a breakfast bar and a piece of licorice that he had stashed in his field jacket along with other emergency rations, so that he would have something that passed for breakfast, as neither the PX nor the mess hall were open at 0430. He saw Caceres come down and talk to Ascencion. Then the supply sergeant appeared with a couple of C-rations for Caceres to take with him out to the Jeep. When he saw this, the Lieutenant joined him, and climbing in the Jeep, told him to crank it up and turn the five-ton heater to its highest setting. Once the Jeep was pretty warm, they sat there and waited for the S-4. At five minutes till 0500, Graham appeared in the courtyard. He had been in the S-4 office tending to some last-minute reports for Major Reardon. His Jeep and driver had been outside the S-4 office, so he rolled up, looked at the Lieutenant, and then gave him the signal to follow him as he pulled out of the Kaserne.

They drove northeast past Frankfurt to Darmstadt. The weather began to get appreciably worse as they climbed into the more mountainous wooded areas. The snow was falling faster, with small flakes reflecting in the headlights as they made the turn for Aschaffenburg. About three hours into the trip, everyone was practically frostbitten. After Aschaffenburg, they cut across the terrain on smaller roads and eventually picked up the main road into Bad Kissingen and the main Kaserne, where the Second Squadron of the 11th Armored Cavalry Regiment was headquartered.

They eased up to the gate to the Kaserne at sunrise. The gate guards asked for identification, which everyone in the Jeeps showed them before they were allowed to pass through. The Border Cavalry Regiments were the Army's hard school. They had a real-life mission to patrol continuously and defend against encroachment by the East Germans, Russians, and Czechs. Here, and in the Second Armored Cavalry Regiment to their south, as the British said of the Northwest Frontier during the Raj, you soldiered well or not at all. One thing was certain; the regiments got the best soldiers. If a soldier made a mistake or messed up in some fashion, he would be shipped back to one of the line divisions, behind the Cavalry Regiments, where if he was bad enough, he would eventually end up in the 8th Infantry Division. The worst of the worst would end up in Mannheim so they would be near the Coleman Barracks Confinement Facility, where soldiers awaiting full court martial, were jailed.

The S-4s and support platoon leaders from the divisional units had battle positions behind the Second Squadron, so the Regimental S-4 and the Squadron S-4 briefed the officers on the locations of the various types of supplies stockpiled, and the main supply routes that would be used to provide fuel, ammunition, and food to the frontline elements. These locations were also important because when the frontline cavalry regiments were shot up and decimated, they would need to fall back through them to pick up enough fuel, ammunition, food, and parts to allow their withdrawal to the rear areas to reconstitute before being sent back out to screen for the withdrawing maneuver units, or provide flank security, as necessary.

When the briefing was over, the Squadron Assistant S-4 told the assembled officers to follow him from location to location to see where the supply points were. This would take a couple of hours since there was so much snow on the roads. After that, the officers would be free to visit their own battle positions and the supply areas for their battalions. At noon, they were to reassemble at the Squadron Headquarters for an out briefing and ask any questions they might have. When he got back to the Jeep, he asked Caceres if he had eaten anything. Caceres told him, "Yes, sir, I found some coffee in the headquarters at the Duty NCO's desk. Then I added C-rations cocoa to it and heated up a can of ham slices on the engine manifold that I ate with cheese, crackers, and a John Wayne bar, so I am okay." The Lieutenant thought to himself, *This kid is a real trooper and we're going to get along well.*

By that time, the Lieutenant had been to the border several times. The only real difference now was the depth of the snow. This and the fact that it was even colder than ever. As they drove from location to location, he and Caceres kept the five-ton truck heater on high and bundled themselves up underneath the poncho liners and their parkas. When they were through with the tour and the debriefing at Squadron Headquarters, all of the officers were released to go back to their home garrisons. Going out the door, Graham caught the Lieutenant's attention and told him to have his driver follow him and his driver to the guesthouse Maria for a hot meal, so they would have something warm in them before they began the trek back to Mannheim in the brutal cold.

They slipped over the snow into the courtyard of the guesthouse. There they all had hot tea or coffee, and hunter soup in large bowls with fresh bread, meat, and cheese. When they were thoroughly warmed, the Lieutenant paid for his meal and his driver's as well. Then they all bundled up in their Jeeps to drive back through the forest, toward the Rhine. It was bright and sunny, almost pleasant, except for the breeze that was blowing through the Jeep. The Lieutenant did his best to stay warm and follow along on the one to two hundred and fifty thousand map sheet that he had cradled in his lap.

With the sun descending rapidly in the west, the temperature dropped dramatically and it became frigid. The good news was that by now they were practically out of the Odenwald, on the main highway that led down out of the mountains to Mannheim. It was a little after 1900 when they arrived back in garrison. The Lieutenant and the S-4 released their drivers to do maintenance on their vehicles and find something to eat in the mess hall as best they could. Before he left, the Lieutenant got Graham's attention and reminded him that he was going to have to leave the next day at 1500, to get to Brigade for duty on Friday night. The S-4 looked at him and asked him, "Didn't you just have Brigade Duty?"

"Yes, but it was weekend Brigade Duty. So now I am on the roster for weekday Brigade Duty. Friday counts as a weekday, so I am going back for another dose." Graham just shook his head and chuckled a little bit because it was the standard indoctrination for a new officer in the brigade.

With the drivers and Jeeps headed back to the motor pool, the Lieutenant folded up his map case and poncho liners as best he could before trudging back to his apartment. After climbing the stairs slowly, trying to warm up, he turned the corner and saw that, unfortunately, there was no one in the kitchen. So, he would be on his own. He walked down to his room and got out of this cold weather gear. Then he decided that he would heat up a frozen dinner in his toaster oven, and make some hot water to go with it, before trying to take a shower. Once he had some food and hot water in him, he decided to make a cup of cocoa and read a little bit, before setting his alarm, and crawling in bed. It was colder than usual in the room, so he actually crawled under the large red blanket that was his comforter from VMI, before he pulled a poncho liner over that and his head, and went to sleep.

HANDY MAN

The next morning, the alarm went off too soon and the Lieutenant slowly crawled out of bed. He splashed some cold water on his face to wake up before he slipped into his PT gear and zipped up the warm-up jacket over his throat. Then he put on a blue watch cap and his leather gloves with wool liners. Unfortunately, this was as warm as he was going to get, before settling in with the company for morning PT. This was followed by a hot cup of instant coffee, a breakfast bar, and a shower. Then he jogged back over at 0900, to do an arms room inspection and inventory for Boris, which involved reading and calling back the serial numbers of every machine gun, rifle, grease gun, and pistol in the company. Since there were over one hundred and eighty soldiers in the company, it took him and the armorer almost two hours to complete the full inventory and fill out the report to the Company Commander. After he finished, he trudged over to the S-4 office, where he picked up the draft of his in-progress review, before going to see the Executive Officer.

Once he reached Battalion, he walked past the Duty NCO to the

XO's office and knocked on the door. When he was told he could come in, he advanced to three paces before the desk and saluted Reardon, who returned it, and told him to take a seat beside him. When he had done that, the XO asked several questions that made almost no sense to the Lieutenant, since they really didn't seem to have a lot to do with the mission of uploading munitions out of the bunkers in war time. Nevertheless, he answered, as best he could, and the Executive Officer took his draft report and told him that he would include it in his more detailed report to the Brigade S-4 and the Brigade Executive Officer. With that, the Lieutenant could sense that it was time to go, so he rose, saluted, and left the office as quick as he could, because he didn't want to give the XO an opportunity to drag him back, for more questioning.

On his way back to his office, he decided to have lunch in the mess hall, since he was going to go on duty, for the whole night shortly. When he walked through the door, he saw Sergeant Wilson and Lieutenant Matthews, his old buddies from Bravo Company, sitting in the far corner. He paid for his meal, took his tray over to their booth, and asked if he could join them. They both smiled and said certainly, so he sat down and tried to catch up on all the latest gossip in his old unit. For one thing, Sergeant Ramsey was preparing to leave the battalion to go back to the States, to a tank battalion at either Fort Knox or Fort Hood if he could get there. He had decided to leave to find a spot where he might have a better shot at making First Sergeant.

The word was that the new Bravo Company Commander was so bad that the Battalion Commander had canvassed some of the other units, and the First Sergeants of the line companies to see if they could locate the man he needed to take over as First Sergeant. Apparently, they had convinced an E-8 by the name of Angel to come to lead Third Armor Division for Mannheim. From all accounts, he was a hardcore, no bullshit individual, who was expected to do well. The downside was that this dashed Sergeant Robb's hopes of becoming the Bravo Company First Sergeant anytime soon. So, it looked like he was going to try to get back up to Battalion, in the operations shop, although, he was probably going to be asked to stay in B Company since he was such an excellent platoon sergeant.

When they were done talking about Bravo, they looked at the Lieutenant and in unison asked how he was doing. He told them that he was surviving as the Support Platoon Leader, which undoubtedly was the single most thankless job in the whole battalion. It had some upsides, in that you got to have a Jeep and you got to play pirate a lot. But other than that, he couldn't recommend it. After hearing this, Lieutenant Matthews and his compatriot smiled and asked him if he wanted to come to their platoon "Fasching" party.

HOLLYWOOD NIGHTS

The Lieutenant looked at them quizzically. He had heard the word "Fasching" before but really had no idea what it meant. Lieutenant Matthews assured him that it was something like Mardi Gras. It ran from New Year's Day until the night before Ash Wednesday. During that time, Germans literally partied around the clock, drinking heavily and, in some instances, became amorously engaged with as many people as possible. They told him that in particular, if he went clubbing, he should wear several neckties and every girl he had sex with would cut off one and keep it as a souvenir. The Lieutenant questioned them with his eyes and was assured by Wilson that this was the case. Germany, in the dark and cold dead of winter, took on, in some ways, the characteristics of a fraternity party. The Lieutenant thought for a little bit and then told them that he had Brigade Duty Friday night and was looking forward to Battalion Duty starting Sunday morning, so he wouldn't be able to stay very late at all on Saturday. But if they gave him the address, he would come over, at least for a while, to hang out with the soldiers of Second Platoon, who always seemed to have a good time. Since he finished eating, the Lieutenant stood up and excused himself from the table, dropped the tray off at the window to the kitchen, before heading over to his office.

Once he had settled in at his desk, he called down to the motor pool to see if Sergeant McNabb was available to discuss the training plan. The Assistant Platoon Sergeant told him, "No, Sergeant McNabb left before lunch on a delivery run and isn't expected back until the end of the day." The Lieutenant swore silently and then reassured himself that at least someday he was going to catch up with that man and they were actually going to do something constructive.

Since he didn't have anything better to do, he walked over to the S-4's office and asked him if he had anything for him to do. The S-4 looked at him and then shoved a pile of papers over to him. "Take a look at these, lieutenant. These are the orders for the tactical exercise without troops, scheduled for February. All of the battalions' headquarters vehicles, and the Jeeps and drivers for the executive officers and Company Commanders are participating. We'll move out in support. The Scouts and Mortars are also going to do training in conjunction with the exercise. The tank platoon leaders will draw Jeeps from Combat Support Company's ground surveillance radar and red eye sections, so that the maneuver units of the battalion, or its leaders anyway, will be in the field together. Our sister battalion and the infantry battalion will also be in the field with us. Each battalion will take a mess truck and trailer with it, to feed the members of the battalion. Alpha Company has been tasked to provide the cooks, mess truck, and trailer to support us."

"The exercise itself will run from Monday, February eleventh to Friday, the fifteenth," He continued. "That is usually the coldest part of the year, so expect to need every bit of your winter gear with you. Before the actual exercise, the officers, including you, will visit all of the assembly areas, refueling points, and trash points in the brigade's area. We'll do the reconnaissance next week. It will start with a mock alert early in the morning and then roll through the night and the next day, to cover the maneuver area. We will not have mess support, so bring some marks to eat at German guest houses or meat markets while we're moving. And be sure and bring some extra money to be able to feed your driver, although each of you will be issued two C-rations per day to take with you." The Lieutenant shuddered to himself at the thought of eating cold C-rations in the dead of winter but figured that he could heat them up the Cav way, putting them on the engine manifold of the Jeep, until they were hot to the touch.

He asked Graham if he could take the orders to his desk to review them for a while, as he had to leave about 1500 to go pick up a Jeep and driver, and draw a sidearm. He checked over the map sheets and it was immediately apparent that the trash points were located in roadside stops. That meant there would be small trash cans there, but no place big enough to put the collective trash he would have from a kitchen trailer. He marked that up as something to ask the Executive Officer about when he had the opportunity.

As 1500 approached, he took all of the orders and maps back to the S-4, who secured them in his safe. Then the Lieutenant told him that he needed to take off for Brigade Duty. He picked up his steel helmet and web gear, before walking over to Headquarters Company. When he got there, he went downstairs to the orderly room and allowed himself at least one cup of good hot coffee, before asking Ascencion to have the armorer come down to issue him a weapon. When the armorer arrived, he led the Lieutenant down to the arms room and told him he was simply going to have to issue him one of the tanker's pistols, since the Lieutenant's weapon under the Table of Organization and Equipment was an M-16 rifle. The Lieutenant thought about this but then realized that support platoon leaders had to range far and wide across the countryside and needed the additional effective firepower that a rifle could bring.

HAND OF FATE

With the holster and a pistol belt around his waist, he secured the lanyard to the webbing and made sure the forty-five was loaded with one round in the chamber and a full clip. With another full clip and one clip, less one round, in the ammo pouch. Then he went back to the First Sergeant's office and asked him who his driver was going to be. Ascencion chuckled and told him, "You're going to get your driver, Lieutenant. Caceres is the newest Jeep driver in the company, so just like you, he drew the duty, too." The Lieutenant smiled because he liked his driver a lot and knew that he would be there on time.

When he was back up in the courtyard, he saw his Jeep pull up with Caceres at the wheel. He asked him if he had ever driven for a staff duty officer at Brigade before. Cassara shook his head, "No." The Lieutenant briefed him on where they were going and what he would be expected to do. The weather was a little warmer, so as Caceres drove him over to Brigade to meet the Adjutant, they had a chance to visit a little more in depth. He learned that Caceres was not only from Puerto Rico, but from a particularly violent part of the island that was crime ridden. Caceres had grown up to be a "hard boy" who could take care of himself. He never went anywhere without at least two knives on him. One was a long combat knife, slipped into his boot on the right side. The other was a smaller, switchblade that he kept in his pocket. The Lieutenant chuckled a little to himself, because when he was a cadet and had to travel cross country on Greyhound buses, and would arrive in inner-city bus stations in the dead of night, he had always separated his cash, with some in the bottom of his boot, some in his breast pocket and some in a waist pocket with two knives: one in his boot and one in the pocket where the cash was. He knew that he and Caceres were going to get along just fine.

As they pulled into Brigade, the Lieutenant showed Caceres where he could park his Jeep and then led him into the headquarters. He took him down to the Duty NCO's office, where the NCO, a Staff Sergeant from 2/13 Infantry named Timms, took charge of Caceres, telling him where he could sleep and rest, and also giving him some tasks to do around the headquarters to keep busy. As the Lieutenant watched Caceres go to work, he smiled again, because the kid reminded him of the Corsican member of the Union Corse in the *Dogs of War*, who was Cat Shannon's personal bodyguard and enforcer.

When the Lieutenant walked up to the Adjutant, he seemed a little perturbed because the Lieutenant was running a few minutes late. But then he brightened immediately when he realized that he had just briefed this particular young officer a few short days ago, so he wouldn't have to go into any great detail to explain what the duties were or show him where the code books were for flash traffic, and where he could hang out. With the short briefing under his belt, the Adjutant told him to take his seat at the duty desk and wait for any further instructions or tasks that the brigade commander might have for him before the day wound down. As usual, there were no specific instructions, and by 1730, the Headquarters was practically a ghost town, until the Adjutant left a little before 1800, leaving the Lieutenant fully in charge once again.

Once he heard the Adjutant's car pull away, the Lieutenant walked down to the Duty NCO's office and told him and Caceres that they could go ahead and take their dinner break. The two of them decided that they would eat with the Infantry Battalion. As they walked away, the Lieutenant was happy that he had had a full lunch, as he went back to the Duty NCO's Office and fished out a couple of pieces of licorice to eat while he waited for their return.

While they were away, he retrieved his copy of the *Odessa File* that he had brought with him and sat back on the couch to read a few pages. After they returned, he went back to the duty office, and just as he sat down, they received their first flash traffic message. He immediately picked up the receiver on his phone, and the Duty NCO did the same at his desk. He transcribed the message, Timms read it back, and they transcribed it again before we went to the S-2's office. With the keys to the door and the safe in hand, he opened, both got out the code book and broke the message. Fortunately, it was a Friday night, and the message was a training message once again.

After he placed the code book back in the safe and locked it as well as the door to the S-2's office, the Lieutenant walked over to the Duty NCO's desk and told him that he would be going back to the Adjutant's office, where there was a sofa where he could try to get some sleep, before the long hours of the deep night. Once he was settled in, he read a few pages of his book to relax before fading out to sleep until a little after midnight.

Then, true to form, at a little after 0100, he received another flash traffic message. He proceeded to break it down and realized that once again, the brigade was not going on alert, but it was only a test. Once he was done, he went back to the duty office, and logged the message in, next to the one he'd received previously. With that done, he decided to step outside to wake up a little bit and enjoy some of the fresh air rolling up from the river. After ten minutes, he realized that he was chilled to the bone and went back into the headquarters building to find a seat next to a heater.

When he was thawed out, he rustled through the pockets in his field jacket and took out a couple of pieces of licorice to eat. He finished chewing, thought for a minute, and then took out four more pieces of licorice and walked down to the Duty NCO's desk, where his driver was sleeping on the sofa and asked Timms if he would like some. Then he woke Caceres up just enough to give him the last two pieces. Both of them enjoyed the treat, particularly since they were not expecting it from a Lieutenant. He told him that he had practically lived on the stuff at White Sands and knew that sometimes, late in the day, it just hit the spot.

Not expecting any more messages, he decided to walk back to the duty desk and sit down in an easy chair in the corner of that room. There, he read a little more, before falling asleep hard, sitting up, by 0600, he could hear Timms making fresh coffee for the brigade staff members, who would be coming in to work for at least part of Saturday morning. When the smell of fresh coffee filled the air, he immediately woke up. He slipped down to the Duty NCO's office and congratulated him on his coffee making skills before drinking a cup of the fresh, wonderful stuff out of a mug that was part of an odd assortment next to the pot.

Finally, at 0700, the Adjutant came in to relieve him and prepare to brief the new duty officer who was just coming in the front door. The Lieutenant told the Adjutant that there were no problems and he logged out, as the new duty officer signed in. When that was done, he called for Caceres and they hopped into the Jeep to make it back to Sullivan Barracks as fast as they could. The drive through the training area was extremely cold because there was still snow on the ground and the wind was up.

As soon as they skidded to a stop in the courtyard of Headquarters Company, he released Caceres to get breakfast and walked down the stairs to the arms room, to sit down with the armorer. Once there he cleared, field-stripped, and cleaned his pistol, before turning it back in. Once he was done, he told the armorer to be sure and have it ready for him to pick up the following morning at 0600, since he had battalion duty on Sunday. The armorer just looked at him and shook his head, trying to figure out how one officer could get back-to-back duty as regularly as this Lieutenant did. He could tell from the quizzical expression on the armorer's face, what he was thinking, so the Lieutenant just told him, "I'm the newly assigned lieutenant and I'm single, so, I'll be pulling a lot of duty until I can get back to the States and get married." The armorer smiled at that. Then the Lieutenant quickly wheeled around, picked up his things, and headed back to his BOQ room to take a hot shower and crash.

When he woke up, he made some hot water for instant coffee while he took another shower and shaved, and this time dressed in civilian clothes. Once he had some coffee and a breakfast bar in him, he walked slowly down the hallway to the communal kitchen. There was no one there, so he looked in the refrigerator to see if there were any leftovers he could take advantage of. Unfortunately, the BOQ dwellers had cleaned everything out, so he decided to slip over to the snack bar. At that moment, he also remembered that he needed to check with the tailor to see whether or not it could pick up the blue suit he had ordered, to be made in England and sent back to him so that he could have something to wear at the rehearsal dinner, since he didn't have a decent civilian suit.

As he walked around the BOQ, he made sure to avoid getting too near the corner of the building where his notorious fellow armor lieutenant liked to throw out furniture and household goods. Not seeing anything coming his way, he continued across the parking lot to the PX. Once there, he headed back to the tailor booth where the gentleman sold suits. Unfortunately, his had not arrived yet, so he wasn't able to pick it up. But he was able to snag a cheeseburger before going back to his room and hanging out. Lieutenant Matthews had given him the address, and he'd written it down in his notepad, so he looked up the address in the ESSO map book that he had bought at the gas station and found the location of the evening's party.

With nothing better to do, he listened to the radio for a while. He was able to pick up the scores from a couple of sporting events back in the States. It seemed like he'd been in Germany forever, and he had completely lost touch with football and basketball. As he sat and relaxed, he thought a lot about his girl and what she was probably doing or would do in the States once it was past noon. On that happy thought, he completely lost consciousness from lack of sleep and passed out again, coming to at a little before 1900.

BOOGIE WONDERLAND

By that time, he decided that he might as well go ahead and drive out on the economy where Matthews' platoon was gathering. Once he was in his car, he put the map sheet on the passenger seat beside him and took off through a light snow, to a group of houses and apartments, on the edge of Viernheim to the north. As he drove around, at first he was concerned that he wouldn't find the place, but as he turned the corner in pitch black, he saw a large number of cars parked in the street and he could hear the forgettable strains of music by *Earth, Wind and Fire* and *Kool and the Gang*, spilling out into the street. He made his way up the stairs to the apartment that the noise was coming out of, and walking through the door, immediately recognized Matthews, Wilson, and a couple of the tank commanders that he remembered.

Once inside, he realized that he, and a fairly young German girl, were the only Caucasians in the entire room. He was a little surprised at first but then remembered that Second Platoon was the all-black platoon for Bravo Company, and he shouldn't really have expected anything else. There were a large number of single black soldiers, men and women, having a marvelous time. And then, his buddy, Charles Matthews grabbed him and told him he should have a drink before the festivities began. He walked over to the makeshift bar and made himself a bourbon and seven, then sipped a little of it before standing next to Matthews, who was truly the life of the party: he was the MC, disc jockey, and spark plug, all rolled into one. He was getting ready to go sit down and simply enjoy his drink when Matthews snatched his arm and dragged him in front of what had been set up as a karaoke microphone and told him to simply follow his lead.

PLAY THAT FUNKY MUSIC

After Charles had the attention of the room, he pulled him to the center of the room, put on some background music, and broke into "We've got a real mutha fuh ya, and if you look, you will discovah that it's the real mutha fuhya yeah." The Lieutenant couldn't help but join in to the general applause of everyone there and it was hilarious. With that done, Charles queued up *Shake Your Groove Thing*, so the Lieutenant decided to retire from his erstwhile soul singing career and hung out in the corner with a couple of the tank commanders.

The one that he was standing next to appeared to be with the German girl, and he asked him how long they had been together. The buck sergeant told him only a couple of weeks. Before that, she had been the girlfriend of another one of the sergeants in the platoon, who had broken up with her because she had a lazy eye and every time he was screwing her, that eye would wander off like she was looking at another brother. It bugged him so much that he had finally given up. Her newfound boyfriend, however, was not put off by that in the least, and the two of them got along marvelously.

Since he had to get up early in the morning to go on duty by 0700 and spend the whole day at Battalion, he made his goodbyes. Charles looked at him with a wry grin and said, "Well, I'm glad that you got the duty and not me because I'm going to be up partying all night long." The Lieutenant just chuckled and wished him well as he headed out the door, into the cold mist that surrounded everything, to the strains of *We Are Family*.

Chapter 32
With a Little Help from My Friends

When he got back to his room, he quickly undressed before setting his alarm for 0600. He was dead tired, so he passed out as soon as he had crawled under his poncho liner on top of his bed. When the alarm went off the next morning, he looked over to make sure that the time was right before he crept out into the cold. He put some water in his hot pot, shaved and took a short shower, before coming out to enjoy what was becoming his standard morning meal: a breakfast bar, licorice, and Nescafe. When he felt almost human again, he got dressed and packed up his poncho liner, a book, the parka, and cold weather headgear, before stepping out into the hallway. As he headed toward the door, the entire building was completely quiet. It was almost like he was Charlton Heston in *The Last Man on Earth.* Taking a deep breath, he said a silent prayer that there would be good coffee in the orderly room when he got to Headquarters Company.

Sure enough, his prayers were answered. As he walked through the door into the cavernous black walls of HHC, he could smell the aroma of strong coffee coming up the stairs. He quickly went down and poured a cup. Ascencion was already in, so he sent his runner to get the armorer, so the Lieutenant could draw a sidearm and prepare to take over the battalion for the rest of the day and into the night. Once the armorer arrived, the Lieutenant drew his pistol and loaded it as usual. Then he thought, since he had about thirty minutes to spare, he would slip over to the mess hall to see if the cooks could make him anything to eat, as he was early. Sparks was in and had his assistant making scrambled eggs, so he served him some with toast and hash browns. When he was through eating, he finished another cup of coffee, before scurrying over to see the outgoing duty officer and take his place at 0700.

As he walked into Battalion, he was happy to see Sergeant Aparicio, the HHC Maintenance Sergeant, who had been in the PSP with him. He asked, "Are you pulling duty with me tonight?"

"Yes, sir," said the sergeant.

The Lieutenant smiled and said, "Great – I just hope we don't have a night at the PSP like we had last time we were on duty together." At this, the sergeant also smiled and shyly said "Yes, sir," again. Then the Lieutenant turned and told him he was going down to sign in. Aparicio just nodded and turned back to reading the SOP.

The Lieutenant walked down the hall and thought to himself, Aparicio was a cool hand and quiet, so they got along well. The outgoing duty officer was the new Alpha Company Platoon leader, Lieutenant Heinrich. He was tall and slender, and an interesting man, but not very personable. He looked haggard after a wild Saturday night. He was as good as dead on his feet and hastened down the hall as soon as the Lieutenant signed in, without even saying goodbye. After he was gone, the Lieutenant went to see Aparicio and sent him out to get breakfast. He told him, "I'll make my first check when you get back."

As soon as the Aparicio was back, good to his word, the Lieutenant slipped on his field jacket and stepped out into the gloom of Mannheim in the early morning. With the big ring of keys in one hand, and his pistol in the other, he started making his arms rooms checks, beginning with Charlie Company, then Alpha, Bravo,

Combat Support, and finally, the black hole of Headquarters Company. It was Sunday morning. The troops were asleep for the most part, so he didn't have problems with anybody, as he signed his check sheets and moved through the billets. When he got back from his checks, he told Aparicio, "Please get me up by noon," because he didn't want to sleep all day. Then he picked up his book and poncho liner and went back to the Adjutant's office, to try and get some sleep on the sofa. As he lay on the couch, he thought about his girl. She would probably be going to church sometime today. Then thought it was good that he was pulling all of his bad holiday and weekend duties now, before she arrived in country.

After what seemed like a couple of hours, but was actually longer, he woke up and realized it was almost as cold inside the office as it was outside the window. The cold and the mist had seeped into everything and it seemed that it slowed him down, but he fought off the lethargy and walked back to the duty desk. He asked Aparicio if all was quiet. He said, "Yes, sir, but I am getting hungry. I let you sleep a little later because you look pretty ragged."

The Lieutenant shook his head in agreement and then said, "Thanks. Go get lunch; I'll take over until you get back. And take your time – I owe you." Aparicio laughed and shot out the door to try and get to the mess hall before the serving line shut down.

After lunch, Aparicio took his seat by the duty phone and the Lieutenant moved over to sit at his duty desk and read a little. In the middle of the afternoon, they got a flash traffic message. It turned out to be a training message, so he logged it as such and then walked down the hall to stretch his legs before going into the Adjutant's office to rest, since Aparicio was a steady hand and it was quiet. He fell asleep and when he woke up, he was bored. It was late in the afternoon, so he went down to Aparicio and told him that he was going to take an early dinner since he hadn't eaten lunch, and that he would release him for dinner as soon as he returned.

He walked down to the mess hall and did not see any of his usual acquaintances, so he asked the Assistant Mess Sergeant to come over and sit for a bit so he could see how things were coming along. The sergeant said, "Everything's fine and we are all doing pretty well. We are getting ready to send the Alpha Company mess trailer to the field, in about a week to cover the reconnaissance of the maneuver area for the upcoming exercise." The Lieutenant told him that he had briefed on that mission but he didn't know that a mess trailer was going out for a one-night stay. The sergeant then said, "The XO got with Sparks and they want the trailer and cook out there to make sure everything works. They also want to make sure that the cooks practice in advance of the brigade exercise, without troops, when they will just have the officers engaged in the exercise to feed." The Lieutenant just shook his head, to signify that he understood before standing up to excuse himself and heading back.

He walked into the duty office, released Aparicio for dinner, and sat at the NCO's window. He spent his time watching groups of soldiers as they came in from the streetcar station, after spending the weekend downtown in the bars, and other entertainment establishments off P-Street. When Aparicio returned, since it was peaceful, the Lieutenant snatched his keys and jacket, drew his pistol, and proceeded to make his early evening arms rooms checks.

Upon his return, they sat opposite each other and told stories about things that happened to them in their earlier assignments. Aparicio doubled over laughing when the Lieutenant told him about the evening in El Paso he spent with his best friend, who was now in Mainz, eating a dinner of wild game with their boss, a crazy Cav captain, in a house full of live rattle snakes that served as a field expedient burglary prevention system. After that, they read back issues of magazines and tried to stay awake.

About three hours later, around 2200, they had a training flash message. After he logged it, he sat down at the duty desk to read a little and wait until just before midnight, then he told Aparicio that he was going to make his last arms room checks and then come back to sleep because he was exhausted. Grabbing the keys and his pistol, he bundled up against the cold and started running to Headquarters

Company. Then he went backward through the units so that he ended up in Charlie Company. The troops there were pretty easy going so he finished his final check and said good morning to the CQ before he rushed out the door and double-timed back to Battalion.

By now, the cold was sapping his energy, so he was starting to get sleepy again. Once he was out of his cold weather gear, he headed for the Adjutant's office and told Aparicio as he passed him to wake him up by 0300, because if bad things happened, they usually happened at that time. He wrapped up in his poncho liner with his parka for a pillow and passed out on the sofa. He woke up a little later and looking at his watch, realized it was 0200. He felt like he hadn't been asleep at all and was just curling up again when he heard Aparicio running down the hallway. He told the Lieutenant to come immediately because they had flash traffic.

TONIGHT'S THE NIGHT

The Lieutenant blearily walked down to the duty desk, and listened over the phone as the codes came in. Once he had the message written down, he grabbed the keys to the S-2's office and safe and went downstairs to break the message. As he broke down the codes, he was shocked a little that it was an alert message. And when he had decoded the entire thing, he realized that they were on full alert.

He immediately locked the code book back in the safe and ran up the stairs screaming at the NCO to start calling the First Sergeant duty roster, to bring the NCOs in because they were on alert and needed to be able to roll within two hours. He ran down to the Adjutant's office, pulled the call roster from the Adjutant's desk, and got on the phone immediately. He called the battalion commander first, followed by the executive officer and the operations officer and then down the line of primary staff officers who would call in the secondary staff officers. He then called all the Company Commanders. The Company Commanders in turn called in to their units to make sure that the First Sergeants were there, or on their way, and ordered the CQs to call all of the platoon leaders or send runners to bring them in if they were in the BOQ without a phone.

Within twenty minutes, the entire headquarters was sheer pandemonium, with Jeeps and drivers appearing out of nowhere. Officers scrambled to get on their gear, as the S-2 tried desperately to get out the CEOIs so that all of the officers could have their radios on the correct frequencies. As soon as the Adjutant arrived, the Lieutenant asked him if he could take over so he could get over to the S-4 office and get into his LBE and take over command of his platoon. The Adjutant asked if he had called all of the officers on the alert roster. The Lieutenant saluted and said, "Yes, sir, I have." The Adjutant nodded and saluted in return.

The Lieutenant signed out, turned on his heel, and then shot out the door, to pick up his things and run over to his office. When he got there, he was happy to see that Graham and Sergeant Dryer were already there and loading the S-4 track to go to the field. Once he had his load bearing equipment and helmet on, he took off at a dead run to the motor pool where his drivers were cranking up their trucks. Sergeant McNabb was not in yet, so he ordered his fuel trucks to move to positions in the street just outside of the motor pool gates where they could top off any tanks or other vehicles that needed diesel or mogas on their way out to their maneuver area alert positions.

After handling that, the Lieutenant was running back to the S-4 office, when it occurred to him that he needed to turn in his side arm to the armorer in Headquarters Company and draw his service weapon, the M-16 rifle that was assigned to him. So, he changed directions and ran across the courtyard to HHC. Once there, he flew down the stairs and tried to get to the arms room; it took him about ten minutes to work his way through the line of enlisted men trying to draw their weapons. Before he could get to the window, he had to wait for the Headquarters Tank Section who were drawing the heavy weapons so they could lug them up to the motor pool. Although he missed his tanks, he was happy that he wasn't humping the machine guns anymore, because the tank sergeants already looked dead on their feet.

He put his head around the corner and caught the eye of the armorer. When he turned, he told the armorer that he wanted to turn in his pistol, so he cleared it and handed it to him along with the clips and holster. Then he waited for his M-16 and enough magazines to put one in the rifle and fill the ammo pouches on his LBE. Once that was done, he ran up the stairs into the courtyard where he made sure again that the rifle was on safe, pulled back the charging handle, and sent the bolt home to put a round in the chamber. Then he ran back to his office at port arms. When he was through the door, he went back to his desk and put his poncho liner on top for use later. Everyone else from the S-4 shop was working on the track and he didn't have anyone to talk to, so he sat down and took stock of his situation. He was in desperate need of coffee. He hadn't even had a chance to really wash his face, much less shave, but he didn't care because they were off to the woods, and he was starving. Fortunately, he had left some licorice in his field jacket, so he ate a couple of pieces, and waited for his driver to appear.

Before he knew it, the front door burst open, and Caceres came into the office with his rifle at port arms and a knife about nine inches long protruding from the top of his right boot. He came up to the Lieutenant and executed the rifle salute as he announced in heavily accented English that he was present for duty. The Lieutenant returned a hand salute and chuckled a little because he really liked the kid. The Lieutenant grabbed his rifle and gear, following the kid out into the courtyard to load his equipment up in the Jeep. While he was doing that one of the troopers from the Commo Platoon ran over and handed him a CEOI. After that he turned on the radios and dialed them to the correct frequencies. Then he sat down, slapped the side of the Jeep, and said, "Get 'em up, Scout." Hearing that, Caceres hit the gas and they sped down to the motor pool to find Sergeant McNabb.

When they arrived at the entrance to the Motor Pool, the

Lieutenant gave the hand signal, and Caceres came to a quick stop to let the Company Commanders and executive officers drive through to their company area. After they passed, Caceres drove down the left side of the Motor Pool to the eight-ton GOERs and five-ton trucks of his platoon. He got out of the Jeep and turned around just in time to see someone walking up behind him. It was Sergeant McNabb, who had been with the troopers in the arms room making sure that they drew their weapons and had their gear, in keeping with the alert procedures. He told his sergeant that he would be leading the ammo trucks out to the PSP, so he should get the cargo vehicles lined up to follow him out of the motor pool. He also told McNabb to stay back with the fueling vehicles to make sure that all of the tanks, tracks, and wheels were topped off with fuel as they left for their alert positions in the woods. When they were finally done with that, McNabb was to go to Support Platoon's position in the assembly area and wait for the Lieutenant to rendezvous with him when he was finished at the PSP.

It was still pitch dark. The flashing yellow whoopy lights on the vehicles turned rapidly and blinded everyone. Shading his eyes, he saw Sergeant McNabb walking up and down the line, climbing up on the trucks and talking to the drivers over the sounds of the engines, which were deafening. When he had finished, he shined his flashlight signaling to the Lieutenant who signaled to the assistant platoon sergeant in his five-ton cargo truck to follow him, with the rest of the cargo vehicles following his truck.

Once the five-ton was moving, the Lieutenant jumped inside his Jeep, and smacking his right hand on the outside of the door said, "Get 'em up, Scout." Caceres looked over at him, laughed, and hit the gas. Deftly driving down the rows of vehicles he slipped out of the motor pool, onto the road, to the alert position, with the five-ton trucks in the lead and the GOERs following. As they moved out, the Lieutenant pressed the handset on the radio to tell the S-4 that he was en route to the PSP. The Graham acknowledged and told him that Reardon would meet him there, undoubtedly pissed off.

The Lieutenant sighed, as he thought that there were few things in life, he was least likely to look forward to than meeting the executive officer in the pre-positioned ammo storage facility, accompanied by an ammo detail and several drivers who were still buzzed, or worse, high on hash or strung out on heroin. Nevertheless, they were his to work with.

The details and his drivers eventually made it to the PSP. The details were composed of troops who had proved themselves to be problems in the line companies, with a few drawn from Scouts and Mortars who had angered their first sergeants. They sat for an hour, waiting for further instructions. After that, as his fuel trucks pulled into position on the other side of a small clearing, away from his ammo haulers, he decided that he needed to address the troops, so he climbed in the Jeep and had Caceres drive him over to a position centered in front of the trucks and details. The troops were in a rare mood, so he cut to the chase, briefing the men on the SOPs for the platoon and the battalion's goals and objectives associated with loading up the ammunition and mines for use on the border.

When he was through, the soldiers still looked a little spacey, with the pale bluish glimmer of being stoned in the backs of their eyes, but not as unhappy since an officer was paying attention to them and trying to make sense out of their hectic and disorganized morning. When he had finished, he saw the executive officer for the first time, out of the corner of his eye, as he pulled up behind him. They sat in silence for another half hour and then they heard the thunderous and creaking sounds of the tanks pulling into position all around them, mixed with the high-pitched whine of the huge M-88 tank recovery vehicles.

From the sound of it, they were probably dragging some deadline vehicles that still needed to be repaired. They would return them to their companies, if Battalion Maintenance and their own maintenance sections could catch up with them, when they were pulling out for the border lands. By now, it was almost 0600, and of course light rain was starting to fall. The Lieutenant was heartened, however, when the XO took off, followed shortly by a radio transmission from the S-4 informing him that the alert was concluded, and that all the units would move back to their company areas in the motor pool, to check their vehicles and alternate crews for breakfast.

Since Support Platoon only had one radio, the Lieutenant had to make the rounds to tell his cargo drivers and the detail to follow him to the alert position on the other side of the clearing where the fuel vehicles were parked and Sergeant McNabb was resting. He gave them the happy news that the alert was over and ordered them to travel in convoy behind his vehicles, back to the motor pool where Sergeant McNabb would take charge of them, since the Lieutenant was coming off twenty-four hours of duty.

When he arrived back at the Motor Pool, the Lieutenant waited for his trucks to pull into position. Then he turned Caceres and the Jeep over to Sergeant McNabb, so he could use it if he needed to. The Lieutenant then told McNabb that he was going to have to go get some sleep, since he'd been on duty twice, Friday and Sunday, and hadn't slept hardly at all because of the alert early Monday morning. When he was done with that, he walked over to the S-4 office and visited with Graham, informing him that he really needed to get some sleep. Graham nodded and told him that of all the things that made him happy about being married, one near the top of the list was that he was off the back-to-back weekend and holiday duty roster for Battalion and Brigade. The Lieutenant saluted, turned, gathered up the stuff from duty that was on his desk, and left at a jog for Battalion to turn in his CEOI in the S-2's office. From there he made his way to his company arms room to turn in his rifle and ammo before hurrying past the orderly room and jogging back to his BOQ room for a long hot shower and a full day's sleep.

I HEARD IT THROUGH THE GRAPEVINE

He did not wake until late in the in the afternoon, then he slipped on a pair of jeans, and stuck his head out of his door to see if anyone else was in on a wet, dreary Monday evening. He thought he could hear signs of life, so he walked down the hallway and stepped in the kitchen. There he found Buzz and Bigfoot shooting the shit, so he sat down to catch up on the gossip in the building. Apparently, everybody was exhausted after the weekend, so Buzz took command and decided that pizza was the order of the day. "Damn," said the Lieutenant, "that sounds great!" and he ran down to his room to retrieve two six-packs of Becks and two five-mark pieces to chip into the community pizza fund.

Buzz and the Lieutenant each opened a bottle and Buzz put the rest in the refrigerator. He nodded to Bigfoot, who headed out the door, but stopped dead when he caught Tom going down the hall and hit him up for ten marks as well. Then, with sufficient funds in hand, he took off for the little converted trailer that sat on the edge of the Kaserne, across the street from the Class VI store that made the best pizza the Lieutenant had ever eaten in his life. It was even better than the New York style pizza from Tarzia's in Lexington, Virginia, that had kept him going when he was a cadet. This place was actually owned and operated by real Italians, who fled the crappy economy in their home country to make their fortune Germany, much like the Greeks – the last vestiges of two great empires. The plan was working well. They were always busy and the location, less than thirty yards from the liquor store meant that a GI just had to pick up his beer or booze and snacks before crossing the street to get up a pizza or two to make a complete meal.

When Bigfoot left, Buzz brought the Lieutenant up to date on the current situation. Susan and Chris, at the end of the hall, were frustrated, and still trying desperately to get reassigned to West Point, where they hoped to get married, or something that approached being married, with or without the help of the Army. Captain Axel Boris, downstairs, rarely came out, but apparently, he was involved, as usual, in some sort of drug bust going on with CID and wouldn't share anything else about the operation because of OPSEC. Bernie's latest girlfriend, the friend of Meghan, who was now in Stuttgart, had apparently broken up with him, but was off and on dating Tom Storch, which kept him from being completely depressed, although he was still hopelessly hung up on Elizabeth. The Lady Godiva girl across the street discovered the 5-68 lieutenant with the alcohol problem, and the two of them were apparently commiserating on again, off again. And, given their volatile natures, they tended to get involved in fairly heated arguments in one BOQ or the other. Fortunately, the furniture throwing lieutenant lived upstairs and across the hall from him, so he did not have to listen to any of it.

The Lieutenant brought Buzz up to speed on the joys of going on alert in the middle of the night, after being on and off duty over seventy-two hours. Buzz shook his head in wonder and smiled, then chuckled, saying, "The good thing about dentistry is we never go on alert."

The Lieutenant told him, "Thank your lucky stars. Most parts of the alert were boring, but with winter exercises coming on, the snow and ice could easily lead to fatalities since tanks are just sixty-ton toboggans." Buzz looked at him incredulously like he was crazy. But the Lieutenant just flashed him a hussar's grin and added, "Look on the bright side, at least now I have a Jeep. That means I can stay warmer and drier, and I don't have hump the machine guns up to the motor pool, which is a huge plus."

When Big Foot got back with the pizzas, the smell of sausage, ham, olives, and warm cheese permeated the entire building. He and Buzz made plates and grabbed beers before sitting down. The Lieutenant did likewise and joined them in the corner. Tom came in, followed by his new lady friend. She was funny and fit right in with them, laughing and telling stories about her arrival in country. Everyone was just feeling good, enjoying the moment.

They had such a good time, and the pizza smell was so enticing that even Axel Boris and Oscar came over to join them. The little dark brown long-haired Dachshund was practically beside himself until he got a piece of sausage and jumped up in Axel's lap. Chris and Susan had decided to stay in for the evening, and Bernie was out on business, so there was plenty of pizza left to put in the fridge when dinner wound down. The Lieutenant finished up his meal and excused himself saying he really needed to get some sleep, since the rumor was that there was going to be another alert and a field exercise probably on Wednesday morning. So, he wanted to get at least one good night's sleep before going out in the woods to freeze for a few nights back-to-back.

SHAKEDOWN STREET

Getting back to his room, the Lieutenant thought about trying to stay up, but finally surrendered to the fact that he was wiped out, and set the alarm for 0530, before crawling under his poncho liner, thinking of his girl, and faded off. At 0500 he came to, turned off the alarm, and decided to get up for a cup of instant coffee to energize himself before getting dressed and running over to the orderly room for a real cup of Jo. It was Tuesday morning and since the wind was brutal, the commander decided on a quick, brisk run of less than two miles. After that, with steam coming out of his mouth and nose, the Lieutenant caught Sergeant McNabb and told him to come by his office after morning formation, so he could discuss the upcoming small-scale exercise with him.

When he walked into the S-4 shop at 0800, Graham waived him into his office and showed him the maneuver plans for the mini exercise and the taskings. It would be a three-day exercise, including a mock alert on Wednesday morning, followed by a road march to the Odenwald. He would be scouting out the assembly area and forward positions for Support Platoon and the battalion trains, as well as the routes back and forth between those positions. By Friday morning, they should conclude the exercise and be road marching back, to arrive mid-afternoon on Friday. To support the exercise, Support Platoon was tasked to provide two five-ton trucks to haul supplies, such as tents for the battalion staff, and one five-ton truck with mogas and another with diesel, to refuel vehicles in the field.

The Alpha Company mess trailer and two-and-a-half-ton truck with a cook and a supply specialist were also going along to provide hot meals twice a day and issue C-rations for lunch during the exercise. One of the five-tons would pull and drop off a "water buffalo" trailer to provide water for cooking and drinking. When the exercise was over, that truck would police up the trash from the mess trailer and the staff area to transport it back to Mannheim to be incinerated. The other would haul back the tents and other equipment for the staff. Both would follow him in his Jeep and the deuce and a half back to garrison. This was a pretty light tasking, so he could leave McNabb behind to supervise the rest of the platoon in the rear and respond to any taskings that came from the other companies, in his absence.

Sergeant McNabb came in at almost 0900 and sat down beside the Lieutenant's desk. He gave McNabb specific instructions to have the fuel trucks, five-tons, and water buffalo ready to go by three in the morning and to coordinate with Sparks to make sure that his mess truck would come with them when he led the convoy to the maneuver area. Since it was bitter cold and he would be in and out of the field for most of the next few weeks, the Lieutenant had pretty much given up on doing any training in the near future and it looked like he would not have a training plan until February. He asked McNabb if he had any questions. The sergeant said, "No," rose, saluted and practically sprinted for the door.

When McNabb was gone, the Lieutenant spent the next couple of hours assembling the maps and maneuver overlays, before fitting them in his large map case so he could locate the routes and assembly areas where he would be with his pirates. After he finished taping the maps together and placing the clear overlay for the start of the exercise inside the case, he spent some time looking at them, and decided that he was going to need something that would let him read the maps in the dark and maybe magnify them a bit so he could better follow the small roads in the villages and the spider trails in the forests. He thought he had seen something like that when he was shopping at the PX, so after lunch there he looked around and finally found a small, black-handled magnifying glass with a little light built in to illuminate what he was looking at. He could put a poncho liner over himself and the map to read it at night with the small light. It would also allow him to warm up from time to time. With his new purchase in hand, he jogged back to his office and told the S-4 that he was going to his quarters to rest and prepare his winter and wet weather gear since they were leaving early in the morning and it would undoubtedly be snowing, raining, or both, in the mountains east of Heidelberg.

By late afternoon, he determined that he was going to need at least one decent meal before going to the woods for three days. So, he decided he would go to the officers club since it was steak night and have what passed for an American meal. On the way back from dinner, he stopped by the PX to buy some Café Mocha, extra licorice, and breakfast bars, just in case. While he was there, walked down to the booth where the English tailor was, to see if his suit was in. It was, so he unzipped the suit bag and looked at it to make sure that it was what he had ordered. Then he zipped it up and signed for it, before he took it to the checkout, along with his other items to pay for them. He walked out into dense fog to his room to go to bed. But once he was there, he couldn't resist trying the suit on. It fit perfectly, so he smiled to himself, because he was one step closer to his wedding.

Chapter 33
All the Tired Horses

The next morning at 0200, the relatively warm darkness of his world was shattered by the crack of a rifle butt pounding against his door and a young soldier screaming for him to get up because he was on alert. The Lieutenant yelled back that he was up, stumbled to the bathroom, washed his face, and shaved quickly, before putting on his fatigues, field jacket, rubber duck, LBE, and steel pot. Then he humped his A-bag, map sheet, and thermos out the door and down the stairs to Headquarters Company. Once there, he stumbled down to the arms room, where he drew his rifle and ammunition before carrying everything back up to the courtyard, where, fortunately, Caceres was smiling and waiting for him. He already had drawn his rifle and had his two knives on him, so he was ready to go to the woods.

The Lieutenant stood by the Jeep waiting on the S-2 NCO who jogged by, throwing him a CEOI that he used to program the radios so he could monitor the battalion command and support nets. Once that was done, he checked in with the S-4, who was already on the radio. He was told to stand by and wait for the command team, that was preparing to lead the line companies when they departed. Before falling in behind Charlie Company's officers with his Jeep and trucks, he walked down the line to check on the maintenance vehicles with a couple of mechanics from Headquarters Company Maintenance and a sergeant from Battalion Maintenance to repair or recover and drag a vehicle in the event that something broke down.

Shea came over the command net to give the order to move out. His driver moved forward and the executive officer followed close behind him. The S-2 and S-3 rolled out next, with the S-4 and Battalion Maintenance Officer trailing them. They were followed by the Company Commanders and platoon leaders in Jeeps. As the last platoon leader for Charlie Company began to move slowly forward, the Lieutenant jumped in beside Caceres, slapped the side of his Jeep, saying "Get 'em up, Scout," and pulled he map case over his lap so he could keep track of their location on the route of march, and bundled up against the cold. Like all convoys, the vehicles in this one sped up and slowed down so there was a constant accordion effect that required the drivers to brake and accelerate in order to maintain the interval between the Jeeps and trucks. For the next hour or so, they rolled slowly through mostly urban areas before turning right and climbing up through the passes to the Odenwald. There was a heavy mist, and a stiff wind blew through the Jeep, so he pulled the flaps of his helmet liner down and pulled his green wool scarf up, to cover his entire face, except for a slit to see out the windshield to the slick road and snow-covered trees that passed silently by him on the right.

There was no radio traffic, but he kept the handset in his right hand in case he had to answer a call or try to through the scarf that made it practically impossible to talk to Caceres who was equally bundled up. Consequently, he used hand and arm signals to tell him to speed up or slow down for the interval and road conditions. After another two and a half hours they arrived at the edge of the maneuver area, where it was truly cold, at the edge of a large windswept valley that was covered in snow. Here the Lieutenant had his first experience with black ice, that was essentially fog, frozen on the blacktop roads. It was invisible to the eye and extremely slippery. When they began to encounter this ice, all of the vehicles reduced speed so much that they crept forward even slower than walking in order to avoid sliding off the road.

From the few road signs he could see and the terrain, the Lieutenant realized that in about five kilometers, he would need to be prepared to turn off with the S-4 into the assembly area for the battalion trains, which consisted of the vehicles following behind him. About half an hour after making the turn toward the forest, the S-4 signaled from his Jeep and the track followed him into the woods. They wound down some narrow trails into a small open area, where he signaled for all of the vehicles following to take up a lagger position, encircling the field. When the S-4 track stopped, the Lieutenant had Caceres pull up and combat park beside it. Then the Lieutenant jumped out of the Jeep and slipped on his parka against the cold, before walking down the line, signaling to each driver where his vehicle should be parked. Once he had all of the Support Platoon vehicles in place, he signaled for the Battalion Maintenance vehicles to pull up beside them. Then he walked down to talk to the battalion motor officer because he assumed that he would want to place his vehicles where he needed them to best accomplish his mission.

Once that was done, he went back to report to the S-4. He found him talking with the executive officer, who had a map case set out on the hood of his Jeep pointing out the positions to be checked out. When they were done, the XO took off for the tracks that made up the Battalion Command Post. As he sped away, the Lieutenant walked up to the S-4 and asked him where he was to place the mess trailer and water buffalo. The S-4 told him that he needed to find some high ground, away from any possibility of spillage of fuel getting underneath the trailer or near the water to avoid contamination.

The Lieutenant walked for about a hundred meters along a spider trial that branched to the left and uphill from the S-4's track. He stepped into a small clearing, and having determined that it was suitable, walked quickly back down the trail where he motioned for the two-and-a-half-ton truck with the mess trailer and five-ton truck with the water buffalo to follow him to that position. When the mess truck and trailer were set up and the water buffalo was in position, he sent the five-ton back to the Support Platoon vehicles in their laager and went over to introduce himself to the ranking cook, Specialist Five Leach.

Leach was white, with a sparse mustache, who smoked continuously and was as thin as a rail, probably from Appalachia as so many soldiers were. The Lieutenant asked how long he had been in the battalion. Leach said, "Almost two years, sir," and looked away to signal that he was not interested in small talk. As he watched him, the Lieutenant realized that Leach was extremely nervous and a little jumpy. At first, he thought it might be withdrawal, but he did not look like he had been using, so the Lieutenant just decided to stand back out of the way and let him and his assistant set up.

The Lieutenant smiled a little as he watched Leach get to work with a will, because he was not the first chain smoking, nervous white cook the Lieutenant had observed. In fact, the first time that he assumed command of a company at Fort Knox during the Christmas holidays, a little over two years earlier, he had to take charge of a cook who had got into an altercation with the Military Police at his quarters and was sandwiched in between two stretchers in the street when he pulled up in his car.

It was cold and almost midnight on that Christmas night two years ago, so he asked the officers if he could sign for the cook after they finished their report and accompany him back to his apartment to counsel him. After the police let the cook go and drove off, he asked the cook to lead him to his home so they could visit. They entered the apartment. The Lieutenant took a seat on the sofa, and the cook introduced him to his wife, who was a Korean national. When offered, he took a glass of iced tea and asked the cook what he liked to do to unwind. Oddly enough, he liked to cook, so the Lieutenant stayed for a shrimp dinner and discovered that the root of the problem was his wife's homesickness. When he was through, he went back to his quarters and back to the company the following day. Shortly thereafter, the trooper re-enlisted to go back to Korea, and that was the last that the Lieutenant saw of him.

With the field kitchen set up under way, the Lieutenant asked Leach when they could expect their first hot meal. He told the Lieutenant that it would be four or five in the afternoon, but that he had several boxes of C-rations that he could heat up in hot water if that would help. The Lieutenant went back to the S-4 and relayed that information to him. Graham transmitted it to the executive officer who then sent it out on the battalion net to all of the officers in the field, informing them that if they needed something to eat, they could come to the battalion trains, and pick up hot C-rations for now, and a meal later.

That being done, the Lieutenant told the S-4 that he planned to do reconnaissance of the various battle positions, and refueling, rearming, and feeding stations, he would be expected to man when the full exercise was in effect. He asked Caceres if he wanted a C-ration, and the Puerto Rican only smiled. Apparently, he had already secured enough for the two of them, so he cranked up the Jeep, and they headed off into the forest so the Lieutenant could check out the positions on the maneuver map. The Lieutenant figured that he needed to cover at least a third of the positions that day and the rest on the second, as he would have to get everything ready to be driven or dragged back to Battalion on Friday, the final day.

Once they were moving, the wind whistled through the Jeep. They were soon chilled to the bone. At the first position, they drove along a narrow track leading into it, and the Lieutenant quickly determined, based upon his Reforger experience the previous fall, it could be travelled by a 113 or a Marder, but nothing larger. As usual, the operations officers were overly optimistic about the ability of the platoons to execute the plan. Having determined that, he and Caceres moved along to evaluate a couple more of the positions.

Then the Lieutenant checked the map to locate a nearby hamlet with a market where they could buy some hot cider or tea to warm up.

After driving slowly through the village, they located a small bakery and meat market where they were able to get cups of hot tea and some bread, mustard, and cheese that was much better than cold C-rations. After paying, they climbed back in the Jeep and covered up the side windows to cut the wind. Caceres cranked the five-ton heater up on high. They used the Lieutenant's Swiss knife to make sandwiches, and they ate in comfort and silence for a little more than five minutes. Then they drove back out into the forest and a light snow that was just beginning to fall.

The Lieutenant checked his map and directed Caceres deeper in the woods and up a steep trail to the next position that was not only higher but snowier as well. He had Caceres pull slowly forward and they entered a clearing covered in deep snow. It looked like it could hold the vehicles from the battalion supply trains if he parked some of the lighter wheeled vehicles in small notches in the trees that lined the spider trials branching out from the clearing.

After another hour, the Lieutenant decided that they needed to call it a day and head back to the battalion trains where his trucks and the mess trailer were parked. After they made their way to the black top highway that led back, he checked the trash pickup points that appeared on the overlay for the Brigade Operations Order and discovered that they were nothing more than medium sized metal waste baskets at small rest areas. He knew in a minute that there was no way they would be sufficient to handle his mess trailer on his small exercise, much less the larger one with tracked vehicles scheduled for later in the month.

When they got back to the battalion trains, the Lieutenant and Caceres drove up to the S-4 track so he could check in with Graham. When he walked up the ramp into the track, he was surprised to find another officer, a captain, sitting in the webbing on the right side. After a short introduction, he learned that Captain Dolan was newly arrived from the Armor Advanced Course at Fort Knox, after an initial three-year tour at Fort Hood in a tank battalion, where he had been a platoon leader, assistant S-3, and tank company executive officer. He was going to take Graham's place since the S-4 was rotating back to the United States. The Lieutenant frowned a bit because he knew that Graham was hoping to stay in the battalion. and take a line company for a year and a half before going back to the States. Unfortunately, the Army's assignment officers in the Military Personnel Center had realized, a little too late, that they had sent so many officers to Germany in the flurry of activity last summer, there weren't enough to fill slots stateside. So, they were not extending any officers in Germany, but instead, forcing them to return to CONUS.

From the moment he walked in, he realized that he liked the new officer who asked him where he was from. When he said "Hays, Kansas."

Dolan immediately smiled and said, "I'm from Salina." That was about ninety miles east of the Lieutenant's hometown. Although he didn't know this officer, since he was a West Pointer based upon the ring that the Lieutenant noticed when they shook hands earlier, they did have the High Plains and their prairie hardships in common. Once the introductions were over, Graham explained that he would be the S-4 through the upcoming winter exercise. Then, upon their return to garrison, he would out process and Dolan would take over his duties.

The Lieutenant congratulated Dolan on joining the battalion, then he turned to Graham and told him about the inadequate trash points and the need for new larger ones to support the coming exercise. The S-4 listened to him politely, then shrugged his shoulders and told him that he would pass his concerns along to the S-3. When they were through, Dolan turned to him and told him that when he took over as S-4, the colonel would appoint him mess officer, so that the cooks in garrison and in the field would be under his command. The Lieutenant nodded in assent and then turned to go back out into the cold. When he was out of the track and walking to the mess trailer, he shrugged his shoulders and shook his head. He had been mess officer before in two different training companies and it was by far the most thankless job in the battalion. This, coupled with being the Support Platoon Leader, was certainly going to make his day, every day, for the next few months.

When he got to the mess trailer, he climbed the few steps up to the kitchen and asked Leach what he was preparing. The cook told him he had canned beef that he was cooking with noodles. This, together with some bread that he had brought from garrison and canned fruit for dessert, made the whole meal. The Lieutenant nodded his head and asked what other meals he had planned. In particular he wanted to make sure that the cook would not, under any circumstances, serve hot dogs, since Shea could not stand them.

Apparently, he had a bad experience in Vietnam when he was in the Eleventh ACR, the Black Horse Regiment, and the smell of hot dogs triggered his "flashbacks" so they were forever banned. The Lieutenant asked the specialist if he was aware of that. Leach told him that he had been cooking in the battalion for a couple of years, so he was well aware of that problem, and he had stocked the necessary foods to provide a variety of hot meals, in the coming days: spaghetti, Swiss steak, and ham. The cook said that he could also prepare the standard breakfast items, - hash browns, toast, eggs, and bacon, and that should see them through.

As the Lieutenant started to leave, he had a second thought, so he turned and asked the cook if he had any Vollmilch: cartons of milk that were pasteurized with radiation and did not have to be refrigerated to stay fresh. Leach nodded and told him that there were a couple of cases in the front of the trailer. Then he walked over, picked up a case, and brought it to the Lieutenant, who took out half a dozen cartoons to carry back to his Jeep. When he had climbed in, he placed them in the back, behind his seat and told Caceres to go ahead and move out.

They drove along the road through the assembly area, looking for the truck drivers from his platoon. When he rolled up to a truck, he got out and asked the driver if he had had anything to eat lately.

Usually, every one of them said, "No." So, he went back to his Jeep and grabbed a couple of breakfast bars and a carton of milk for each and gave it to them. At first, they looked at him a little funny, but then he showed them the label on the side of the bar, that read, "One bar and a glass of milk makes a complete meal," and said, "So there you go." After that, they chuckled and thanked him, as he moved down the line, freezing. When he was done, he went back to the S-4 track to check on things and Caceres put poncho liners over the windows of the Jeep to keep warm.

He checked in with Graham and Dolan, who were on the radio getting reamed by Reardon for some minor infraction of the SOP, or a maintenance issue. The Lieutenant just smiled, thankful that he was not immersed in the conversation. He backed up and sat in the right rear corner of the track, relaxing until the royal ass chewing was over and they could all get back to doing something constructive. As soon as the radio went silent and he put down the handset, Graham asked him when the food would be ready so the officers could go down to the kitchen and get some hot food. He told him that he had checked with the cook a little over a half hour earlier and he was setting up the chow line so he should be ready by now. Graham nodded and picked up the handset to inform the XO, who would announce it to the battalion's officers. Even before the radio went silent, the Lieutenant was out of the track and in his Jeep heading for his truck drivers to send them to eat. Since the sun was setting and the temperature was dropping rapidly, they were in need of a hot meal to warm up if they were to get any rest.

When Caceres pulled up to his trucks, the Lieutenant jumped out and went to each one, climbing up on the running boards to talk to the drivers. Being veterans of the field, they already had their A-bags unpacked and sleeping bags out and were wrapped in them. When they heard that dinner was ready, they pulled on their cold weather gear and parkas. They looked like a squad of misguided Eskimos at first but picked up the pace when they caught the scent of beef and noodles. With that done, he checked on the mechanics to make sure they had the word. They were already on the march when he had walked about halfway to the maintenance area. He turned back to the Jeep so Caceres could drive them back to the kitchen trailer. When they got there, he had Caceres get a plate to eat, while he manned the radio. After he was done, Caceres came back with a cup of hot water to make some C-ration cocoa, and the Lieutenant took the opportunity to get a meal himself. It was pretty bland, but hot and filling; that was what counted in the field.

He finished eating, and walked back to the Jeep and climbed in with a cup of hot water to make some hot chocolate. When he was done stirring in the powder, he took a sip and then rifled through his A-bag until he located some John Wayne bars he had stashed from gunnery in December. He pulled out two and gave one to Caceres who smiled and thanked him. He finished the bar and the cocoa. It was getting colder, and the wind was coming up, so he had Caceres turned the Jeep, so it was tail to the wind and backed up to a large evergreen to use as a windbreak. Then they bundled up in their parkas, with a poncho liner over them and additional poncho liners strung up in the windowless doors of the Jeep to cut the wind and retain the warmth from the five-ton heater that was straining to beat back the cold. Having spent months in the field in the desert and now Germany, he was not about to sleep on the ground. In New Mexico, there had been an endless assortment of creeping creatures that would crawl up on you, and at night, there were vehicles moving around that could easily run right over you. After almost two years, he had developed the ability to sleep sitting up and looking over at Caceres he could see that he had the same skill. He had been a tank gunner and was used to traveling and resting in that position; he was already asleep.

WASTED DAYS AND WASTED NIGHTS

The Lieutenant was used to getting up by 0500, but with the cold he came awake before 0330. He crawled out of the Jeep and walked over to the mess trailer to see if the cooks were up. They were, with their burner units on high and the trailer itself was actually warm and comfortable. He stepped into it and asked if they had any coffee. They did, so he got a cup for himself and one for Caceres. He took them back to the Jeep, where he and his driver sat together in the darkness and drank their coffee, thinking of friends, family, and warmer and happier places. When he was done, the Lieutenant went back to inspect the trailer and see what breakfast looked like. Leach was very good at his job and had everything under control. So, the Lieutenant went back to his Jeep, and using hand and arm signals motioned for Caceres to take him back to the S-4 track, where he reported in. The crew had pulled out the three-sided canvas extension that made a tent-like structure behind the track and both the outgoing S-4 and the incoming S-4 were sleeping on cots. The track driver was curled up in a sleeping bag, on the seats in the track, oblivious to the world.

The Lieutenant stepped inside, looking for a sign of life. As he walked beside his cot, Graham came out of his sleep and asked the Lieutenant how his people were doing. He told him that the cooks were up and he could relay to the XO that breakfast would be ready at 0600. He also told him that he planned to recon all of the remaining assembly and supply locations that were further out. He would hit them all, with a break for lunch in a village with decent food. Then he and Caceres would spend the night freezing, before getting up early to organize the convoy and road march back to garrison on Friday.

After breakfast, Caceres and the Lieutenant drove from battle position to battle position and an assembly area after that, to become familiar with the terrain. There was one position that was particularly far out of the way. They went deeper and deeper into the woods and the roads became narrower and narrower, until they found themselves on an earthen dam that was covered in ice, with water and ice on both sides. The Lieutenant dismounted and walked slowly in front of the Jeep to test the surface, until they reached a small clearing that they inspected from the Jeep to see how many vehicles could be parked there. He concluded that it could hold his platoon, or a tank company, but it would be far too dangerous to use it in the winter, as you would not be able to get a track to it. So, he marked it on his map and dismounted to lead Caceres slowly and carefully back over the ice bridge.

They drove as quickly as possible to the nearest highway. By the time they reached the black top, "white ball" road, it was past noon so they descended into the valley and stopped in the first village they came to for lunch. It was cloudy, with a light drizzle falling. The civilians did not get many GIs in the old forest and were not welcoming, so the trip took on a gloomy atmosphere. Finally, they found a Gast Haus that was open, so they dismounted and walked in to get something to eat. The proprietor was the first friendly face they had seen so far, and his mood brightened more when the Lieutenant slipped out of his rubber duck and he saw the German last name on his field jacket. The Lieutenant conferred briefly with Caceres and then used his best restaurant German to order two plates of curry wurst with fresh bread, French fries, cold salads, and hot coffee. The Lieutenant asked for the bill to pay in advance and build goodwill, rounding the tab up to the nearest whole mark, as was expected, and they took their seats at a small table, facing the door.

The owner's wife, who apparently was also the cook, brought out the coffee first, followed by the meals and he thanked her in German. As usual, the food was marvelous and the curry and coffee warmed them up. As they got up to go, he left a five-mark piece on the table as an extra thanks. The owner, who was watching them, smiled, and they hurried out the door to fire up the Jeep and its heater for the drive back to the assembly area, as stinging sleet began to fall. When they arrived, having seen the places that few tourists would ever see, he had Caceres pulled up beside the mess trailer so he could check in with Leach. So far, he was not only doing well, he was enjoying cooking. The Lieutenant asked if he could get a thermite container of hot water to take to the drivers to see if they wanted to make coffee or cocoa to warm up. He made the rounds, dispensing Vollmilch and breakfast bars along with the water to his troops and the mechanics as well. Then he and Caceres stopped by the S-4 track to catch up with Graham before eating dinner and settling in for the night.

They made it through the night, huddled over the heater, in their cocoon of parkas and poncho liners. Before first light appeared on the horizon, he had Caceres drive to the trailer through several inches of new snow. He was delighted to find fresh coffee in a thermite container and breakfast almost ready. He took of the black leather shells of his gloves and stuffed them in the right pocket if his field pants before pouring a cup of "Go Juice" for himself and Caceres. Once they had slammed a quick breakfast of toast, bacon, and eggs, he and his driver went to work with a will, because this would be his busiest day so far.

After calling Graham on the radio to let him know that breakfast was ready and he was riding herd on the drivers, he checked in with the fuel trucks and had them moved to the road so they could refuel Jeeps with Mogas and tracks and trucks with diesel. Then he checked the cargo truck that was being loaded with the tentage and other equipment for the battalion staff and then that the trash truck was loaded with trash from the staff and kitchen and finally, that the kitchen trailer was packed up and ready to go back to Mannheim. He did this while receiving a nearly incessant stream, of less that helpful and sometimes contradictory "advice" from the XO. He could practically hear the S-4 rolling his eyes over the radio as Caceres drove as fast as he could, with tires spinning through snow and ice to get from position to position.

When the kitchen trailer and his trucks were assembled, he slapped his right hand of the outside of the Jeep's door and said, "Get 'em up, Scout." Caceres laughed through the scarf that was wrapped around his face, slipped into first gear, and took off. The Lieutenant used hand signals to direct him to the far side of the assembly area, where they looked for the Battalion Maintenance Officer, Lieutenant Parsons, who was also newly arrived from a tank battalion in Texas, having been caught up late in the emergency transfer of officers to Europe to reinforce NATO. They talked for a little bit, and despite the fact that he was a West Pointer, the Lieutenant liked Parsons a lot. He stood about five foot six, apparently hated the cold, had a healthy dose of cynicism, and his nickname from the Military Academy "Conan," described him perfectly. The Lieutenant laughed and told him that he was happy to find a kindred spirit. Then he got back to business and asked if he was ready to roll. Parsons said he would be in about ten minutes, as he had to get a track with a broken quill shaft hooked up to a recovery vehicle to drag it back to garrison. The Lieutenant nodded and told him to pull his Jeep up behind the Lieutenant's convoy when he was ready, then he would inform the S-4, who would pass the report up to the XO and the command group.

The Lieutenant and Caceres then drove back to his small convoy to make sure they were ready. It was a good thing that they did so because as they pulled up, his radio came to life to inform him that they were all going back early so it was time to move out. He turned to Caceres and signaled him to drive down the line to let the drivers know that they would be leaving shortly. When he was done, he saw that Parson's was ready, and got back to the front of his platoon, just in time to pull in behind Charlie Company for the trek back.

By now it was almost 1000. It was clear and beautiful, but windy and colder than anything he had experienced before. He and Caceres were both wrapped up in cold weather gear and their ponchos and scarves, so they looked like a couple of olive drab mummies as the vehicles cautiously made their way over the snow and ice-covered roads to the west. They finally reached Darmstadt and the highway after about two and a half hours. The Lieutenant was practically frozen, even with his gloved hands and the radio's handset under the poncho liner that was draped over the heater and his lap. Caceres was also suffering and put his right hand in front of the heater once in a while to thaw out, so he could drive.

They turned south with the convoy and left the Wald and its

black ice behind, so they sped up and made better time on the clear roads back to Mannheim. They were moving along well and came up to the roadside stop that was the brigade's trash point. He signaled for the trash truck to pull over and offload the bags while the convoy continued. The point was only a small barrel and as anticipated, was completely covered, so the Lieutenant just shook his head, and when the truck was emptied, he led it back to join the convoy again, ahead of the BMO's group. Then he and Caceres raced back to their original place in front of the support vehicles.

When they pulled through Kaserne's gate, he motioned to Caceres to pull into the motor pool and drive back to his platoon's area. Once there, he released the cook, driver, kitchen truck, and mess trailer to Alpha Company. Then he dismounted and went into the Headquarters maintenance bays. After locating Sergeant McNabb, he turned his returning troops over to him and told him that he would see him on Monday. McNabb saluted and headed out to the platoon area, while the Lieutenant told Caceres to drive up to Battalion so he could turn in his CEOI to the S-2 NCO before they went to the S-4 shop.

When he was done at Battalion, he had Caceres drive over to his office. As they pulled up, the Lieutenant wrestled his A-bag out from behind his seat and stuffed in his cold weather gear. Then he grabbed his rifle and released Caceres, who had been passing the time sharpening his boot knife with the whet stone he carried in the lower left pocket of his field jacket. As Caceres jammed the Jeep into gear, the Lieutenant walked slowly into the S-4 office, since he was numb from the waist down, from the cold and sitting practically motionless during the long drive.

While he was getting his gear and rifle laid on the desk, and hanging up his TA-50, map case, and steel pot on the coat tree near it, Graham came in behind him and asked how he was. The Lieutenant replied, "Just fine; warming up and getting my circulation back. How about you?"

Graham shook his head a little and muttered, "I just finished with the XO's after-action ass chewing, but it went pretty well for me since I'm leaving. He asked Captain Dolan to stay back to go over some of the issues in greater detail, so I took the opportunity to release my driver, bail out, and go home to get some rest. I strongly suggest that you do the same."

The Lieutenant felt relieved and said, "You don't have to tell me twice," as he shouldered his A-bag and picked up his rifle. When he reached the door, Graham added, "I'll be preparing a report over the weekend. See me first thing Monday so we can go over it before I have to meet with the XO and the other staff officers to fine tune things, for when we go out for the larger exercise."

The Lieutenant nodded and said, "Will do," before walking quickly to his company arms room, where he cleared his rifle, to turn it, and the full magazines he had been issued, in before humping his A-bag up the stairs and out into the foggy near darkness of late afternoon.

Back in his quarters, he slid the A-bag onto the floor in his bedroom and separated out the clothing he needed to take to the cleaners, putting it in one of his two laundry bags from VMI that was hanging from the door. When that was done, he took the waterproof bag and cold weather gear out and laid them in the living room to dry. Then he stripped off what was left of his clothes and tossed them in the laundry bag before turning on the shower. Once the water was as hot as he could stand it, he climbed in and let it run over him for at least ten minutes, to thaw out.

Chapter 34
Hot Stuff

When he had feeling back in his limbs, he climbed out and dressed in a t-shirt, jeans, sneakers, and V-neck sweater. He was still chilled but no longer numb, so he decided to have a German beer to warm up. He was still stiff, so he slowly stalked down the hallway to the kitchen. There was no one there. He walked in, took a bottle of Spaten out of the refrigerator and opened it before he sat down in the corner and relaxed. He had been there for about five minutes, rubbing his hands and arms, and savoring the beer before Buzz walked in with a bag full of groceries. It was a cold Friday night, and Buzz had decided that it was time for chili.

He immediately set to work, throwing ground beef, diced onions, and several other items in a large pot on the stove. Buzz had spent some time at Fort Hood, so he made it Texas style, with no beans, but lots of chili powder, diced pickled jalapeños, and tabasco, for an extra kick. Buzz was a maestro at work; and he smelled and tasted the concoction regularly to make sure it was coming along well before he ran back to his room and came back with a quarter of a small flask of Kentucky bourbon and a cup of cold coffee that he added, along with a bottle of beer. The Lieutenant looked at him quizzically and Buzz said, "Chuck wagon chili. Six Bs: beef, beer, bourbon, black coffee, brown sugar, and bacon, but I don't have the last two, or you'd really be in for a treat." Then he turned down the heat to let it simmer before opening a beer for himself and sitting down with the Lieutenant to ask how his mini vacation had been.

The Lieutenant told Buzz that he spent most of the time frozen and slipping on snow and ice. But other than that, it was not a bad experience for a Cav outing. He got to see some of the heavy forests, that were beautiful covered in snow. Buzz just chuckled, and told him, "I like to travel, but not in a Jeep with no side windows, and a frigid wind whipping through it, for days on end." The Lieutenant agreed that it was not the ideal tourist experience and he was more than happy to be back in garrison, to spend an evening with good food and company. Then he asked if Elizabeth would be joining them. Buzz said, "Yes, she is running late because she has a couple of students who need extra help. I expect her in the next half hour, or at least that is the plan."

As Buzz finished, they both took a drink and just enjoyed the smell of the chili that had permeated the building and got the attention of Tom from across the hall, who came into the kitchen, saying, "Hello," and grabbing a beer before taking his usual place at the end of the table. He looked a little morose, so the Lieutenant asked him how things were going. He answered, "Not very well. I had a falling out with Cathy, the transportation lieutenant I was seeing, so I'm back at square one." He seemed pretty broke up so the Lieutenant decided to change the subject, and asked him, "Do you have any field duty coming up?"

"No, not at least for a month, but when the weather gets a little better, I will have to accompany a large shipment of ammunition to Grafenwoehr in anticipation of the spring tank gunnery. That will take some time."

The Lieutenant nodded and said a prayer of thanks that he was not going along on that expedition, since Graf supposedly was even worse than Baumholder, then he added, "I'll be out on another winter exercise in a little more than a week, and after that, if I don't get myself killed, and the old man doesn't cancel my leave again, I'll go back to Kansas for a short stay and a wedding before coming back here. I look forward to seeing you and introducing my girl to you when you get back from Graf." Tom smiled and said that sounded good to him and maybe he would have a new girl himself by then. The Lieutenant wished him well and they made a toast, clinking their beer bottles together, as the sound of a door slamming echoed down the hall.

They saw movement in the doorway. Looking up at Captain Pavel, the Lieutenant asked him how he was doing. Pavel held up his hand as he made his way to the refrigerator to liberate and open a beer, then sat down heavily and took a swig and a deep breath. He looked like hell. He was pale and sallow, with glassy eyes and a stubble beard. After he had recovered a little, he took another drink and then said, "Just fine. I have been out since Tuesday, before your exercise, running continuously, from position to position for the last four days, keeping up with the scouts and mortars as they moved, and checking on the ground surveillance radars and stinger sections too. But don't worry. As soon as I get another beer and some of that chili in me, followed by a double shot of Jägermeister, I'll be back among the living."

He chuckled a little and the Lieutenant laughed with him "Life in the Cav, sir: a case of Cs, ten gallons of water and a pat on the ass?"

"Exactly LT," said Pavel. "The problem with CSC is that you have all of these separate elements: four different platoons, that are constantly changing positions, and being checked by Battalion and higher headquarters, so you spend your entire time running in circles trying to make the staffs happy."

The Lieutenant nodded in understanding, saying, "Being a newly minted Support Platoon Leader, I get it."

Pavel nodded and then added, "I've spent the last three hours locating and shepherding back a GSR team in a Jeep with a radio that went out early in the morning. A couple of new kids. It never occurred to them that they could call Battalion from a pay phone for instructions, so I had to trace them back to their last position and follow them back. Where do they find these people? Anyway, we'll have training session with the First Sergeant Monday and believe me, we won't have this problem again."

The Lieutenant asked if he had been able to get any German food along the way. "Just cold meat, bread, and cheese with a tube of hot mustard," he replied, "so I'm famished."

They spent the next quarter of an hour telling stories and catching up on the BOQ gossip, until at last Buzz gave the pot a final stir, and announced that the chili was ready. The Lieutenant looked over and asked. "Do we have any ketchup and crackers?" Buzz immediately shot him a look, in the words of Harry Flashman, like a Turk at a christening. The Lieutenant quickly realized that he was committing a Texas chili sin, but followed up with, "I was raised on hillbilly chili, and it reminds me of home." So, Buzz relented and pulled a bottle of Heinz out of the refrigerator and a box of cheap saltines down from the cupboard, as he said, "Help yourself." They rose to fill their bowls in turn, and the Lieutenant added crushed crackers into his with a large shot of ketchup, before opening another beer and sitting down with his compatriots and diving in.

As he was finishing his chili, Elizabeth came through the doorway, crossed the room, and gave Buzz a kiss and a hug, before heading over to the stove. She turned slightly, looked in the refrigerator, and finally found an open bottle of red wine that looked like it could work. Then she poured herself a glass and carefully placed it on the table before helping herself to a large bowl of chili and sitting down to join the conversation. When she saw him and Captain Pavel still practically frozen, she asked where they had been. Pavel told her that they had been east of Mannheim, in the Odenwald, in the snow for several days. She looked at both of them like they were crazy, and then asked, jokingly, if they had taken any time to ski. They both immediately said no, and they weren't planning on going skiing anytime soon because they had enough snow to last them for a long while. After enjoying another bowl of chili, the Lieutenant felt warm enough to maybe sleep, so after washing his dishes, he excused himself and went back to his room to take another hot shower before diving into bed and sleeping till almost noon the next day.

GOOD DAY IN HELL

When he woke up, he looked over at his alarm clock to check the time, before he looked around the room, to make sure he was really in his quarters. He got up, dressed, and headed down the hall and the stairs to the basement and the Autovon phone, to call Father Piska, to confirm their meeting that evening. He was the Catholic Chaplain who served the units in Mannheim. Piska had been in the Army for over twenty years, with a couple of tours in the combat zone, and although he was a full colonel, he was unassuming and easy to talk to. He had been meeting the Lieutenant on Saturdays when he was in garrison for pre-marriage classes, as required by the Catholic Church. While at first it had seemed odd to be getting marriage training from a man who was never married, the Lieutenant soon discovered that he knew a lot about the challenges in the relationship as he had seen about everything that could go right and wrong after twenty years ministering to soldiers. Also, since the Lieutenant was a Protestant, he learned a lot about that faith as well as married life.

The Lieutenant liked the priest a lot. They usually went out to dinner for their meetings, so it was perfect for both of them and the Lieutenant looked forward to getting out to see more of the surrounding villages. His favorite place was a small hotel in Viernheim that had excellent Wiener schnitzel and even better Hungarian goulash with lots of paprika to heat it up. After climbing the stairs, he looked in the kitchen, but not seeing anyone, went back to his room to eat a quick lunch of a breakfast bar and a Coke while he wrote a letter to his girl about his latest trip to the woods and anticipation of his coming back to see her soon.

As evening fell, he washed and dressed, with the warm light gray ski jacket he picked up in a German department store. He made his way out and warmed up his car, with the heater on high, before driving over to the chapel where the priest had his office. He parked and walked to the door where Father Piska met him. Then they turned and climbed into the priest's Mercedes sedan. Piska took off like a rocket for the Autobahn. The priest asked if goulash sounded good and the Lieutenant answered, "Yes," even before the question was finished. They both laughed. Then as they drove to the restaurant, the priest asked quite a few more questions about what the Lieutenant had been up to since their last meeting. When the Lieutenant told him that he had been on an exercise in the forest east of Heidelberg, the priest asked how it went. The Lieutenant told him just fine but with his new job as Support Platoon Leader, he would expect to be busier than ever, which was good because it made the days go by faster.

Over dinner they visited a little about the final arrangements for the wedding that were being handled by his girl back in Kansas. When they were done eating, they had a brandy for dessert and the chaplain drove him back to drop him off beside his car. This was the last session, so the Lieutenant thanked him. Then the chaplain told him he was really looking forward to meeting his wife when she got in country because she sounded like a fascinating person. The Lieutenant assured him that she was, then saluted, turned and trudged through light snow to his little Volkswagen.

THE HUSTLE

The next few days were simply monotonous, consisting mostly of short runs in the morning, staff work at his desk, and the never-ending quest to keep Sergeant McNabb around long enough to develop a lesson plan, although it would be too cold to do anything until after his wedding. As luck would have it, one Wednesday morning, the Adjutant called and told him to come to Battalion. When he got there, he was handed orders to appear on Friday for another court martial as a member of the jury and due to his rank, probably the jury foreman. Given the short notice, the Lieutenant left immediately to get his greens into the cleaners so they would be pressed before he had to wear them on Friday.

He got through the next couple of days with no major incidents. He tried his hand at making spaghetti in the kitchen for his BOQ family one Thursday night. It turned out fairly well, although he was a novice, and Buzz had to come in and keep him from making a complete disaster. Everyone was in good spirits and looking forward to warmer weather in the spring.

The next morning, he appeared in greens at the Kaserne where the JAG and MPs were stationed. The jury was composed mostly of second lieutenants, along with the Lieutenant, a captain who was foreman, and one sergeant major, because the accused was an enlisted soldier. The trial was a fairly simple case of theft. It was presented before lunch and the jury came to a guilty verdict by 1330. After he was dismissed, he drove by the S-4's office and after finding out that there was nothing pressing, drove back to his quarters, He was looking forward to a weekend off, without duty somewhere.

CUT THE CAKE

Although it was Friday night, he decided to eat a frozen dinner, rest in his room, listening to the radio, and reading a little. When Saturday came, and he finally had some time to himself, he decided that he would drive over to Heidelberg to see the sights around the university there. On Sunday, he thought he would drive up one of the narrow two-lane roads into the Odenwald and visit the little villages along the way. He had a lot of fun going up into the mountains, and back. On Sundays, the civilians would visit their little bakeries for strong coffee or tea and beautiful desserts: tortes of all types, fruit covered pastries, and croissants. Dressed like a German student, he was well received and was able to converse a little with the younger folk, and the wait staff in particular, who were surprised that he was an American soldier and had a lot of questions about the United States. He had a hard time, based upon the American TV shows, convincing them that the country was not full of oil wells and cattle ranches as appeared on Dallas, or Indians and buffalo as they saw on older TV and movies. At first it seemed odd, but then he realized that Americans had the same misconceptions about Germans, in that not all of them wore lederhosen or carried around steins of frothy beer. The Lieutenant enjoyed the Sunday socializing and as the sky darkened, bid his new friends farewell before heading back to his quarters for a good night's sleep.

FLY, ROBIN, FLY

The next week went by pretty fast. His platoon was due for quarterly maintenance and given the number of vehicles, maintenance would take ten days to two weeks before they were all done. The NCO's and drivers would handle this with little guidance from him. Nevertheless, there was always plenty to do, since he had been alerted to the fact that he would be taking over as mess officer after the winter exercise. He needed to spend some time with Sparks and the cooks to get to know them; watching them, and asking them what they needed, and spend some time with his drivers to check on their needs as well. By Wednesday morning, he had made substantial progress but was called down to Battalion for a meeting with the S-4 and the executive officer to go over deficiencies in supply and transport, over and over again, for the next two mornings. The only bright spot was that toward the end of the week, he was able to purchase the airline tickets to fly from Frankfurt International through Heathrow to the States to get married and bring his new wife east with him to spend the next couple of years in Europe. As he made his purchases, reality set in and he began to believe that it was actually going to happen, so he decided to return his library books, well in advance of his departure.

During the next couple of weeks, the S-3 shop issued new plans and orders for the winter exercise. It would now consist of more vehicles and officers than the earlier TEWT. This outing would last for five days, from early Monday morning on February eleventh through the evening of Friday the fifteenth. Brigade also convened a full-scale Pegasus exercise, which was fun, because sitting in a room with map boards was a lot easier than running around the forest in a Jeep, freezing his tail off, followed by half a dozen trucks, slipping through the snow, with no radios. He was also tasked to attend training on his nuclear, biological and chemical duties, since that job had followed him from Bravo Company to Headquarters Company, and for the first time they had a chemical officer assigned to the battalion who was eager to train and be part of the unit. He was a fun loving second lieutenant who made the training easy and as lighthearted as possible, given the subjects.

He took time to check with the Adjutant to make sure that his leave was properly recorded at Battalion and connected with Master Sergeant Fykes to confirm he could still drop him off at the airport and then pick him, and his bride, back up once they returned. He had decided he would take off on February twenty-eighth, which was a Thursday. This gave him a week to catch up on the arrangements that his girl had made for their March eighth wedding date.

BLUER THAN BLUE

The following week there was little for the Lieutenant to do but check the status of his vehicles that were in maintenance and attend meetings. The depression and boredom of Germany in winter had finally hit him. Although he hadn't realized it before he got there, Germany, being so far north was almost like being in Alaska. The darkness was oppressive, lasting from mid-afternoon until late the following morning. January moving into February was practically impervious to happiness, very much like the "Dark Ages" at VMI that he had endured four times in the mountains of Virginia. This was why the German's Fasching holiday was filled with loud music, bright lights, and booze. The lack of light would get to you otherwise.

PRE-ROAD DOWNS

Finally, on a Thursday afternoon, with the weekend of February ninth approaching, he received a copy of the new order and overlays for the small-scale exercise from the S-3 NCO. It would begin with the usual 0300 alert on Monday. Each tank platoon would be represented by one tank commanded by the platoon leader, accompanied by the Company Commanders and executive officers in their Jeeps, supported by battalion maintenance and supply. CSC's Scouts and Mortars would go out in their tracks. Ground surveillance radars and air defense would be in their Jeeps. All of them would depart in advance of the line companies and the staff.

He ran into Captain Pavel in the hall of the BOQ on Friday afternoon after he had stuffed his maps and new overlays into his large map case. Pavel was on his way out to dinner but stopped and asked the Lieutenant if he was ready to get out into the woods again. The Lieutenant just laughed and said, "Probably about as much as you are." Pavel just chuckled and replied, "True enough, but at least I'll get a good meal tonight before I have to inspect all my folks tomorrow to make sure they are ready for the long haul."

With that, he rushed down the stairs and through the door into the foggy night, while the Lieutenant went back to his room for a TV dinner. He liked Pavel, even though he was the battalion snitch, the eyes and ears of the commander; he was good at training, and incredibly funny.

Next day, he drove up to Nick's for dinner through a dense, clammy fog. He enjoyed it thoroughly and told Nick that the next time he saw him he would probably be bringing his wife in, although he might come in for dinner after this next winter exercise, if he could. Early Sunday, after a quick trip to the PX, the Lieutenant got with Buzz and handed him a large pot roast and vegetables so that they could have a good hot meal that night, since he would be rolling out early the next morning. The Lieutenant also donated a bottle of red wine, that was not his favorite but would go well with dinner. That evening was a lot of fun. Most of the residents came down to dinner. The Lieutenant and Pavel regaled them with tales of what their adventures were going to be like. Elizabeth was there, of course, and she had lots of questions. She was fascinated by most things military and about maneuvers in particular. Nevertheless, by eight in the evening, the Lieutenant excused himself to get some sleep. He set his little alarm for 0300 and prepared for the worst.

Chapter 35
Déjà Vu

True to form, a little before three, he woke to the harsh sound of pounding on his door. He hurried out of bed and screamed that he was up to the soldier who was tasked to bring him back and told him he could go. Then he shaved, dressed himself, and hurried down the stairs. He had packed his A and B bags with his thermos, a big bag of licorice, and as many breakfast bars as he could fit in. Bracing himself, he shouldered the A bag and carried the B bag into the mist to Battalion and his office. When he arrived, he was heartened to see that Graham and Sergeant Dryer were already there. He said hello and set the bags down, then strapped on his LBE and helmet, slung his map case over his shoulder and grabbed both bags before hurrying out the door.

Stepping into the cold, he was delighted to see that Caceres already had the Jeep parked in the courtyard with the engine running and the heater on full blast. The ever-cheerful driver stepped out and saluted before helping him wrestle the bags into the back of the Jeep. Then they climbed in and drove down to Headquarters Company. When they got to the entrance, he hurried through the big wooden door and down the stairs to the arms room to draw his rifle and four magazines, before pounding back up the stairs to the courtyard where he made sure the M-16 was on safe, slammed in a magazine, locked and loaded the first round into the weapon, and wedged it between the seat and the transmission. Then he motioned to Caceres to drive down to the motor pool where his convoy of support vehicles was preparing to move out.

He had more vehicles now, due to the addition of tanks to the exercise. He had two 8-ton fuel GOERs that were particularly hard to move over black ice. Each had two drivers so they could spell each other. He spoke with them to make sure they had a warm place to sleep since their cabs were right over the engines and there was very little space to stretch out. They told him that they had made arrangements to sleep in tents with the company maintenance section coming along to support the staff vehicles.

He then walked down the line to the mess and cargo trucks. The cargo drivers had all been with him on the prior exercise, so he stopped and visited briefly with each one. They joked a little about the cold and the fun they were going to have. Then he moved over to check the mess trailer. He had the same cook and mess truck driver as well so he spoke with them briefly before sliding back into his Jeep so Caceres could drive him over to check in with the Battalion Maintenance Officer, Lieutenant Parsons, aka, "Conan." They bullshitted for a while since they hadn't seen each other very often at work and Conan lived on the economy.

Heading back down to the Support Platoon area he saw Sergeant McNabb slipping into the Headquarters Company maintenance bays. He had just arrived and was getting settled in behind his desk. He had not drawn a rifle since he was not going to the field and rose slowly to salute the Lieutenant as he walked in the door. They spoke briefly and McNabb told him that he was sending the new assistant platoon sergeant, a white E-5 named Brady, who had just arrived from Fort Dix, with an additional five-ton cargo truck and driver, due to the larger amount of trash that would need to be hauled back after this exercise.

Since they had not discussed this change or the arrival of a new NCO earlier, and the Lieutenant really did not want another vehicle to keep up with, he was not happy but was not in the mood to get in an argument at four in the morning. So, he thanked McNabb, who told him not to worry as he was staying back to make sure the rest of the platoon stayed out of trouble while he was in the woods. The Lieutenant saluted and trooped the line down to Brady's truck where he made sure that he and the driver had their weapons and gear, and instructions to follow the other five-ton cargo truck in the convoy.

When he got back to his Jeep, he crawled in and swore that if anything, it was getting colder. Rubbing his hands behind the heater, he looked around and then signaled for Caceres to drive over to the staff vehicles so he could coordinate with Graham. The S-4 told him there were no changes to the operations plan and that they would be taking off as soon as the S-2 issued the CEOIs. While he was still standing beside Graham's Jeep, the Battalion Supply Sergeant, Dryer, peered out of the back of the S-4 track and said, "Lieutenant, after watching you on the last exercise, I know you'll be running day and night, freezing the whole time. You need to get that thermos full of hot coffee right now to get ahead of the crappy weather."

The Lieutenant thanked him for the good advice and told Caceres to drive up to the mess hall so he could go in and fill the thermos with steaming hot coffee. When that was done, they pulled around to the edge of the courtyard, and ran into the S-2, who tossed him the CEOI that he had been waiting for. He put the lanyard around his neck, opened it, and tuned the radios to the command frequency and the logistics frequency. That way he could tell what the battalion was doing by listening in and communicate over the logistics frequency to coordinate the movement of food, fuel, and simulated ammunition.

With the yellow lights of the vehicles flashing, the entire area looked like an amber version of Dante's Inferno. They were distracting and annoying, so the Lieutenant was happy to hear Shea's order to "move out" come over the command net. Pavel and his scouts were first to go, followed by the mortars, air defense, and GSR platoons. The battalion command and staff group was next, followed by the three line companies, each with two Jeeps for the CO and XO, a track for the First Sergeant and three tanks, one for every platoon leader.

He pulled in behind Charlie Company with his small convoy, followed by Battalion Maintenance and their noisy 88s. As usual, the last vehicle was an ambulance track from the Medical Platoon. This was because civilians often failed to accurately judge the speed of military convoys at night and would run into the rear of the trailing vehicle. An armored personnel carrier could withstand the impact and that way medics were on the scene right away to administer first aid, which was of great importance.

ON AND ON

As they pulled out of the housing area toward the Autobahn, he and Caceres encountered hard pellets of sleet flying into the Jeep. The Lieutenant was glad that, based on their last experience, they both had on their field pants and shirts, field jackets, rubber ducks, and parkas. The wind whipped past his right side, so he slipped his map case over his lap and down the right side of his leg to fend off the moisture. He and Caceres both had their wool scarves wrapped around their heads to cover everything except slits for their eyes. This left a small portion of his right cheek exposed. The Lieutenant would never forget these trips over the remainder of his three-year tour. He ended up with frost bite on that cheek, which would be with him for the rest of his life.

When they reached the foothills of the mountains, the convoy slowed, and then slowed more, until it was crawling along, slower than a man could walk. After that, they inched along, barely moving over the frozen snow and black ice. Due to the reduced speed, they did not reach the maneuver area until it was almost noon. The staff and line companies had taken up their initial positions, so the

Lieutenant with his trucks moved into their assembly area, followed by the BMO and his vehicles. The Lieutenant made sure that his people were in position and taken care of. Then he released the five-ton with the tentage and supplies for the staff to help set up and return.

They pulled away. He walked the area and checked with Leach to make sure there would be a hot meal ready for delivery at the end of the day. The cook assured him that it would be ready, so he had Caceres drive him over to the S-4 track to check in with his boss. The original plan had not changed so he would lead a truck loaded with thermite containers full of food forward to the line companies where the First Sergeants would pick them up and feed their troops, then return the empty containers for transport back to the cook. Since his soldiers were spread far and wide, the CSC First Sergeant picked up their food directly from the cook. So did the junior sergeant in the operations shop who borrowed the S-3 Air's Jeep to make the run. The staff officers themselves came by the trailer when they could, with their drivers since their schedules were unpredictable.

He and Caceres pulled up at the trailer to pick up the food. The Lieutenant went to the driver of the truck that would be hauling the food and his assistant and told them to make sure that they ate before they had to deliver it, otherwise they probably wouldn't be fed at all. The same thing went for Caceres, who made himself a plate and devoured it in record time so he would be ready to go when his officer slapped the side of the Jeep. As for the Lieutenant, he just needed more hot coffee. With that and a couple of breakfast bars, he figured he would make it.

The first night went as well as could be expected. At first light, the Lieutenant was surprised to see a first lieutenant with a Jeep and a driver pull up beside him. The visitor was an experienced Support Platoon Leader from another battalion in the division. He was there to act as a controller during the maneuver, grade the Lieutenant on his performance, report back to Brigade, and give him advice to help him in his job. The Lieutenant liked the man and he seemed to be a decent sort, so they got on well, swapping stories, standing in the freezing snow, moving to try to keep warm before hopping in their Jeeps for the day's maneuvers.

The line companies and scouts were moving quite a bit, but since it was the first full day of the exercise, Conan and the Lieutenant only had to move their vehicles once during the day. Unfortunately, it was getting colder by the minute and the wind was coming up so that by evening it was brutally frigid. After they returned from delivering the evening meal, he and his controller friend decided, as night fell, to go back their Jeeps up to some heavy evergreen trees that would block the wind. This gave them the advantage of the trees and the backs of the Jeeps as insulation against the full force of the wind. They also put poncho liners up in the open doors and cranked their heaters up on high, to try to stay warm as they monitored the radios. As usual, the Lieutenant preferred to sleep sitting up for two reasons: first he did not like being on the ground with tanks and other vehicles running amok, and second, he never knew when something would happen, resulting in a frantic call over the radio, requiring an immediate response to avoid disaster.

SLEEP'S DARK AND SILENT GATE

As he sat rubbing his hands together over the heater, he realized just how crippling the European winter could be. The damp sea air coupled with the wind blowing off the north German plain simply made life miserable and wore you down. He looked over at Caceres and was happy to see that he had finally gone to sleep. For himself, he was determined to stay awake as long as he could, before waking him to keep watch. After a little over three hours, he was just too exhausted and felt himself slipping away so he nudged Caceres, who nodded at him. The Lieutenant gave him a thumbs up with his left hand and promptly passed out.

Before he knew it, without warning, he suddenly felt like he was going to throw up. He had a sickly-sweet taste in his mouth reaching deep into the back of his palette and his breathing was labored. Caceres was unconscious. Looking out the windshield, he could see that heavy snow was falling and swirling around him. Then he realized, looking back, that snow had drifted in behind the Jeep all night due to the trees, and the vehicle was full of carbon monoxide. He shook Caceres awake, yelled at him to move forward, tore down the poncho liners, and threw his door open. Then he ran over to the other Jeep and dragged his controller and driver out into the open so they could both breathe as much fresh air as possible. The other lieutenant and his driver immediately realized what had happened, and they pulled their Jeep up and out of the way from the trees, which made them colder but much safer.

Visibly shaken, they drove over to the kitchen trailer to get fresh hot coffee from the cook and rinse the awful taste out of their mouths. As soon as breakfast was ready, the Lieutenant rounded up his truck driver and assistant to have breakfast with Caceres before they drove out to the small assembly area behind the line companies where the first sergeants picked up hot food for their troops. After that was done and they were back in the battalion's large assembly area, the Lieutenant's radio crackled to life on the command channel. Since they were halfway through the exercise, they had reached the point where the line companies would start to fall back, simulating the active defense. This meant that the trains would have to leapfrog back, at least once a day for the next two days.

The remainder of Wednesday passed without any major problems, other than the snow that continued until nightfall. The next morning was Thursday and they would be leaving Friday. Unfortunately, the cook was practically out of rations because everyone in the field was eating more than anyone had planned. The Lieutenant checked with the S-4 and he was ordered to have Leach and his assistant drive to a nearby Kaserne in Hanau to pick up enough rations for dinner that night and breakfast on Friday. The two of them left the trailer with the assistant Support Platoon Sergeant to issue C-rations to anyone who needed them and moved it to a new assembly area when the trains were ordered to move. The Lieutenant explained to Leach where they would be when they moved. He remembered it from the earlier exercise and took off through the snow on his mission.

At noon, the S-4 gave the Lieutenant the order to round up and move the trains to the new assembly area. With the ice and snow, and a broken-down tank slowing them down, it took almost two and a half hours to fall back ten kilometers and set up again. As 1500 approached, Leach pulled in and hurriedly set up with his assistant to cook the evening meal. He had managed to pick up enough canned chicken and vegetables to make chicken and rice, with fresh bread and canned fruit. It would have to suffice. Now that food, which was the Lieutenant's main job, was under control, he covered up in his Jeep and napped until Caceres tapped him on the shoulder to let him know that chow was ready. They left to make their delivery to the line units, and when they pulled back up to the mess trailer in the darkness, the cook motioned to the Lieutenant that he wanted to talk to him.

RENEGADE

The two of them walked over to the tree line, behind the ration truck. There the cook told him that on the trip when he was coming back with the rations, he was driving and hit and killed a small German roe deer. "Sir, this is a real problem. The German forstmeisters are very protective of their deer and hunting rights. I was in Korea for my first tour. We'd be better off running over a kid outside Seoul. We have to get rid of this deer."

"What do you suggest?"

The cook looked up and said, "Sir, the best thing I can think of is to field dress and cook it. My wife is German and I cook with her at home. I picked up some mushrooms, onions, and spices plus a little brandy to make hunter sauce so we can have a good meal tonight. Then I'll shred and chop the rest and mix it with hamburger to serve cream chip gravy on toast for breakfast. As for the rest of the deer, I'll put it in a trash bag and keep it on my trailer. Then when we get back to Mannheim, I'll put it in with the trash from the mess hall that will go to the German incinerator downtown. That should take care of it."

"Very well," said the Lieutenant, "we'll get the evening meal out and when everything quiets down, you can go to work."

He and Caceres drove over to the S-4 track and parked. He went in to see Graham, who was out with the XO, and found Dryer holding down the fort. Things were under control, so he told Dryer that he and Caceres were going to take a nap and asked him to wake them if anything came up. Late in the afternoon, around 1600, Dryer woke them up so they could lead the ration truck forward. They made their delivery, and by the time the first sergeants brought back the empty thermite containers, the wind had come up and was blowing through the Jeep with a vengeance, as they made their way back to the Mess Trailer.

Upon his return, he stopped by to see how Leach was doing. The cleaned body of deer was cooking marvelously, while the head and the remainder of the carcass stared at him, with a razor-sharp butcher knife resting beside it. He quietly told the cook to get him when it was done, before joining Caceres in their home away from home. The driver went to sleep and the Lieutenant waited for the cook. With all the running back and forth, leading the trains, convoys, and mess trucks, living on breakfast bars, coffee/cocoa, and licorice, he was exhausted and famished. When Leach waived at him from the side of the trailer, he woke Caceres and told him what was going on. The two of them made plates of venison with mushroom gravy, accompanied by fresh bread. They thanked Leach and enjoyed every bite, while they watched him cut all to meat off the bone, before he diced it finely to go in the breakfast gravy. When they were done eating, they slipped over to the Jeep and settled in. Caceres took the first watch so his officer could get at least some sleep before they had to make the trek back to garrison.

The assembly area fell quiet. The wind died down, as did the precipitation, so where they were located it was almost pleasant, as he fell asleep. A little after 0100, Caceres shook his left shoulder to wake him up. He gave a thumbs up with his right hand to acknowledge that he was on watch. His right foot had fallen asleep, so he moved it back and forth to bring back his circulation and took a deep breath, to let it out like smoke in the darkness. Then he realized that he needed to relieve himself, so he grabbed his rifle, pulled off his poncho, climbed out of the Jeep, and stalked off into the woods about twenty yards. When he found a likely spot, he propped his rifle against a tree, opened his entrenching tool, and dug a cat hole. Then he unbuckled his LBE and hung it on a branch.

RED NECK FRIEND

When he had finished doing his business, he stood up and pulled up and fastened his pants. Suddenly he sensed something behind him and felt the sharp edge of a knife against his throat. He stiffened and then froze as a slurred voice said, "I'm going to kill you. You're a rich, candy-ass college boy and I hate college boys." He recognized the voice; it was Leach. He glanced down and recognized the knife as well. It was the knife used earlier to field dress the deer. With his rifle out of reach, he had no weapons available, so he decided to remain calm. From the back of his mind, he thought he recognized the cook's accent, and so he asked, "Where are you from?"

Instinctively, the cook answered, "Tennessee."

The Lieutenant thanked his lucky stars and said, "My momma's from Tennessee. Erin, near Belle Valley, northeast of Shilo and about thirty miles from Fort Campbell. You know it?"

"Yes, sir."

"I'm a hillbilly just like you. I was raised on hoe cakes and buttermilk, just like you. And biscuits with black-strap sorghum molasses, just like you. And ham hocks, pinto beans, cold kraut, and shoe-fly pie, sour cream, and raisins with cracked pecans. I came up from nothing just like you."

Then Leach softly said, "Yes, sir," and as the knife fell back from his neck, the Lieutenant asked, "So hillbilly to hillbilly, what the hell is going on?"

Leach told him that he was constantly being tasked to go to the woods and his wife was upset all the time. He was angry about it and drank the brandy left over from cooking the deer. But mostly he'd been told to do it. The Lieutenant asked him, "Who?"

The cook replied, "The Mess Sergeant. Sparks sends me out to the woods because I won't deliver drugs for him."

The Lieutenant then asked him, "How about McNabb?" Leach looked away and fell silent.

Having his answer, the Lieutenant told him, "Never mind. I'll fix this. This never happened. You get some sleep. Get up in a couple of hours, drink coffee, and make breakfast."

The cook looked at him and asked. "How are you going to fix this, sir?"

The Lieutenant just smiled and said, "Don't you know? Officers are magic. And to quote that great American film, *Animal House,* 'You can't spend your whole life worrying about your mistakes."

STAYIN' ALIVE

After sending Leach back to his trailer, the Lieutenant strapped on his LBE, picked up his rifle, and checked to make sure that he had a round in the chamber and it was off safe, just in case. Then he walked back to the Jeep and climbed in beside his driver who was snoring softly, fast asleep. When Caceres woke up, he could tell that the Lieutenant's mood was different, so he asked him what was wrong. The Lieutenant briefly relayed what had happened and then told him not to tell anybody. Caceres looked at him and as his eyes narrowed, before he whispered, "If anyone comes after you again, sir," he motioned as though making a slit from the naval to the throat. When the Lieutenant looked at his driver quizzically, Caceres simply said, "That's the way things are done in my neighborhood."

Chalking that up as a good thing to know, the Lieutenant told him, "I can only hope it doesn't come to that. I'll take care of it." Then the Lieutenant grabbed his thermos and got out of the Jeep to make sure that the cook and his assistant were up, getting chow ready. After downing a cup of coffee to wake up, he filled his thermos and made his way back to the Jeep so Caceres could get some breakfast with the truck driver and prepare to deliver the morning meal. He was not hungry himself, as he was still full of adrenaline, so he made mocha with a C-rations cocoa packet poured in his thermos and sat thinking. One thing was for sure, he would never be caught again without at least a Swiss knife in his pocket, and when he got back, he was going to attach the jet-black K-bar combat knife that he had picked up from a Third Cav first lieutenant for fifteen bucks at the end of a poker game in El Paso, to his LBE. From that day on, he would always have nine inches of steel at his side in the field. Besides, it was perfect for cutting open cases of C-rations and you could eat with it in a pinch.

MORNING SIDE OF THE MOUNTAIN

At first light, the Lieutenant and Caceres took the five-ton truck with rations and the large fuel GOERs forward to be led to the companies by the first sergeants who returned the empty containers for transport back to the mess trailer as soon as the tanks and tracks were topped off.

When he and Caceres, along with the truck and the GOERs, arrived in the assembly area, he waved the truck on to meet the cook, and he had Caceres drive him over to see the S-4. Lieutenant Graham told him that he was heading back early with Captain Dolan, the command team, and the operations officer for a meeting at Brigade, so Sergeant Dryer would get the track back, but he would, in essence, be the S-4 and responsible for getting all of the support vehicles back to post. Plus, he was now the Mess Officer and since Brigade had not identified any new trash points, he would have to get a truck right away to load and drop of the refuse from the exercise at that small roadside stop he had used earlier. As Graham and Dolan rode away, the Lieutenant saluted and then shook his head. Behind him Sergeant Dryer chuckled a bit and then said, "That's the Army, LT!" The Lieutenant looked over and just said, "Indeed," before hustling back to the Jeep so Caceres could drive him over to Sergeant Brady to coordinate the move.

When he pulled up, he told Brady to have his driver climb into the rations five-ton. That truck would follow the Lieutenant to the mess trailer where they would pick up the trash bags. Meanwhile the sergeant would drive his truck to the staff area to pick up the staff's tents and other field equipment from Headquarters Company. He would also take the five-ton diesel and mogas trucks along to refuel the staff vehicles and then escort them to the entry to the highway where the road march would start so he could top off all of the Jeeps that were left in the area for the road march home. The Lieutenant let the NCO pass on the orders.

When the five-ton and GOERs pulled up behind him, he directed Caceres to the mess trailer. The cook's helper and the trooper with the five-ton driver loaded the trash bags into the truck. While they were doing that, the Lieutenant checked in on Leach, who appeared to have recovered from the night's activities. He pointed to a smaller bag of trash that he had set aside and nodded. The Lieutenant nodded back, and seeing that the truck was loaded, walked over to his Jeep, climbed in, slapped the door, and said "Get 'em up, Scout." Caceres, who had been sitting, seething, and visibly sharpening his larger knife, placed it and his whet stone back in the sheath on is right side, slammed the Jeep in first gear and took off like a rocket. As they pulled out, he saw the yellow rotating lights of the battalion commander and staff Jeeps move out of the trees and onto the highway. They disappeared quickly since they could move a lot faster than the tracks.

When the Lieutenant got to the roadside stop about ten kilometers away that was the designated trash point, Caceres pulled over by the only available container: the small trash can the Lieutenant had earlier warned battalion and brigade about. The bags from the first exercise were still there so the Lieutenant climbed out and directed the troops to offload the new bags and stack them next to the older ones. When they were finished, he picked up the handset and called Dryer to tell him that they were on the way back, and Caceres drove him to the assembly area as fast as he could.

Once he was back, he saw the yellow lights flashing on the battalion's tracks and began assembling his small convoy before coordinating with battalion maintenance. He stopped to visit with Conan a little when they heard the assistant S-3 come over the command net to give the order to move out. The crackle of the radio transmission ended, and the creaking of tracks and thunder of diesel engines filled the forest. Then, just as they had come into the woods, the Scouts departed first, rolling out in the lead, followed by the battalion staff tracks, mortars, air defense, ground surveillance radars and, finally, the line companies. He recognized the palm trees on the turrets of Charlie Company's tanks, so he pulled in behind them before he called Conan on the radio's logistics net to let him know that he was under way. He stepped out and signaled to the cook, who pulled out with the mess trailer, followed by Brady's truck hauling the water buffalo, and the rest of the support vehicles.

When they were all underway, he jumped in and slapped the door of the Jeep. Caceres was off like a shot. It brought to mind the words from *American Graffiti,* "Peel out, I just love it when guys peel out." They quickly made their way up the convoy to their normal place, well in front of the mess trailer. When they reached the highway where his fuel trucks were waiting, he told the drivers to fall in behind the last support platoon vehicle. When they rejoined the convoy, he signaled to Caceres to get back in front of the cook's truck and radioed Sergeant Dryer to let him know they were all underway and to check in to see if there were any issues. Apparently, everything was under control, so he and Caceres bundled up against the cold and settled in for the trip down from the ice in the mountains to the lowlands where they could speed up.

BALLAD OF EASY RIDER

The trip back to Manheim was uneventful, except for an Alpha Company tank that blew an engine on the way down and had to be dragged by an M-88. Fortunately, that was the BMO's problem, and he had great mechanics who had it hooked up in no time, so that tank and medic track in the rear were able to catch up and keep the show on the road. When they got back to garrison, he released Leach to Sparks. Then he went to see Sergeant McNabb, who, oddly enough, did not seem happy to see him. He turned the drivers and their vehicles over to McNabb with orders to get the tanks and other vehicles refueled and added that he would see him first thing Monday morning.

With that done, he had Caceres drive him to the S-4 shop. As he walked in, he ran into the S-2 NCO who was picking up the CEOIs. The Lieutenant removed the lanyard from around his neck and turned it over to the sergeant, who had just retrieved Dryer's as well. Then he removed his LBE and placed it on a hanger in the coat rack at his desk and put his helmet on the shelf on top of it. He was dead tired and preparing to leave when he noticed light under the S-4's office door. He walked up and knocked on the door. The new S-4, Dolan, told him to come in. After a week in the woods, without a shower, shave, or decent meal, and very little sleep, the Lieutenant looked like hell. Dolan told him so. The Lieutenant thanked him for the compliment and asked if there was anything he needed to do before he left for the day. Dolan told him, "No."

The Lieutenant then added, "Based on my observation, the Mess Sergeant only had one cook and trailer capable of operating in the field. After years of neglect and almost no training due to fuel constraints, I have serious doubts about the fitness of the other four trailers as well as their equipment. On Monday, I plan to have Sparks pull all of the trailers out on the parade ground and cook the battalion's meals on them until his cooks are proficient. In addition, I will need a copy of the mess TOE in order to have a full field layout and inventory of the mess equipment." Dolan nodded in agreement.

The Lieutenant said, "Great, I'll see you first thing Monday, after PT," and saluted. Dolan returned it, so he turned and walked out to his desk, picked up his cold weather headgear, and stepped out into the courtyard where Caceres was patiently waiting for him, guarding his rifle and spare mags.

He jumped into the Jeep and had Caceres drive him over to the entrance to Headquarters Company. Once there, he grabbed his rifle and cleared it. Then he practically ran down to the arms room where he signed in the rifle, mags, and ammo. Then he stopped by the orderly room where he let Christian, the Assistant First Sergeant, know that he was leaving before he pounded up the stairs and out the front door. Caceres looked over at him when he came back and told him, "Sir, you're beat. Let me drive you to your quarters so you don't have to hump your A and B bags over there."

The Lieutenant agreed, so in less than five minutes they were out the gate and in front of his BOQ. He pulled his bags out and placed them on the sidewalk. Caceres saluted him and drove off for the motor pool. As usual, he was grinning from ear to ear, because he really liked driving a Jeep and apparently liked the Lieutenant as well. The Lieutenant returned the salute, shouldered his A bag, and carried his B bag up the steps to his door.

Practically stumbling into his apartment, he put both of his bags down on the floor. There he sorted out his dirty clothing to put into one bag that he carried back out to the street and over to the laundry and tailor shop, where he dropped them off. Once that was done, he made his way back to his quarters, where he set the remainder of his gear out to dry in the living room before taking a long hot shower. By now, he realized that he really hadn't slept for five days, and he was so tired that he wasn't even hungry. Looking through the refrigerator, he finally found a frozen roast beef dinner to heat up in his toaster oven.

When that was finally heated up, he ate it and crawled into bed, to sleep until noon the next day.

His room was filled with gray light when he woke up. He remained in bed for several minutes, reassuring himself that he wasn't dying from carbon monoxide or being threatened with murder. Taking a deep breath, he slid out of bed and headed to the bathroom to get ready. Then he dressed in his German clothes and made his way down to the kitchen, where he was overjoyed to find a coffee maker with a half full carafe and a couple of German pastries that were up for grabs. He poured a cup of coffee and after taking a sip, wrapped both of them in a paper towel and carried them back to his room where he enjoyed them and spent the afternoon reading and warming up, before heading out for dinner.

Chapter 36
Mexico

As the sun went down, he walked out the doors and down to the curb where the Sirocco was parked. Once it had warmed up, he pulled out and made his way through the traffic circle to head to the highway to Viernheim and Nick's. He enjoyed the short drive up to the gasthaus, and played the song "Mexico" by James Taylor, as he made his way through freshly fallen snow. Oddly enough, it made him feel a little warmer, recalling his time on the border across the Rio Grande from Juarez. He pulled up outside the restaurant and drove around to the side of the building, where he parked and pulled down his turn signal lever to the left, to leave his left turn signal lights blinking since the road was so narrow.

When he walked in the gasthaus, Nick looked up and smiled, before waiving him to his usual place at the bar. The Lieutenant ordered a pilsner and a bowl of hunter soup, along with his usual dinner. A soon as Nick came back with the beer and soup, the Lieutenant took a large drink of the Kulmbacher Mönchshof and then turned his attention to the soup; it was delicious and comforting. Nick left him to himself until he brought back the grilled pork loin and French fries.

As he set those down, he remarked that the Lieutenant looked worn out and asked him where he had been. The Lieutenant told him that he had been on an exercise in the Odenwald, east of Frankfurt. It had been brutally cold and he had not been able to get any real sleep or a full meal for almost a week. Nick enjoyed hearing about his time in the field, since he ran the hotel and restaurant almost single handedly, he could not get away. Like most Greeks, he was adventurous and liked to travel, so he really enjoyed hearing about the wanderings of his customers.

He finished eating and told Nick he was going to forgo the Metaxa and head back to his quarters because he was tired. He paid his bill and went out the door in his ski jacket into a light snow. The drive back to the BOQ was a little tricky because the roads were slick, but he made good time and climbed the steps into the building in a good mood.

Once he reached the second floor, he started past the communal kitchen and pulled up. He could tell from the smell that it had been a pizza night, so he looked in the refrigerator for his favorite food: cold leftover pepperoni pizza. He was still hungry, so he picked up a piece and ate it as he walked down the hall to his room. After he opened the door, he washed his hands and sat down to write his girl a two-page letter, catching her up to date on his more recent travels, without mentioning the problems with carbon monoxide or the cook. He told her that he planned to call her the next week, and that he was looking forward to flying out on the twenty-eighth. Secretly, he was also glad that he had changed the beneficiary on his SGLI to protect her and give her a future in case anything happened to him.

A HARD DAY'S NIGHT

The next morning was Sunday, and ironically sunny, so he decided to drive over to Heidelberg for brunch at the officers club. According to Buzz and Pavel, it was fantastic. He made the twenty-kilometer drive, listening to a Pablo Cruise tape, until he pulled up to the gate guard on duty at the entrance to the parking lot and was waived through after he showed his ID. It took a while, but he finally found a spot for his car and walked into the club building. He showed his ID to another guard who was checking them for security purposes. He knew that Heidelberg was a hotbed of terrorist activity and home to the Red Army Faction, so he appreciated the efforts.

Brunch was all he'd been promised. Every kind of Army breakfast, along with double Screwdrivers and Bloody Marys. He had never developed a taste for tomato juice, so he stuck to a Screwdriver, coffee, and a full breakfast. The wonderful thing about Heidelberg was you had an opportunity to visit with officers who would come in from all over the world. He hoped that he might run into a Brother Rat or VMI alum but was not so lucky on this trip. When he was through, with the rest of the day ahead of him and good weather, he decided to explore downtown Heidelberg. It was a college town, so there were coffee shops, sandwich shops, record stores, and bars up and down the main street, as well as a large variety of young people from other countries in Europe, if not the world. He visited with some of the students in broken German and better English before heading back to Mannheim in the middle of the afternoon.

SUITE: JUDY BLUE EYES

For winter, it was a sunny and warm day. He took the Autobahn and had just hit eighty miles per hour, when he saw an older, little, white VW Beetle broken down on the side of the road with a flat tire. A small blonde girl, with bright blue eyes was standing beside it, looking forlorn. She reminded him a lot of Elizabeth, who had the same frame and eyes, so he pulled over and asked if she needed help. She did not have a spare, but lived nearby, in Karlsruhe, and accepted his offer to drive her home.

Once they were on the highway, he tried to talk with her but she stayed quiet. To break the silence, he slipped in a cassette tape of ABBA music and settled in. The trip only took twenty minutes and she had him pull over in a quiet residential neighborhood. She said, "Danke," and walked quickly for about a block to a set of stairs where an older man, her father no doubt, was waiting. They spoke briefly and the Lieutenant could tell that her father was unhappy because she had taken a ride from an American, as evidenced by his green USAREUR license plates. Picking up on the hostility, he knew it was time to go, so he pulled up into a three-point turn and made tracks for Mannheim.

OUTLAW BLUES

When he parked in front of the BOQ, it was still warm so, on a whim, he decided that he would try doing something he had not done in a long time before he got to Germany. He would try to make soup for dinner. He hustled over to the PX to pick up some stew beef from Ireland, mushrooms, onions, bell peppers, tomato paste, canned vegetables, and French bread. He already had a couple of bottles of red wine that would go well with it in his room, so he picked them up and carried everything back to the communal kitchen. By now it was almost 1600. He was determined to give it his best shot, but he was pretty unsure. He'd even managed to burn water, trying to make spaghetti a couple of years earlier at Fort Knox. So, he was worried to say the least. Fortunately, after two minutes or so, rattling around the kitchen, banging pots and pans, Elizabeth heard him and came across the hall to rescue him.

She took over for him browning the beef in olive oil. When it was done, she added water, tomato paste, and the other ingredients. Then she proved herself a wizard with the variety of German and Indian spices available in the cabinet above the stove. Once it had come to boil, she reduced the heat to a simmer and let the aroma of beef, onions, peppers, paprika and curry powder fill the air. Once she had the situation in hand, he asked if she wanted a glass of wine. "Certainly," she told him, so he poured them both a glass of red and then sheepishly asked if it was all right for him to slice the French bread. She nodded as she sat down at the table, so he did that and set it out on a large platter with butter, to await the arrival of their fellow diners. Just as he finished doing that, Buzz walked in the door, poured himself a glass of wine and joined Elizabeth. The two were soon engaged in intimate and private conversation, so the Lieutenant sat at the far end of the table and thought of home.

Evening was fast approaching, and he could hear small pellets of sleet, driven by a strong wind, bouncing off the glass of the windows, and felt a cold draft from the windows beginning to fill the room. The onset of the freezing evening and the wonderful smell of hunter's soup bubbling in the pot that wafted down the halls, soon attracted the other residents who walked down, laughing and chatting, to join them. As they came in, Elizabeth took charge and directed all of them to get their soup, wine and bread, and take their places at the table to join her, Buzz, and the Lieutenant for dinner. Sharon and Chris were in high spirits, visiting with Bernie, who, for a change, was unaccompanied except for the faint smell of hash, and Tom Storch, who had also been away earlier in the week, inventorying ammunition and weapons stored in stock points throughout the area. Bigfoot was also flying solo and sat down by Buzz with a bowl of soup and a bottle of Beck's from the refrigerator, as he was not fond of red wine.

Lieutenant Pavel was the last resident of the floor to come in. He looked haggard and worn, since he had spent almost four weeks in the field, due to the fact that he had spent those days that he was not involved in the exercises supervising skills test training with the Scout Platoon and the other platoons that made up Combat Support Company. He also passed on the wine, grabbing a beer and a bowl of soup with a couple of pieces of bead before he slipped in beside the Lieutenant, thankful to be indoors, warm, and dry.

When everyone was eating and chattering, the Lieutenant stood and thanked Elizabeth for making supper and pulling his fat out of the fire. Everyone clapped for her and she smiled broadly before taking a bow and retaking her seat, snuggling up next to Buzz. As soon as he had finished eating, the Lieutenant stood up to place his dishes in the sink. Seeing this, Buzz spoke up, volunteering to do the dishes. The Lieutenant thanked him and bid farewell to everyone before making his way back to his room.

COURAGE AND SHUFFLE THE CARDS

He climbed on the twin bed and pulled his well-worn VMI comforter over him, up to his chin. Lying there, he soaked in the warmth of the room, and thought to himself, *Tomorrow will be a red-letter day...my first time taking on the "heroin trade," even if indirectly.* He also decided, that, since the chain of command might not be too supportive, to adopt the motto of Lola Monez, the heroine in a novel by Fraser that he read when he was much younger: "Courage and shuffle the cards," before finally falling asleep.

The next morning after PT, in the snow, he caught Sergeant McNabb and told him that he needed to see him in his office at 0930 sharp. After getting dressed, he had breakfast before visiting with Sergeant Sparks. He told him that he needed to see him at 1000 sharp, in his office. The Mess Sergeant looked at him sullenly, out of the corner of his left eye. Unphased, the Lieutenant reminded him that he was now the Mess Officer and that Sparks now reported to him. This brought him up short and he said, "Yes, sir," before heading back to his office, between the kitchen and the serving line.

HELLO, GOODBYE

Once he was back in his office, he said, "Hello," to Caceres who was seated at the small desk by the door, normally occupied by Specialist Christopher, the supply officer's driver, who was on a detail that day. Caceres, who was filling in for him, smiled and went back to his usual pastime of sharpening his knives. When the S-4 came through the door from Battalion, the Lieutenant asked if he had a copy of the table of organization and equipment for the Mess Platoon, to include the trailers. Captain Dolan said, "Sure." Then they went back in his office and the S-4 handed a large computer printout to the Lieutenant who took it back to his desk. Once there, he looked down at the extensive list and smiled a little to himself.

At 0930, he looked up to his left and saw that Sergeant McNabb was coming through the door, on time for a change. He was impeccably dressed in starched fatigues and highly shined boots, looking just like a recruiting poster. And the Lieutenant thought that he looked a lot like the armorer in his first company, who turned out to be a serial killer. Somehow, it seemed appropriate. McNabb saluted, and after returning it, the Lieutenant motioned for him to take a seat, to the left of his desk. Then he looked at him intently and said, "You know, sergeant, I've never fired anybody in my life. But today, I'm firing you. You have twenty-four hours to find yourself a new job. I know you have several connections in the line-haul, and supply and transport battalions, so I'm sure you can make that happen. And maybe you can find an NCO there who wants to come work for me. This much I do know: if you're not gone within twenty-four hours, I will go with the S-4 up to the Sergeant Major and make sure that you're out of here. And since you haven't worked for me for ninety days, I can't rate you." He could tell that McNabb was thinking fast, listening to Caceres' knife on the whet stone, and weighing his options. Then before the Lieutenant could say another word, he jumped up, saluted, and was almost running out the door. The Lieutenant never saw him again.

A HARD DAY'S NIGHT

At 1000, the Mess Sergeant came in and the Lieutenant, after exchanging salutes, beckoned him to take the seat beside his desk. He looked closely at Sparks, who was dressed in short-sleeved kitchen whites. He was a large man with massive shoulders. His eyes were bloodshot, and watery: the telltale signs of heavy smoking, and alcohol, or hash, or both, but apparently not heroin as he had no needle marks on his arms. The Lieutenant looked intently into the sergeant's eyes and in a low measures voice, said, "Sergeant, it has come to my attention that you are continually sending the same cook and trailer to the field for every exercise. Why is that?"

"He's the only man I have that is trained to cook in the field, Lieutenant."

The Lieutenant looked at him, thought for a few seconds, and then said, 'I'm heartened by that." The sergeant stared at him quizzically, so the Lieutenant continued, "That means we have to train your cooks. Everyone of your cooks. And that means we'll pull all of the trailers out onto the parade ground, end to end. And you will cook every meal for the battalion outside on those trailers until I'm happy."

The bleary-eyed sergeant looked out the window to the snow filled courtyard and said, "In the snow, sir?"

"Yes. I just came back from a week in the snow. Your people have to be able to cook in the snow, so we will train in the snow."

Sparks turned back to the Lieutenant, mulling over the situation, and said, "Sir, we can't do that."

"Why not?"

"Because we don't have all of the equipment."

"I'm even more heartened by that," said the Lieutenant, as he pulled the Table of Organization and Equipment from his desk. "You personally signed for all the property on this list, didn't you?" The sergeant hesitated, thinking about lying for a minute, as the sound of Caceres' largest knife, scraping on his whetstone, filled the room. Finally, he said, "Yes, sir, from Headquarters Company."

"Then we'll do a full field layout tomorrow at 1000. Twenty-four hours from now I will inventory every pot, pan, knife, fork, and spoon. Anything missing will be surveyed against you and the money taken out of your pay to cover it. But something tells me you have enough connections to find everything."

The sergeant watched him dully, until he got the message. Finally, what passed for realization sank in, and he nodded his head in resignation. The Lieutenant then simply looked at him and said,

"You're dismissed." Sparks rose, saluted, turned, and hurried past Caceres to the door. When he was gone, Dolan came out of his office and asked him how things were going. The Lieutenant told him that he was going to inventory the Mess Platoon's equipment, shut down the mess hall, and the meals would be cooked in the trailers, until all of the cooks were fully trained. The S-4 looked at him quizzically, and apprehensive at the same time, so the Lieutenant chimed in, "Don't worry, officers are magic. On top of that, I'm a graduate of Major Charles F. Riley's finishing school for wayward boys in the Third ACR. It's going to happen."

The next morning, once he was dressed, he ate some licorice and breakfast bar with a cup of instant coffee before heading over to his office for the daily meeting with the S-4. When they were done, he bundled up against the cold, grabbed a clipboard and the TOE list, and walked around the building to the parade ground, which was still covered in two or three inches of snow. The Lieutenant was happy to see that all five kitchen trailers were set up end-to-end, with tarps and equipment laid out in front of them.

Beginning promptly at 1000, he inventoried all of the equipment with the Mess Sergeant. When they were done, it looked like Sparks was short a couple of serving spoons, so he immediately dispatched the nearest cook to retrieve them from the mess hall, where they had been used to serve the troops breakfast. When they were laid out on the tarp in front of the trailer where the Lieutenant was standing, all of the equipment was accounted for, so he congratulated Sparks and told him that since he was Mess Officer, he did not have to pay surcharge for his meals and he could expect to see him at least twice daily until the cooks were trained. "Based upon what I've seen, I imagine that you will have everybody trained in a week or so."

The Mess Sergeant said, "Yes, sir, that's right." With that, the Lieutenant saluted, the sergeant returned it, and he headed back to his office.

OUTLAW MAN

When he walked in the S-4 shop he saw that Dolan's driver was back at the desk by the door. Then he noticed a Sergeant First Class, a little shorter than him, with sandy hair, grey eyes, and a shit-eating grin, standing by his desk. As he approached, the sergeant immediately came to attention and saluted him. He returned it and simply asked, "Who are you?"

He replied, "Sergeant Conley, sir." The Lieutenant signaled for him to sit beside the desk. He then quickly explained that he was formerly assigned to a line-haul battalion, and that he had decided to swap out with Sergeant McNabb, so he could come over and take charge of a platoon that was stationed in Benjamin Franklin Village, close to where he and his wife lived.

Once in a while, a commissioned officer will run across a sergeant who is a smooth operator. Conley was such an operator, and the Lieutenant sensed it immediately. He asked how long he would need to put together a training plan for the drivers. Conley quickly responded that he needed to assess all of them and get the vehicles through with the ongoing maintenance, but he should have everything ready within a week. Upon hearing his, the Lieutenant began smiling inwardly. Then he said, "Excellent. As you may know, I'll be on leave for the first three weeks of March, and after that, we go to Baumholder for gunnery, so I need you to get to work immediately." No sooner had he finished than Conley rose, saluted, and disappeared out the door. Apparently, his new sergeant was old school, a self-starter, like the drill instructors in his first battalion, who preferred to do their jobs with a minimum of officer interference.

When Sergeant Conley was gone, the Lieutenant looked over his right shoulder and asked Sergeant Dryer what he thought. Dryer thought for a couple of seconds – then smiled and said, "Sir, I think he is exactly what you need."

"Thanks," said the Lieutenant, "I'm going to grab a quick lunch with my cooks, run some errands, and try to call my girl back in

Kansas. I'll be back before evening formation. After that, I'll get dinner in the mess hall. See you in the morning." With that, he walked out the door. He made his call to his girl and caught her up on his travel schedule, before they hung up so she could get some sleep. He returned to Headquarters Company as the First Sergeant was calling roll. When the company was dismissed, he made his way to the mess hall where he checked in with Sparks to see how things were going. The sergeant, while not enthusiastic, was performing, so things were looking up. After he ate, he caught the eye of the Leach, the cook who had tried to kill him earlier. Leach was taking utensils and empty pans from the steam table out to the trailers. The cook smiled faintly before looking away. The Lieutenant didn't say a thing and turned to walk out the door, thanking the cooks on the serving line as he left.

With Sergeant Conley on board in Support Platoon, the Lieutenant focused his energies on the mess hall. As promised, he ate two meals a day there and followed up with Sparks each time to make sure that he was training the cooks. On the whole, they were pretty sullen, but he could tell that they were becoming more adept and comfortable cooking on the trailers, so he was feeling good, and looking forward to getting his drivers trained once gunnery was done.

LIFE IN THE FAST LANE

The next weekend went smoothly. He checked on his cooks and had breakfast and dinner with them on both Saturday and Sunday. Finally, after PT and breakfast on Monday, he asked Sparks to meet him in his office at 0900. The Mess Sergeant arrived promptly and it appeared that he had been sleeping better. The Lieutenant asked him, "Based on your experience, are your people trained enough to cook in the field?"

The sergeant said, "Yes, sir!" without hesitation.

The Lieutenant told him, "Good. You and your platoon are released to store the trailers and cook in the dining facility. But I want you to put together a duty roster to rotate cooks and trailers into the field for the upcoming exercises." Sparks agreed immediately, rose, saluted, and hustled out the door.

When the sergeant was gone, the Lieutenant went over, knocked on the door, and when beckoned, came in, saluted, and reported back to the S-4 that the cooks were trained. He also explained that he had a new platoon sergeant in Support Platoon and that he expected a training schedule the following day. At first, the S-4 looked at him in amazement. But the Lieutenant assured him that it was true, before thinking to himself that he was in the company of another West Pointer, who had probably never seen a VMI officer at work.

The following day, Sergeant Conley visited with him after morning PT. They agreed to meet at 0900, so the Lieutenant got dressed, grabbed a couple of breakfast bars, and jogged back to his office so he could check in with the S-4 and be on time for his first real conference with his new Platoon Sergeant. Good to his word, Conley came through the door and saluted beside the Lieutenant's desk at exactly 0900. After the Lieutenant returned the salute, Conley sat down and opened a folder that he had carried in his left hand.

The Lieutenant looked at the first sheet of paper: it had a matrix of the vehicles in the platoon, showing their maintenance status and the dates when they had, or would complete, maintenance. The second sheet had a matrix of the drivers and the vehicles that they were trained to operate. The third and final sheet was a thirty-day calendar with training tasks and names of drivers set out in detail. As he looked it over, he said, "It looks like we have drivers who cannot operate a manual transmission, is that right?"

"Yes, sir."

"How is that even possible when they graduated from AIT?"

"Sir, there is not enough time allocated in AIT to train all of them on a standard transmission, so they get assigned to vehicles based upon their experience, mostly before they came in service. Besides, all of the GOERs are automatic."

"So that means we have to train them here?"

"Yes, sir, it does."

"Thanks," said the Lieutenant, "it looks like you have your work cut out for you and you need to get busy."

"Yes, sir," said Conley, "I do," as he rose and saluted. The Lieutenant returned it, and before he could say another word, Conley disappeared out the door on his way back to the motor pool.

Chapter 37
Already Gone

That night he decided to have dinner at the officers club, since it was steak night, and he did not feel like driving anywhere in the cold, misty evening. He smiled when he came through the door because he saw his classmate, Mad Dog, and his girlfriend, who was a little blonde captain, from the MP Brigade. He walked across the room and asked if he could join them. They both said, "Sure," so he sat down, ate, and brought them up to date on the wedding plans, or what he knew of them.

Since he was tied up overseas, his girl and her family had made the arrangements and as far as he knew, he just had to show up, get married, and get back to Germany with her. The little blonde just laughed and said, "You will be a lot more involved in everything than you think. When is she due in country?"

"The fifteenth of next month and we will be living in the BOQ until we can get married quarters."

"That," she said, "will be both a surprise and an education. I look forward to meeting her; she sounds interesting."

"That she is," he added and then excused himself to get some rest and a few things packed for the trip. They wished him a safe trip, and he took off to drive back to the "Q" in a slow cold rain.

The next day was the day before he was to leave, and he could feel both excitement and relief in that he was getting ready to make his escape from the battalion to finally get married. He met with both the S-4 and Sergeant Conley to go over his schedule and tasks to be completed in his absence. After lunch, he confirmed his ride to the airport with Master Sergeant Fykes and left early to get some last-minute banking done, and purchase traveler's checks before he made his way over to the Adjutant's office to sign out on leave.

After that he stood for evening formation and checked out with the First Sergeant and Sergeant Coley before walking to his quarters. There he heated up a TV dinner and had a bottle of beer to go with it. Once he finished eating, he packed everything except his shaving kit and traveling clothes: jeans, the German long-sleeved shirt, and ski jacket. By then, night was falling fast and he was restless, so he decided to set his alarm for 0400 and try to sleep.

GOODBYE YELLOW BRICK ROAD

As soon as the alarm clattered, he slipped out of bed and put water in his hot pot so he could enjoy a cup of instant coffee and a breakfast bar, before making his bed and cleaning the place up a little for his bride. When he was done with that, he finished packing, locked the door, and carried his luggage down to the curb exactly at 0600. Good to his word, Fykes pulled up at the curb when he reached the sidewalk. He was dressed in starched fatigues, driving a dark gray Ford Taunus sedan, and motioned to the Lieutenant to put his suitcase and suit bag in the back seat, before taking a seat in front beside him.

They had a nice visit and the sergeant made sure that he had the Lieutenant's arrival time and gate number so he could pick him and his new wife up when they got in. Sergeant Fykes, like Conley, had been around and was fully integrated into the German economy so, in short, he could take care of anything and everything that might come up. He was a veteran of the Autobahn as well, so they flew down the highway, arriving at Frankfurt International in less than forty minutes. When they pulled up to the departure doors, he saluted Fykes, who wished him well, saluted, and accelerated away for the drive back to Mannheim before he knew it. The Lieutenant made his way past the wine, luggage, and sex shops, and Polizei carrying submachine guns. Germany never ceased to amaze him, and he shook his head a little, before using his green ID card to get through customs.

SPANISH HARLEM

Shortly after boarding, the plane made the red-eye short hop to London's Heathrow, where he was treated to the usual bad English food that passed for breakfast; he regretted not bringing some licorice. A little less than thirty minutes later, the jet took off to the west, and he took a deep breath, before letting it out slowly. He couldn't believe that he was finally going to marry the girl who was his destiny. Relaxing, he started to drift off to sleep. But he could swear before he lost consciousness, that he heard the ghost of Cat Shannon softly whistling "Spanish Harlem" a few rows back.

He thought to himself that he had finally made it and after five months or so as a pirate in Support Platoon, he could get back in a line company as executive officer and then pull the last few months of his extended tour on staff at Battalion or maybe Brigade, before getting back to the States and law school. Little did he know that upon his return, he and his bride would face obstacles, tragedies, and challenges that dwarfed anything he had seen so far in the Pacesetter Battalion of the Grey Lions Regiment, call sign Pagan. To survive, he would have to embody the spirit of his division: he would become the Pathfinder.

THE END

. . .

About the Author

Robert Lee Manteuffel was born and raised on the high plains of western Kansas. His family provided officers to the armies of Europe for over seven centuries, amassing numerous awards, including the Pour le Merite, and Knight's Cross with oak leaves, swords and diamonds. On May fifteenth 1977, he graduated first in his class from the Virginia Military Institute, and was inducted into the Kappa Alpha Order. Commissioned as armor lieutenant, he served in basic training battalions and, while attached to the Third ACR, tested five of the first seven M-1 tanks ever built. While waiting the requisite two years for law school, he was involuntarily transferred to USAREUR in 1979, when the Shah fell in Iran and the Soviets massed on the Afghan border. In Germany, he served as in a tank battalion, where he took command of a company as a first lieutenant. Upon his return, he attended the US Air Force Institute of Technology for a master's degree in Operations Research, Strategic and Tactical Sciences, to become the mirror image of a Soviet operations officer. After three years as a computer gamer using simulations for weapons engineering and force design, he was separated from the Army and went to law school at SMU as a Hatton W. Sumners Scholar. He practiced as a Texas trial lawyer for over thirty years with a modicum of success.

About the Publisher

TACTICAL 16

Tactical 16 Publishing is an unconventional publisher that understands the therapeutic value inherent in writing. We help veterans, first responders, and their families and friends to tell their stories using their words.

We are on a mission to capture the history of America's heroes: stories about sacrifices during chaos, humor amid tragedy, and victories learned from experiences not readily recreated — real stories from real people.

Tactical 16 has published books in leadership, business, fiction, and children's genres. We produce all types of works, from self-help to memoirs that preserve unique stories not yet told.

You don't have to be a polished author to join our ranks. If you can write with passion and be unapologetic, we want to talk. Go to Tactical16.com to contact us and to learn more.

All of Tactical 16's books are available on our online bookstore, T16Books.com. Visit it today to see more books from our selection of authors and to find a new adventure to read!

www.ingramcontent.com/pod-product-compliance
Ingram Content Group UK Ltd.
Pitfield, Milton Keynes, MK11 3LW, UK
UKHW020128290926
15471UKWH00003B/361

9 781966 41304